T0214633

IFIP Advances in Information and Communication Technology 539

Editor-in-Chief

IFIP – The International Federation for Information Processing

IFIP was founded in 1960 under the auspices of UNESCO, following the first World Computer Congress held in Paris the previous year. A federation for societies working in information processing, IFIP's aim is two-fold: to support information processing in the countries of its members and to encourage technology transfer to developing nations. As its mission statement clearly states:

IFIP is the global non-profit federation of societies of ICT professionals that aims at achieving a worldwide professional and socially responsible development and application of information and communication technologies.

IFIP is a non-profit-making organization, run almost solely by 2500 volunteers. It operates through a number of technical committees and working groups, which organize events and publications. IFIP's events range from large international open conferences to working conferences and local seminars.

The flagship event is the IFIP World Computer Congress, at which both invited and contributed papers are presented. Contributed papers are rigorously refereed and the rejection rate is high.

As with the Congress, participation in the open conferences is open to all and papers may be invited or submitted. Again, submitted papers are stringently refereed.

The working conferences are structured differently. They are usually run by a working group and attendance is generally smaller and occasionally by invitation only. Their purpose is to create an atmosphere conducive to innovation and development. Refereeing is also rigorous and papers are subjected to extensive group discussion.

Publications arising from IFIP events vary. The papers presented at the IFIP World Computer Congress and at open conferences are published as conference proceedings, while the results of the working conferences are often published as collections of selected and edited papers.

IFIP distinguishes three types of institutional membership: Country Representative Members, Members at Large, and Associate Members. The type of organization that can apply for membership is a wide variety and includes national or international societies of individual computer scientists/ICT professionals, associations or federations of such societies, government institutions/government related organizations, national or international research institutes or consortia, universities, academies of sciences, companies, national or international associations or federations of companies.

More information about this series at http://www.springer.com/series/6102

Zhongzhi Shi · Cyriel Pennartz
Tiejun Huang (Eds.)

Intelligence Science II

Third IFIP TC 12 International Conference, ICIS 2018
Beijing, China, November 2–5, 2018
Proceedings

Springer

Editors
Zhongzhi Shi
Chinese Academy of Sciences
Beijing
China

Tiejun Huang
Peking University
Beijing
China

Cyriel Pennartz
University of Amsterdam
Amsterdam
The Netherlands

ISSN 1868-4238   ISSN 1868-422X (electronic)
IFIP Advances in Information and Communication Technology
ISBN 978-3-030-13167-8   ISBN 978-3-030-01313-4 (eBook)
https://doi.org/10.1007/978-3-030-01313-4

This Springer imprint is published by the registered company Springer Nature Switzerland AG
The registered company address is: Gewerbestrasse 11, 6330 Cham, Switzerland

Preface

This volume comprises the proceedings of the Third International Conference on Intelligence Science (ICIS). Artificial intelligence research has made substantial progress in certain areas to date. However, the deeper understanding of the essence of intelligence is far from sufficient and, therefore, many state-of-the-art intelligent systems are still not able to compete with human intelligence. To advance the research in artificial intelligence, it is necessary to investigate intelligence, both artificial and natural, in an interdisciplinary context. The objective of this conference is to bring together researchers from brain science, cognitive science, and artificial intelligence to explore the essence of intelligence and the related technologies. The conference provides a platform for discussing some of the key issues that are related to intelligence science.

For ICIS 2018, we received more than 85 papers, of which 44 papers were included in this program as regular papers and four as short papers. We are grateful for the dedicated work of both the authors and the referees, and we hope these proceedings will continue to bear fruit over the years to come. All papers submitted were reviewed by three referees.

A conference such as this cannot succeed without help from many individuals who contributed their valuable time and expertise. We want to express our sincere gratitude to the Program Committee members and referees, who invested many hours for reviews and deliberations. They provided detailed and constructive review reports that significantly improved the papers included in the program.

We are very grateful for the sponsorship of the following organizations: Chinese Association for Artificial Intelligence (CAAI), IFIP TC12, China Chapter of International Society for Information Studies, Peking University, and supported by Beijing Association for Science and Technology (BAST), Shanghai Association for Science and Technology (SAST), Beijing Association for Artificial Intelligence (BAAI), CIE Signal Processing Society and Institute of Computing Technology, Chinese Academy of Sciences. Thanks to Professor Jinwen Ma, who served as chair of the Organizing Committee, and Dingsheng Luo as secretary general.

Finally, we hope you find this volume inspiring and informative.

August 2018

Zhongzhi Shi
Cyriel Pennartz
Tiejun Huang

Organization

Sponsors

Chinese Association for Artificial Intelligence (CAAI)
China Chapter under International Society for Information Studies

Organizers

Peking University

Support

IFIP Technical Committee 12
Beijing Association for Science and Technology (BAST)
Shanghai Association for Science and Technology (SAST)
Beijing Association for Artificial Intelligence (BAAI)
CIE Signal Processing Society

Steering Committee

Deyi Li   President of CAAI, China
Jianxian Lin   Peking University, China
Yanda Li   Tsinghua University, China
Bo Zhang   Tsinghua University, China
Ruqian Lu   Chinese Academy of Sciences, China
Wen Gao   Peking University, China
Guangjian Zhang   Chinese Association for Noetic Science, China

General Chairs

Yixin Zhong   Beijing University of Posts and Telecommunications,
                        China
Jay McClelland   Stanford University, USA
Xingui He   Peking University, China

Program Chairs

Zhongzhi Shi   Institute of Computing Technology, Chinese Academy
                        of Sciences, China
Cyriel Pennartz   University of Amsterdam, The Netherlands
Tiejun Huang   Peking University, China

Program Committee

Guoqiang Bi, China
Yongzhi Cao, China
Jie Chen, China
Zhicheng Chen, China
Bin Cui, China
Shifei. Ding, China
Gordana Dodig-Crnkovic, Sweden
Elias Ehlers, South Africa
Yongchun Fang, China
Jiali Feng, China
Hongguang Fu, China
Xiao-Shan Gao, China
Yang Gao, China
Xiaozhi Gao, China
Huacan He, China
Tiejun Huang, China
Zhisheng Huang, The Netherlands
Licheng Jiao, China
Gang Li, Australia
Hongbo Li, China
Yu Li, France
Yujian Li, China
Tianrui Li, China
Zhong Li, Germany
Zhouchen Lin, China
Chenglin Liu, China
Feng Liu, China
Mingzhe Liu, China
Tian Liu, China
Chenguang Lu, China
Dingsheng Luo, China
Xudong Luo, China
Jinsong Ma, China
Jinwen Ma, China
Shaoping Ma, China
Shimin Meng, China
Kedian Mu, China

Miec Owoc, Poland
He Ouyang, China
Gang Pan, China
Germano Resconi, Italy
Paul S. Rosenbloom, USA
Chuan Shi, China
Zhongzhi Shi, China
Bailu Si, China
Andreej Skowron, Poland
Sen Song, China
Nenad Stefanovic, Serbia
Huajin Tang, China
Lorna Uden, UK
Guoying Wang, China
Houfeng Wang, China
Pei Wang, USA
Wenfeng Wang, China
Xiaofeng Wang, China
John F.S. Wu, China
Chuyu Xiong, USA
Hui Xiong, USA
Guibao Xu, China
Chunyan Yang, China
Dezhong Yao, China
Yiyu Yao, Canada
Jian Yu, China
Yi Zeng, China
Xiaowang Zhang, China
Yinsheng Zhang, China
Zhaoxiang Zhang, China
Zhisen Zhang, China
Chuan Zhao, China
Yang Zhen, China
Yixin Zhong, China
Fuzhen Zhuang, China
Xiaohui Zou, China

Organizing Committee

Chair

Jinwen Ma   Peking University, China

Secretary General

Dingsheng Luo   Peking University, China

Vice Secretary General

Xiaohui Zou   Sino-American Searle Research Center

Local Arrangements Chair

Shuyi Zhang   Peking University, China

Finance Chair

Wenhui Cui   Peking University, China

Publication Chair

Tian Liu   Peking University, China

Publicity Chair

Yongzhi Cao   Peking University, China

International Liaison

Kedian Mu   Peking University, China

Abstracts of Keynote
and Invited Talks

Progress Toward a High-Performance Brain Interface

Andrew Schwartz

Department of Neurobiology,
University of Pittsburgh

Abstract. A better understanding of neural population function would be an important advance in systems neuroscience. The change in emphasis from the single neuron to the neural ensemble has made it possible to extract high-fidelity information about movements that will occur in the near future. Information processing in the brain is distributed and each neuron encodes many parameters simultaneously. Although the fidelity of information represented by individual neurons is weak, because encoding is redundant and consistent across the population, extraction methods based on multiple neurons are capable of generating a faithful representation of intended movement. A new generation of investigation is based on population-based analyses, focusing on operational characteristics of the motor system. The realization that useful information is embedded in the population has spawned the current success of brain-controlled interfaces. This basic research has allowed to us to extract detailed control information from populations of neural activity in a way that this can be used to restore natural arm and hand movement to those who are paralyzed. We began by showing how monkeys in our laboratory could use this interface to control a very realistic, prosthetic arm with a wrist and hand to grasp objects in different locations and orientations. This technology was then extended to a paralyzed patient who cannot move any part of her body below her neck. Based on our laboratory work and using a high-performance "modular prosthetic limb," she was able to control 10 degrees-of-freedom simultaneously. The control of this artificial limb was intuitive and the movements were coordinated and graceful, closely resembling natural arm and hand movement. This subject was able to perform tasks of daily living– reaching to, grasping and manipulating objects, as well as performing spontaneous acts such as self-feeding. Current work with a second subject is progressing toward making this technology more robust and extending the control with tactile feedback to sensory cortex. New research is aimed at understanding the neural signaling taking place as the hand interacts with objects and together, this research is a promising pathway toward movement restoration for those who are paralyzed.

Predicting the Present: Experiments and Computational Models of Perception and Internally Generated Representations

Cyriel M. A. Pennartz

Cognitive and Systems Neuroscience Group,
University of Amsterdam

Abstract. The last three decades have witnessed several ideas and theories on brain-consciousness relationships, but it is still poorly understood how brain systems may fulfill the requirements and characteristics we associate with conscious experience. This lecture will first pay attention to the basic requirements for generating experiences set in different modalities, such as vision and audition, given the rather uniform nature of signal transmission from periphery to brain. We will next examine a few experimental approaches relevant for understanding basic processes underlying consciousness, such as changes in population behavior during sensory detection as studied with multi-area ensemble recordings. For visual detection, the primary sensory cortices have been a long-standing object of study, but it is unknown how neuronal populations in this area process detected and undetected stimuli differently. We investigated whether visual detection correlates more strongly with the overall response strength of a population, or with heterogeneity within the population. Zooming out from visual cortex to larger neural systems, we asked how "visual" the visual cortex actually is by studying auditory influences on this system and considering interactions between visual and auditory systems. Finally, we will consider the topic of perception in the context of predictive coding. Predictive coding models aim to mimic inference processes underlying perception in the sense that they can learn to represent the hidden causes of the inputs our sensory organs receive. They are not only about "predicting-in-time" but also about predicting what is currently going on in the world around us – "predicting the present". I will present novel work on predictive coding in deep neural networks, and link the inferential and generative properties of these networks to conscious representations. I will argue that a productive way forward in research on consciousness and perception comes from thinking about world representations as set across different levels of computation and complexity, ranging from cells to ensembles and yet larger representational aggregates.

Brain Science and Artificial Intelligence

Xu Zhang

Institute of Brain-Intelligence Science and Technology,
Zhangjiang Laboratory, Institute of Neuroscience,
Chinese Academy of Sciences,
Shanghai Branch of Chinese Academy of Sciences, China

Abstract. Cognition is the mental action or process of acquiring knowledge and understanding through thought, experience and the senses. The processes can be analyzed from different perspectives within different contexts, notably in the fields of linguistics, anesthesia, neuroscience, psychiatry, psychology, education, philosophy, anthropology, biology, systemic, logic and computer science. So far, we still do not know how many neuron types, neural circuits and networks in our brain. It is important to construct the basis for deciphering brain and developing brain-inspired artificial intelligence (AI). In 2012, Chinese Academy of Sciences started the Strategic Priority Research Program, mapping brain functional connections. This research program tried to set up new research teams for interpreting and modeling the brain function-specific neural connectivity and network. In 2014, we started the Shanghai Brain-Intelligence Project, for translational research and R&D. We tried to map the somatosensory neuron types and their connectivity with single-cell Tech and the trans-synaptic tracers. We were also interested to link Neuroscience and AI development. Our team has produced the deep-learning, neural network processors, and achieved the applications of AI Tech, such as the speech recognition and translation technology, and the bionics of eyes and control system through the physiological, mathematical, physical and circuit models.

Intelligence Science Will Lead the Development of New Generation of Artificial Intelligence

Zhongzhi Shi

Institute of Computing Technology,
Chinese Academy of Sciences

Abstract. The State Council of China issued the notice of the new generation AI development plan in last year. The notice points out that AI has become a new focus of international competition and a new engine of economic development. We must firmly grasp the great historical opportunity of the development of artificial intelligence, play the leading role of intelligence science and drive the national competitiveness to jump and leap forward. Intelligence Science is the contemporary forefront interdisciplinary subject which dedicates to joint research on basic theory and technology of intelligence by brain science, cognitive science, artificial intelligence and others. The presentation will outline the framework of intelligence science and introduce the cognitive model of brain machine integration, containing environment awareness, motivation driven automated reasoning and collaborative decision making. Finally, explore the principle of cognitive machine learning in terms of mind model CAM.

Scientific Paradigm Shift for Intelligence Science Research

Yixin Zhong

Beijing University of Posts and Telecommunications

Abstract. Intelligence science is a newly inceptive and highly complex scientific field, which is the most height of information science while is rather different from the classical physic science. However, the research of intelligence science carried on so far has been basically following the scientific paradigm suitable for classical physic science. The incompatibility between the properties of intelligence science and the scientific paradigm suitable for classical physic science has caused many problems. For overcoming these problems and making good progress in intelligence science, the shift of the scientific paradigm from the one suitable for classical physic science to the one suitable for intelligence science is demanded. What is the concept of scientific paradigm then? What is the scientific paradigm suitable for intelligence science research? What kinds of progresses can be, or have been, achieved in intelligence science research through the scientific paradigm shift? These are the topics in the paper.

Visual Information Processing – From Video to Retina

Tiejun Huang

Department of Computer Science, School of EE & CS,
Peking University, China

Abstract. Visual perception is a corner stone for human and machine. However, the conventional frame by frame video employed in computer vision system is totally different with the spike train on the visual fibers from the biological retina to the brain. This talk will give a background on the challenges for the visual big data processing nowadays, then introduce our works on mapping and simulation of the neural circuits in the primate retina, and a new sensor chip based on the spiking representation, to be potentiality used for machine vision including autonomous driving, robot perception etc.

The Human Brainnetome Atlas and Its Applications in Neuroscience and Brain Diseases

Tianzi Jiang

Institute of Automation of the Chinese Academy of Sciences

Abstract. Brainnetome atlas is constructed with brain connectivity profiles obtained using multimodal magnetic resonance imaging. It is in vivo, with finer-grained brain subregions, and with anatomical and functional connection profiles. In this lecture, we will summarize the advance of the human brainnetome atlas and its applications. We first give a brief introduction on the history of the brain atlas development. Then we present the basic ideas of the human brainnetome atlas and the procedure to construct this atlas. After that, some parcellation results of representative brain areas will be presented. We also give a brief presentation on how to use the human brainnetome atlas to address issues in neuroscience and clinical research. Finally, we will give a brief perspective on monkey brainnetome atlas and the related neurotechniqes.

A Brief Overview of Practical Optimization Algorithms in the Context of Relaxation

Zhouchen Lin

Fellow of IEEE, Peking University

Abstract. Optimization is an indispensable part of machine learning. There have been various optimization algorithms, typically introduced independently in textbooks and scatter across vast materials, making the beginners hard to have a global picture. In this talk, by explaining how to relax some aspects of optimization procedures I will briefly introduce some practical optimization algorithms in a systematic way.

Clifford Geometric Algebra

Jiali Feng

Information Engineering College,
Shanghai Maritime University, Shanghai

Abstract. Turing question: "Can Machine think?" involves the basic contra-diction in philosophy: "Could the material be able to have spiritual?" By it a secondary contradiction chain, that from the general matter to the life, to the advanced intelligence, can be induced. The law of unity of opposites of con-tradiction and the law of dialectical transformation have become the core issues that must be studied in Natural Sciences, Social Sciences, Noetic Sciences and Intelligence Sciences. The space-time position is the basic attribute of when and where things are represented. If take "two different things must not be in the same position at the same time" as the "simultaneous heterotopy" principle or basic assumption, and the exclusiveness between different objects based on "simultaneous heterotopic" can be equivalented the "overtness" of object, then whether a non-zero distance between two different objects is existing? or not, would not only is a criterion for the existence of differences between them, but also can be seen as a source of contradictions between the two. Since the range of displacement of one object, which such that the non-zero-distance between both contradictions could be maintained, can be considered as "the qualitative criterion" for its quality can be maintained, therefore, the law that "the quality of object can be remained, when the range of quantitative change does not exceed, can be expressed as a qualitative mapping from quantities to quality. The (non-essential) differences in the nature of object caused by different distances between two contradictions, can be expressed by the function of degree of conversion from quantity to quality. The movements, changes, developments of contradictory between both objects, and one changes to its opposites and so on, are regulated by the non-zero distance changes that accompany displacement varies with time, as well as changes of qualitative criterion. In mathematics, distance is defined as "the square root of the inner product of two vectors (positive definite)". The inner product is defined, in physics, as the work done by a force vector (or force function) on an object along with the direction of motion. The inner product is an invariant under the coordinate displacement translation and rotation, by which a polarization circle can be induced. When an object moves with a velocity under the action of a force, since distance is defined as the integral of velocity over time, but the integral of force and the inner product of the velocity is an outer product at the polarization vector. Using an application example, the analyzes about the mutual entanglement relation among the three definitions of distance, inner product and work function, and by it the Geometry Product = Inner product + Outer product (Clifford geometric algebra) structure are induced are presented in this paper, and the corresponding attribute coordinate for representation of it is given too. In addition, combining with pattern classification and identification, the relationship and the differences between the theories and methods in this paper and the (Clifford and Capsule) Artificial Neural Network are discussed. It improves a referenceable way for the creation of Noetic Science, Intelligent Science, and Synthesis Wisdom.

Urban Computing: Building Intelligent Cities Using Big Data and AI

Yu Zheng

Urban Computing Lab at JD Group

Abstract. Urban computing is a process of acquisition, integration, and analysis of big and heterogeneous data in cities to tackle urban challenges, e.g. air pollution, energy consumption and traffic congestion. Urban computing connects sensing technologies, data management and AI models, as well as visualization methods, to create win-win-win solutions that improve urban environment, human life quality and city operation systems. This talk presents the vision of urban computing in JD group, introducing the urban big data platform and a general design for intelligent cities. A series of deployed applications, such as big data and AI-driven location selection for business, AI-based operation optimization for power plants, and urban credit systems are also presented in this talk.

New Approaches to Natural Language Understanding

Xiaohui Zou (ID)

Sino-American Searle Research Center

Abstract. This talk aims to disclose the know-how of launching a new generation of excellent courses and to develop the learning environment in which human-computer collaboration can optimize the expert knowledge acquisition. The method is to form a teaching environment that can be integrated online and offline with some technical platform of cloud classrooms, cloud offices and cloud conference rooms. Taking Chinese, English, classical and summary abstracts as examples, human-computer coordination mechanism, to do the appropriate new generation of quality courses. Its characteristics are: teachers and students can use the text analyzed method to do the fine processing of the same knowledge module, and only in Chinese or English, through the selection of keywords and terminology and knowledge modules, you can use the menu to select as the way to achieve knowledge. The module's precision machining can adopt the big production method that combines on the line first, complete coverage and accurate grasp each language point and knowledge point and original point even their respective combination. This method can finish fine processing instantly for any text segment. The result is the learning environment that enables human-computer collaboration to optimize the expert knowledge acquisition. Natural language understanding is only a research field that has great significance to human beings. Digital Chinese character chess, using numbers and Chinese characters as twin chess pieces, with their meaningful combination as a language point and knowledge point, the purpose is to find the original chess soul namely original point. Assume that every sentence, every paragraph, every article has at least one original point. Whether reading comprehension, writing expression, or even automatic recognition, it is intended to clearly highlight the original points, of course, also to list language points and knowledge points. These jobs were originally mainly experts' expertise, now and in the near future, computers will also be able to handle them automatically. Its significance is that this project of this learning environment software based on the National Excellent Courses is already owned by Peking University and that is constructed by using the numbers-words chessboard with the feature of the introduction on the knowledge big production mode for the textual knowledge module finishing. The new approaches obtains a breakthrough from three types of information processing, namely: phenomenon of object, intention, text and its implication on the nature of mechanism, principle, law, and even chaos or Tao, all can be paraphrased in sequence and positioning logic on essential information, linkage function on formal information and generalized translation on content information, under the guidance of language, knowledge, software three kinds of GPS, such as GLPS and GKPS and GSPS.

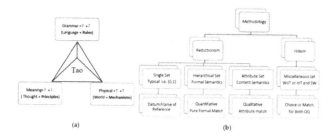

Fig. 1. Model a shows the biggest ontology; Model b shows how formal techniques of set systems link philosophical methodology and scientific methods system.

Fig. 2. Highlight bilingual pairs of Chinese and English arranged separately

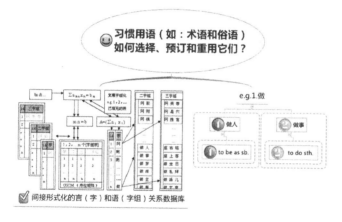

Fig. 3. Three types of bilingual systems as the GLPS and GKPS and GSPS

Neuromorphic Computing: A Learning and Memory Centered Approach

Huajin Tang

Neuromorphic Computing Research Center,
Sichuan University, China

Abstract. Neuromorphic cognitive computing is a new theme of computing technology that aims for brain-like computing efficiency and intelligence. Neuromorphic computational models use neural spikes to represent the outputs of sensors and for communication between computing blocks, and using spike timing based learning algorithms. This talk will introduce the major concepts and developments in this interdisciplinary area from the learning and memory centered perspective, and discuss the major challenges and problems facing this field.

Theory of Cognitive Relativity

Yujian Li

College of Computer Science, Faculty of Information Technology,
Beijing University of Technology
liyujian@bjut.edu.cn

Abstract. The rise of deep learning has brought artificial intelligence (AI) to the forefront. The ultimate goal of AI is to realize a machine with human mind and consciousness, but existing achievements mainly simulate intelligent behavior on computer platforms. These achievements all belong to weak AI rather than strong AI. How to achieve strong AI is not known yet in the field of intelligence science. Currently, this field is calling for a new paradigm, especially Theory of Cognitive Relativity. The starting point of the theory is to summarize first principles about the nature of intelligence from the systematic point of view, at least including the Principle of World's Relativity and the Principle of Symbol's Relativity. The Principle of World's Relativity states that the subjective world an intelligent agent can observe is strongly constrained by the way it perceives the objective world. The Principle of Symbol's Relativity states that an intelligent agent can use any physical symbol system to describe what it observes in its subjective world. The two principles are derived from scientific facts and life experience. Thought experiments show that they are of great significance to understand high-level intelligence and necessary to establish a scientific theory of mind and consciousness. Other than brain-like intelligence, they indeed advocate a promising change in direction to realize different kinds of strong AI from human and animals. A revolution of intelligence lies ahead.

Keywords: The principle of world's relativity · The principle of symbol's relativity · First principles · Thought experiments · Artificial intelligence

Two-layer Mixture of Gaussian Processes for Curve Clustering and Prediction

Jinwen Ma

Department of Information Science,
School of Mathematical Sciences and LMAM, Peking University

Abstract. The mixture of Gaussian processes is capable of learning any general stochastic process for a given set of (sample) curves for regression and prediction. However, it is ineffective for curve clustering analysis and prediction when the sample curves come from different stochastic processes as independent sources linearly mixed together. In fact, curve clustering analysis becomes very important in the modern big data era, but it is a very challenging problem. Recently, we have established a two-layer mixture model of Gaussian processes to describe such a mixture of general stochastic processes or independent sources, especially effective for curve clustering analysis and prediction. This talk describes the learning paradigm of this new two-layer mixture of Gaussian processes, introduces its MCMC EM algorithm and presents some effective practical applications on curve clustering analysis and prediction.

Contents

Perceptual Intelligence

Intelligent Robot

Fault Diagnosis

Ethics of Artificial Intelligence

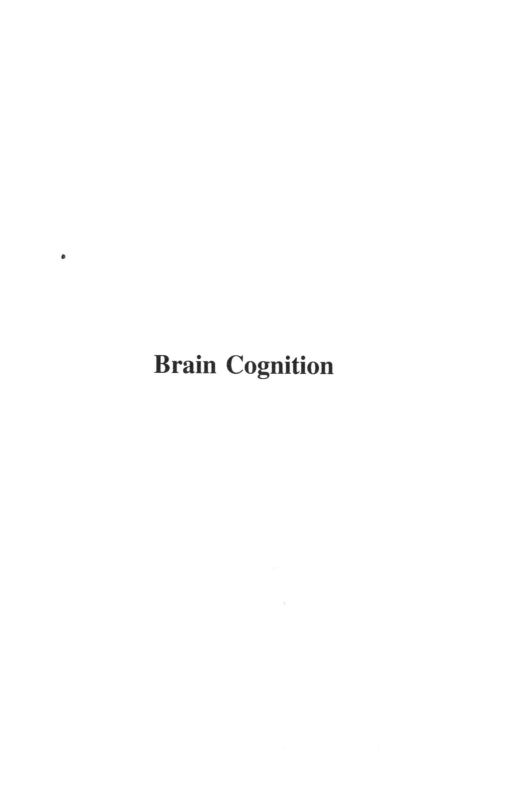

Brain Cognition

Multi-task Motor Imagery EEG Classification Using Broad Learning and Common Spatial Pattern

Jie Zou, Qingshan She[✉], Farong Gao, and Ming Meng

School of Automation, Hangzhou Dianzi University, Hangzhou 310018, China
729904283@qq.com, {qsshe,mnming}@hdu.edu.cn, frgao@126.com

Abstract. Motor imagery electroencephalography (EEG) has been successfully used in the brain-computer interface (BCI) systems. Broad learning (BL) is an effective and efficient incremental learning algorithm with simple neural network structure. In this work, a novel EEG multi-classification method is proposed by combining with BL and common spatial pattern (CSP). Firstly, the CSP algorithm with the one-versus-the-test scheme is exploited to extract the discriminative multiclass brain patterns from raw EEG data, and then the BL algorithm is applied to the extracted features to discriminate the classes of EEG signals during different motor imagery tasks. Finally, the effectiveness of the proposed method has been verified on four-class motor imagery EEG data from BCI Competition IV Dataset 2a. Compare with other methods including ELM, HELM, DBN and SAE, the proposed method has yielded higher average classification test accuracy with less training time-consuming. The proposed method is meaningful and may have potential to apply into BCI field.

Keywords: Electroencephalography · Brain-computer interface · Broad learning · Common spatial pattern

1 Introduction

The brain-computer interface (BCI) is a kind of communication and control system that does not depend on the brain's normal output channels of peripheral nerves and muscles. Therefore, it can be a very helpful aid to the people suffering motor disabilities [1]. A popular paradigm for BCI is motor imagery (MI), i.e., the subjects perform the imagination of movements and the imagined movements are distinguished by the system and translated into computer commands.

Due to the characteristics of non-stationary, time-varying and inter-subject variability, EEG signals are difficult to analyze [2]. At present, the common spatial pattern (CSP) algorithm and its extensions are the most effective feature extraction methods of discriminating different patterns of EEG signals [3]. CSP is designed for two-class BCIs, which completes feature extraction by diagonalizing two covariance matrices simultaneously to construct an optimal filter [4]. Various multi-class approaches to extend CSP have been reported and have been shown to yield good experimental results [5].

© IFIP International Federation for Information Processing 2018
Published by Springer Nature Switzerland AG 2018. All Rights Reserved
Z. Shi et al. (Eds.): ICIS 2018, IFIP AICT 539, pp. 3–10, 2018.
https://doi.org/10.1007/978-3-030-01313-4_1

In recent years, Deep learning has increasingly gained attention in the field of machine learning and artificial intelligence, and has been successfully applied in many engineering problems. Zheng [6] has used deep belief networks (DBN) to complete emotion classification based on EEG and achieve satisfactory results. Tabar [7] has investigated convolutional neural networks and stacked autoencoders (SAE) to classify motor imagery EEG signals. However, most of deep learning networks suffer from the time-consuming training process. Therefore, deep learning algorithms are generally difficult to meet the BCI system with high real-time requirements.

Single layer feedforward neural networks (SLFN) have been widely applied to solve problems such as time-consuming and low accuracy [8]. The random vector functional-link neural network (RVFLNN) is a member of the SLFN that is effectively eliminates the drawback of the long training process and also it provides the generalization capability in function approximation [8]. But RVFLNN could not work well on remodeling high-volume and time-variety data in modern large data, a latest algorithm known as broad learning (BL) has been proposed by Philip [9] in 2017, which aims to offer an alternative way for deep learning and structure. BL is designed for the network through the broad expansion in both the feature nodes and the enhancement nodes, and then the incremental learning approach is developed for fast remodeling in broad expansion without a retraining process.

In this paper, we explore the use of broad learning techniques for MI EEG signal classification. Although the BL method has strong adaptability, direct learning may not be able to extract the essential features of the EEG signal. Considering that CSP can effectively extract discriminatory information in multichannel EEG data associated with motor imagery and BL has fast training speed and good classification accuracy, a combination of them are applied to multiclass EEG classification. The rest of this paper is organized as follows. Section 2 presents briefly the related work consisting CSP and BL, and then gives the details of the proposed method. Section 3 then gives the experimental results and analysis on multiclass MI EEG data from BCI Competition IV Dataset 2a. Finally, the conclusion of this work is summarized in Sect. 4.

2   Methods

2.1   Multiclass Common Spatial Pattern

In binary CSP, diagonalizing two covariance matrices simultaneously that the EEG signal's variance between different modes can be maximized. The one-versus-the-rest based CSP (OVR-CSP) approach computes spatial patterns for each class against all others. It treats one of the situations as a class, and the remaining situation as another class, so that it can be converted into binary CSP for each type of model.

EEG signal matrices X_i $(i = 1, \ldots, C)$ with dimension of N by T, where N is the number of channel, T is the number of sampling points per channel, and C is the number of MI tasks. The normalized covariance matrix for each class of EEG signal is denoted as:

$$R_i = \frac{X_i X_i^T}{\text{tr}\left(X_i X_i^T\right)}, \quad i = 1, \dots, C \tag{1}$$

where X_i^T is the transpose of X_i and tr($\bullet$) represents the trace of the matrix. Sum the covariance matrix to get $R = \sum_{i=1}^{C} R_i$ and eigendecomposition as follows:

$$R = U_0 \Lambda U_0^T \tag{2}$$

where U_0 is the $N \times N$ unitary matrix of principal components, and Λ is the $N \times N$ diagonal matrix of eigenvalues. The whitening transformation matrix is then formed as:

$$H = \Lambda^{-1/2} U_0^T \tag{3}$$

Let $R_1' = R_2 + \dots + R_C$ represent another class relative to R_1, and $S_1 = H_1 R_1 H_1^T$, $S_1' = H_1 R_1' H_1^T$. S_1 can be decomposed as $S_1 = U_1 \Lambda_1 U_1^T$, S_1' can be decomposed as $S_1' = U_1 \Lambda_1' U_1^T$, and $\Lambda_1 + \Lambda_1' = I$, so the relationship of covariance matrix is expressed as:

$$\left(H_1^T U_1\right)^T R_1 \left(H_1^T U_1\right) + \left(H_1^T U_1\right)^T R_1' \left(H_1^T U_1\right) = I \tag{4}$$

The eigenvector corresponding to the first m largest eigenvalues in U_1 is selected to design a spatial filter for the first type of mode. The projection direction can be expressed as:

$$P_1 = U_{1,m}^T H_1 \tag{5}$$

Projection direction P_j of various motor imagery tasks can be calculated using Eq. (5). Sample X is projected with the projection direction in j-type mode. $Z^j = P_j X \in R^{m \times T}$ is the filtered signal, and $Z = [Z^1, \dots, Z^C] \in R^{M \times T}$ is a set of filtered signal where $M = C \times m$. The variance of each component in Z is normalized, and it can be calculated by logarithm as:

$$f_p = \log\left(\frac{\text{var}(z_p)}{\sum_{q=1}^{M} \text{var}(z_p)}\right), \quad p = 1, \dots, M \tag{6}$$

where var(z_p) represents the variance of the pth row component in Z. $F = [f_1, \dots, f_M]$ is the feature vector of the sample, which is a set of normalized components from Z.

2.2 Broad Learning

In this section, we will elaborate on the specific implementation of this algorithm. The input of BL is first mapped to create a set of transferred feature that is the basic part of the enhancement node, and the incremental learning technique is used to update BL dynamically that can achieve satisfactory performance in training accuracy.

Assume the input data set X contains N samples and each sample is m-dimensional. For n feature mappings, each mapping generates k nodes, and feature mappings can be represented as the equation of the form

$$Z_i = \phi(XW_{ei} + \beta_{ei}), \ i = 1, \dots, n \tag{7}$$

where W_{ei} and β_{ei} are randomly generated matrices. The feature nodes are summarized as a set of nodes given by $Z^n = [Z_1, \dots, Z_n]$. Similarly, the m-th group of enhancement node can be expressed as:

$$H_m = \xi(Z^n W_{hm} + \beta_{hm}) \tag{8}$$

And the enhancement nodes are also summarized as $H^m = [H_1, \dots, H_m]$.

Next, the output expression is constructed for width learning as follows

$$\begin{aligned}
Y &= [Z_1, \dots, Z_n | \xi(Z^n W_{h1} + \beta_{h1}), \dots, \xi(Z^n W_{hm} + \beta_{hm})] W^m \\
&= [Z_1, \dots, Z_n | H_1, \dots, H_m] W^m \\
&= [Z^n | H^m] W^m
\end{aligned} \tag{9}$$

where Y is the output of the broad learning algorithm, and $W^m = [Z^n | H^m]^+ Y$ are the connecting weights for the broad structure and can be easily computed through the ridge regression approximation of $[Z^n | H^m]^+$ using Eq. (10). We can update the weight of the model by using the idea of incremental learning algorithm, and the classifier model is also updated.

The pseudoinverse of the involved matrix are calculated by

$$A^+ = \lim_{\lambda \to 0} (\lambda I + AA^T)^{-1} A^T \tag{10}$$

The main idea of the BL algorithm is to achieve the required training accuracy as we expect by increasing the number of enhancement nodes or feature nodes. In this paper, the increases p enhancement nodes can be completed by Eq. (11).

Denote $A^m = [Z^n | H^m]$ and A^{m+1} as

$$A^{m+1} = [A^m | \xi(Z^n W_{h_{m+1}} + \beta_{h_{m+1}})] \tag{11}$$

where $W_{h_{m+1}} \in \mathfrak{R}^{nk \times p}$ is the weight vector, and $\beta_{h_{m+1}} \in \mathfrak{R}^p$ is the bias. Both of these quantities are randomly generated. The new generation of weight and biases is generated by mapping features to the p enhancement nodes.

In [10], RVFLNN used the stepwise updating algorithm for adding a new enhancement node to the network.

The pseudoinverse of A^{m+1} is calculated as

$$\left(A^{m+1}\right)^{+} = \begin{bmatrix} (A^{m})^{+} - DB^{T} \\ B^{T} \end{bmatrix} \tag{12}$$

where $D = (A^{m})^{+}\xi\left(Z^{n}W_{h_{m+1}} + \beta_{h_{m+1}}\right)$.

$$B^{T} = \begin{cases} (C)^{+} & if \quad C \neq 0 \\ \left(1+D^{T}D\right)^{-1}B^{T}(A^{m})^{+} & if \quad C = 0 \end{cases} \tag{13}$$

If A^{m} is of the full rank, then $C = 0$ and no computation of pseudoinverse is involved in updating the pseudoinverse $(A^{m})^{+}$ or weight matrix W^{m}, and $C = \xi\left(Z^{n}W_{h_{m+1}} + \beta_{h_{m+1}}\right) - A^{m}D$.

Then, the new weights are calculated as

$$W^{m+1} = \begin{bmatrix} W^{m} - DB^{T}Y \\ B^{T}Y \end{bmatrix} \tag{14}$$

To this end, the most important work of BL is finished that adds p additional enhancement nodes.

2.3 Our Algorithm

The whole process of this work includes two stages: feature extraction and feature classification. The motor imagery EEG classification system can be described as the following steps:

Step 1: The features of motor imagery EEG signal are mainly distributed in the frequency range of 8–30 Hz, and thus the Butter-worth bandpass filter of 8–30 Hz is used to preprocess the data.

Step 2: The OVR-CSP algorithm is used to extract the features of the filtered training data in Step 1. The projection direction of each filter can be obtained using Eq. (5). The training samples are projected in these projection directions and then Eq. (6) is used to calculate the feature vector F.

Step 3: The extracted feature vector is used as the input of the BL algorithm, and each input vector is mapped as a feature node or an enhancement node, and the weight W between the input and the output of network is calculated using these nodes and data labels.

Step 4: When the training error threshold is not satisfied, we need to increase the number of enhancement nodes to improve the performance of the network, and the weights need to be updated using Eq. (14). By gradually adjusting the value of the weight matrix, the classifier model is approached in the end.

Step 5: For a test sample, its feature is extracted according to Step 2, and then putted it into the BL classifier obtained by Step 4 to get its classification result.

3 Experiments and Discussion

In this section, the experiments on real world EEG data from BCI Competition IV Dataset 2a were performed to verify the validity and practicability of our proposed algorithm, as compared with the other state-of-the-art approaches including ELM [11], HELM [12], DBN [13] and SAE [14].

3.1 Datasets and Settings

The data of BCI Competition IV Dataset 2a was obtained from motor imagery experiment with normal subjects, in which the task was to control a feedback bar (in a screen, according to a cue randomly provided) by means of imagination of several specific movements. It contains data acquired from 9 healthy subjects that execute four-class mental imagery tasks, namely the imagination of movement of the left hand, right hand, both feet, and tongue. Two sessions, one for training and the other for evaluation. Each session comprised 288 trials of data recorded with 22 EEG channels and 3 monopolar electrooculogram (EOG) channels.

Considering the complexity of EEG signal, the training and test samples need to be preprocessed, and the effect of parameters C and s on the results should be considered, and the optimal values of parameters are obtained by using the artificial fish swarm algorithm to iterate 20 times. After determining the $C = 0.12267$ and $s = 0.89107$, the number of feature windows N_1, nodes of each window N_2 and enhancement layer nodes N_3 in the BL algorithm also should be determined. In this paper, the best case is determined as $N_1 = 9$, $N_2 = 8$ and $N_3 = 55$.

After that, the number of enhancement nodes in the BL algorithm is increased with 2 each to improve the training accuracy. For each dataset, the training procedure is stopped when training error threshold is satisfied. The test experiments will be conducted in ten runs on each learning method and the average results are provided.

3.2 Experimental Results

In the experiment of this paper, several algorithms including ELM, DBN, SAE and HELM are selected for comparison. The results are shown in Table 1.

Table 1. Classification accuracy of each algorithm on data set BCI Competition IV Dataset 2a

Methods	ELM	DBN	SAE	HELM	Our method
A01	78.18 ± 0.0015	72.05 ± 0.0241	74.08 ± 0.0173	$\mathbf{81.00 \pm 0.0001}$	79.16 ± 0.0104
A02	47.65 ± 0.0021	46.39 ± 0.0236	45.50 ± 0.0058	48.11 ± 0.0013	49.37 ± 0.0125
A03	78.49 ± 0.0005	$\mathbf{80.07 \pm 0.0052}$	78.78 ± 0.0001	76.93 ± 0.0010	78.44 ± 0.0096
A04	62.88 ± 0.0019	60.97 ± 0.0272	60.28 ± 0.0028	62.93 ± 0.0008	64.99 ± 0.0131
A05	37.78 ± 0.0008	38.19 ± 0	37.94 ± 0.0018	37.57 ± 0.0012	39.20 ± 0.0096
A06	48.97 ± 0.0025	44.27 ± 0.0017	43.83 ± 0.0137	$\mathbf{50.89 \pm 0.0011}$	49.83 ± 0.0113
A07	81.97 ± 0.0010	76.18 ± 0.0181	76.01 ± 0.0077	80.30 ± 0.0011	83.54 ± 0.0123
A08	82.46 ± 0.0013	80.83 ± 0.0247	77.77 ± 0.0027	81.53 ± 0.0010	82.92 ± 0.0048
A09	73.88 ± 0.0071	70.08 ± 0.0307	$\mathbf{79.20 \pm 0.0101}$	73.92 ± 0.0071	75.17 ± 0.0129
Mean	65.81	63.22	63.71	65.90	66.91

In the Table 1, the results showed that our method yielded the best mean testing accuracy. Specifically, our method gained the best mean accuracy on subjects A02, A04, A05, A07, and A08, while HELM performed best on subjects A01 and A06, and DBN achieved the best result on subject A03, and SAE performed best on subject A09. The mean accuracy of our algorithm is 1.1% higher than the ELM algorithm, 3.2% higher than the SAE algorithm, 0.99% higher than the HELM algorithm and 3.69% higher accuracy than the DBN algorithm.

Training time-consuming is used to demonstrate the time complexity of an algorithm. So, Table 2 presents the training time of different algorithms.

Table 2. The training time of each algorithm in the dataset BCI Competition IV Dataset 2a

Methods	ELM	DBN	SAE	HELM	BL
Train time (s)	0.0127	0.0247	0.0137	0.2550	0.0153

Among the five classification algorithms, ELM is the most efficient one, while SAE, BL and DBN are relatively comparable, and HELM is the least time efficient. The training time of the proposed method is less than DBN, and H-ELM. These results show that our method can achieve excellent trade-off between classification accuracy and computational cost.

3.3 Discussion

The proposed method exhibited an excellent performance in both classification and computational efficiency, which verifies the effectiveness of the BL algorithm in EEG signal classification. When compared with the ELM as well as DBN, SAE and H-ELM with deep architecture, BL achieved relatively better performance by its novel strategy of weight updates. Furthermore, our method gained the best mean testing accuracy on subjects A02, A04, A05, A07 and A08, but the individual difference in EEG signals cannot be completely eliminated. For example, the SAE achieved better mean testing accuracy on subject A09. In terms of computational efficiency, the dataset with small sample size is used in this paper to experiment.

4 Conclusion

In this paper, we have proposed a novel multi-task motor imagery EEG classification framework using BL and OVR-CSP. Different from the general methods, the proposed method adopted the width architecture of BL and learned from the input data using the CSP method to extract the essential features of the signal. This framework yielded the best mean testing accuracy in all five methods. It is observed that learning time of our method is smaller than some multilayer architecture of deep learning simultaneously. However, the BL algorithm used in this paper only considers the impact of the increase of the enhancement nodes, and the combined impact of the feature nodes and enhancement nodes is not considered. It may affect the performance of the classifier. Therefore, the follow-up work is necessary to research.

References

1. Schalk, G., Mcfarland, D.J., Hinterberger, T., et al.: BCI2000: a general-purpose brain-computer interface (BCI) system. IEEE Trans. Biomed. Eng. **51**, 1034 (2004)
2. Pfurtscheller, G., Neuper, C.: Motor imagery and direct brain-computer communication. Proc. IEEE **89**, 1123–1134 (2001)
3. Aghaei, A.S., Mahanta, M.S., Plataniotis, K.N.: Separable common spatio-spectral patterns for motor imagery BCI systems. IEEE Trans. Biomed. Eng. **63**, 15–29 (2015)
4. Li, M.A., Liu, J.Y., Hao, D.M.: EEG recognition of motor imagery based on improved CSP algorithm. Chin. J. Biomed. Eng. **28**, 161–165 (2009)
5. Zheng, Y.C., Kai, K.A., Wang, C., et al.: Multi-class filter bank common spatial pattern for four-class motor imagery BCI. In: International Conference of the IEEE Engineering in Medicine & Biology Society, p. 571 (2009)
6. Zheng, W.L., Zhu, J.Y., Peng, Y., et al.: EEG-based emotion classification using deep belief networks. In: IEEE International Conference on Multimedia and Expo, pp. 1–6 (2014)
7. Tabar, Y.R., Halici, U.: A novel deep learning approach for classification of EEG motor imagery signals. J. Neural Eng. **14**, 016003 (2016)
8. Igelnik, B., Pao, Y.H.: Stochastic choice of basis functions in adaptive function approximation and the functional-link net. IEEE Trans. Neural Netw. **6**, 1320–1329 (1995)
9. Philip, C.C., Liu, Z.: Broad learning system: an effective and efficient incremental learning system without the need for deep architecture. IEEE Trans. Neural Netw. Learn. Syst. **29**, 10–24 (2018)
10. Chen, C.P., Wan, J.Z.: A rapid learning and dynamic stepwise updating algorithm for flat neural networks and the application to time-series prediction. IEEE Trans. Syst. Man Cybern. Part B Cybern. **29**, 62 (1999)
11. Huang, G.B., Zhou, H., Ding, X., et al.: Extreme learning machine for regression and multiclass classification. IEEE Trans. Syst. Man Cybern. Part B **42**, 513–529 (2012)
12. Tang, J., Deng, C., Huang, G.B.: Extreme learning machine for multilayer perceptron. IEEE Trans. Neural Netw. Learn. Syst. **27**, 809–821 (2017)
13. Hinton, G.E., Osindero, S., Teh, Y.W.: A fast learning algorithm for deep belief nets. Neural Comput. **18**, 1527–1554 (2016)
14. Tao, C., Pan, H., Li, Y., et al.: Unsupervised spectral–spatial feature learning with stacked sparse autoencoder for hyperspectral imagery classification. IEEE Geosci. Remote Sens. Lett. **12**, 2438–2442 (2015)

From Bayesian Inference to Logical Bayesian Inference

A New Mathematical Frame for Semantic Communication and Machine Learning

Chenguang Lu[(✉)] [iD]

College of Intelligence Engineering and Mathematics,
Liaoning Engineering and Technology University, Fuxin 123000, Liaoning, China
lcguang@foxmail.com

Abstract. Bayesian Inference (BI) uses the Bayes' posterior whereas Logical Bayesian Inference (LBI) uses the truth function or membership function as the inference tool. LBI is proposed because BI is not compatible with the classical Bayes' prediction and does not use logical probability and hence cannot express semantic meaning. In LBI, statistical probability and logical probability are strictly distinguished, used at the same time, and linked by the third kind of Bayes' Theorem. The Shannon channel consists of a set of transition probability functions whereas the semantic channel consists of a set of truth functions. When a sample is large enough, we can directly derive the semantic channel from Shannon's channel. Otherwise, we can use parameters to construct truth functions and use the Maximum Semantic Information (MSI) criterion to optimize the truth functions. The MSI criterion is equivalent to the Maximum Likelihood (ML) criterion, and compatible with the Regularized Least Square (RLS) criterion. By matching the two channels one with another, we can obtain the Channels' Matching (CM) algorithm. This algorithm can improve multi-label classifications, maximum likelihood estimations (including unseen instance classifications), and mixture models. In comparison with BI, LBI (1) uses the prior $P(X)$ of X instead of that of Y or θ and fits cases where the source $P(X)$ changes, (2) can be used to solve the denotations of labels, and (3) is more compatible with the classical Bayes' prediction and likelihood method. LBI also provides a confirmation measure between -1 and 1 for induction.

Keywords: Bayes' Theorem · Bayesian inference · MLE · MAP
Semantic information · Machine learning · Confirmation measure · Induction

---

The original version of this chapter was revised: An error in Equation (31) has been corrected. The correction to this chapter is available at https://doi.org/10.1007/978-3-030-01313-4_51

© IFIP International Federation for Information Processing 2018
Published by Springer Nature Switzerland AG 2018. All Rights Reserved
Z. Shi et al. (Eds.): ICIS 2018, IFIP AICT 539, pp. 11–23, 2018.
https://doi.org/10.1007/978-3-030-01313-4_2

1   Introduction[1]

Bayesian Inference (BI) [1, 2] was proposed by Bayesians. Bayesianism and Frequentism are contrary [3]. Frequentism claims that probability is objective and can be defined as the limit of the relative frequency of an event, whereas Bayesianism claims that probability is subjective or logical. Some Bayesians consider probability as degree of belief [3] whereas others, such as Keynes [4], Carnap [5], and Jaynes [6], so-called logical Bayesians, consider probability as the truth value. There are also minor logical Bayesians, such as Reichenbach [7] as well as the author of this paper, who use frequency to explain the logical probability and truth function.

Many frequentists, such as Fisher [8] and Shannon [9], also use Bayes' Theorem, but they are not Bayesians. Frequentist main tool for hypothesis-testing is Likelihood Inference (LI), which has achieved great successes. However, LI cannot make use of prior knowledge. For example, after the prior distribution $P(x)$ of an instance x is changed, the likelihood function $P(x|\theta_j)$ will be no longer valid. To make use of prior knowledge and to emphasize subjective probability, some Bayesians proposed BI [1] which uses the Bayesian posterior $P(\theta|\mathbf{X})$, where $\mathbf{X}$ is a sequence of instances, as the inference tool. The Maximum Likelihood Estimation (MLE) was revised into the Maximum A Posterior estimation (MAP) [2]. Demonstrating some advantages especially for working with small samples and for solving the frequency distribution of a frequency producer, BI also has some limitations. The main limitations are: (1) It is incompatible with the classical Bayes' prediction as shown by Eq. (1); (2) It does not use logical probabilities or truth functions and hence cannot solve semantic problems. To overcome these limitations, we propose Logical Bayesian Inference (LBI), following earlier logical Bayesians to use the truth function as the inference tool and following Fisher to use the likelihood method. The author also set up new mathematical frame employing LBI to improve semantic communication and machine learning.

LBI has the following features:

- It strictly distinguishes statistical probability and logical probability, uses both at the same time, and links both by the third kind of Bayes' Theorem, with which the likelihood function and the truth function can be converted from one to another.
- It also uses frequency to explain the truth function, as Reichenbach did, so that optimized truth function can be used as transition probability function $P(y_j|x)$ to make Bayes' prediction even if $P(x)$ is changed.
- It brings truth functions and likelihood functions into information formulas to obtain the generalized Kullback-Leibler (KL) information and the semantic mutual information. It uses the Maximum Semantic Information (MSI) criterion to optimize truth functions. The MSI criterion is equivalent to the Maximum Likelihood (ML) criterion and compatible with the Regularized Least Squares (RLS) criterion [10].

Within the new frame, we convert sampling sequences into sampling distributions and then use the cross-entropy method [10]. This method has become popular in recent

---

[1] This paper is a condensed version in English. The original Chinese version: http://survivor99.com/lcg/CM/Homepage-NewFrame.pdf.

two decades because it is suitable to larger samples and similar to information theoretical method. This study is based on the author's studies twenty years ago on semantic information theory with the cross-entropy and mutual cross-entropy as tools [11–14]. This study also relates to the author's recent studies on machine learning for simplifying multi-label classifications [15], speeding the MLE for tests and unseen instance classifications [16], and improving the convergence of mixture models [17].

In the following sections, the author will discuss why LBI is employed (Sect. 2), introduce the mathematical basis (Sect. 3), state LBI (Sect. 4), introduce its applications to machine learning (Sect. 5), discuss induction (Sect. 6), and summarize the paper finally.

2   From Bayes' Prediction to Logical Bayesian Inference

Definition 1.

- x: an instance or data point; X: a random variable; $X = x \in U = \{x_1, x_2, ..., x_m\}$.
- y: a hypothesis or label; Y: a random variable; $Y = y \in V = \{y_1, y_2, ..., y_n\}$.
- θ: a model or a set of model parameters. For given y_j, $\theta = \theta_j$.
- X_j: a sample or sequence of data point $x(1), x(2), ..., x(N_j) \in U$. The data points come from Independent and Identically Distributed (IID) random variables.
- D: a sample or sequence of examples $\{(x(t), y(t))|t = 1$ to $N; x(t) \in U; y(t) \in V\}$, which includes n different sub-samples X_j. If D is large enough, we can obtain distribution $P(x, y)$ from D, and distribution $P(x|y_j)$ from X_j.

A Shannon's channel $P(y|x)$ consists of a set of Transition Probability Functions (TPF) $P(y_j|x)$, $j = 1, 2, ...$ A TPF $P(y_j|x)$ is a good prediction tool. With the Bayes' Theorem II (discussed in Sect. 3), we can make probability prediction $P(x|y_j)$ according to $P(y_j|x)$ and $P(x)$. Even if $P(x)$ changes into $P'(x)$, we can still obtain $P'(x|y_j)$ by

$$P'(x|y_j) = P(y_j|x)P'(x)/ \sum_i P(y_j|x_i)P'(x_i) \tag{1}$$

We call this probability prediction as "classical Bayes' prediction". However, if samples are not large enough, we cannot obtain continuous distributions $P(y_j|x)$ or $P(x|y_j)$. Therefore, Fisher proposed Likelihood Inference (LI) [7].

For given X_j, the likelihood of θ_j is

$$P(X_j|\theta_j) = P(x(1), x(2), ..., x(N)|\theta_j) = \prod_{t=1}^{N_j} P(x(t)|\theta_j) \tag{2}$$

We use θ_j instead of θ in the above equation because unlike the model θ in BI, the model in LI does not have a probability distribution. If there are N_{ji} x_i in X_j, then $P(x_i|y_j) = N_{ji}/N_j$, and the likelihood can be expressed by a negative cross entropy:

$$\log P(\mathbf{X}_j|\theta_j) = \log \prod_i P(x_i|\theta_j)^{N_{ji}}$$

$$= N_j \sum_i P(x_i|y_j) \log P(x_i|\theta_j) = -N_j H(X|\theta_j) \tag{3}$$

For conditional sample $\mathbf{X}_j$ whose distribution is $P(x|j)$ (the label is uncertain), we can find the MLE:

$$\theta_j^* = \arg\max_{\theta_j} P(\mathbf{X}_j|\theta_j) = \arg\max_{\theta_j} \sum_i P(x_i|j) \log P(x_i|\theta_j) \tag{4}$$

When $P(x|\theta_j) = P(x|j)$, $H(X|\theta_j)$ reaches its minimum.

The main limitation of LI is that it cannot make use of prior knowledge, such as $P(x)$, $P(y)$, or $P(\theta)$, and does not fit cases where $P(x)$ may change. BI brings the prior distribution $P(\theta)$ of θ into the Bayes' Theorem II to have [2]

$$P(\theta|\mathbf{X}) = \frac{P(\mathbf{X}|\theta)P(\theta)}{P_\theta(\mathbf{X})}, \; P_\theta(\mathbf{X}) = \sum_j P(\mathbf{X}|\theta_j)P(\theta_j) \tag{5}$$

where $P_\theta(\mathbf{X})$ is the normalized constant related to θ. For one Bayesian posterior, we need n or more likelihood functions. The MLE becomes the MAP:

$$\theta_j^* = \arg\max_{\theta_j} P(\theta|\mathbf{X}_j) = \arg\max_{\theta_j}\left[\sum_i P(x_i|j) \log P(x_i|\theta_j) + \log P(\theta_j)\right] \tag{6}$$

where $P_\theta(\mathbf{X})$ is neglected. It is easy to find that (1) if $P(\theta)$ is neglected or is an equiprobable distribution, the MAP is equivalent to the MLE; (2) while the sample's size N increases, the MAP gradually approaches the MLE.

There is also the Bayesian posterior of Y:

$$P(Y|\mathbf{X}, \theta) = \frac{P(\mathbf{X}|\theta)P(Y)}{P_\theta(\mathbf{X})}, \; P_\theta(\mathbf{X}) = \sum_j P(\mathbf{X}|\theta_j)P(y_j) \tag{7}$$

It is different from $P(\theta|\mathbf{X})$ because $P(Y|\mathbf{X}, \theta)$ is a distribution over the label space whereas $P(\theta|\mathbf{X})$ is a distribution over the parameter space. The parameter space is larger than the label space. $P(Y|\mathbf{X}, \theta)$ is easier understood than $P(\theta|X)$. $P(Y|\mathbf{X}, \theta)$ is also often used, such as for mixture models and hypothesis-testing.

BI has some advantages: (1) It considers the prior of Y or θ so that when $P(X)$ is unknown, $P(Y)$ or $P(\theta)$ is also useful, especially for small samples. (2) It can convert the current posterior $P(\theta|X)$ into the next prior $P(\theta)$. (3) The distribution $P(\theta|X)$ over θ space will gradually concentrate as the sample's size N increases. When $N \rightarrow \infty$, only $P(\theta^*|X) = 1$, where θ^* is the MAP. So, $P(\theta|X)$ can intuitively show learning results.

However, there are also some serious problems with BI:

(1) **About Bayesian prediction**

BI predicts the posterior and prior distributions of x by [2]:

$$P_\theta(x|\mathbf{X}) = \sum_j P(x|\theta_j)P(\theta_j|\mathbf{X}) \text{ and } P_\theta(x) = \sum_j P(x|\theta_j)P(\theta_j) \tag{8}$$

From a huge sample **D**, we can directly obtain $P(x|y_j)$ and $P(x)$. However, BI cannot ensure $P_\theta(x|\mathbf{X}) = P(x|y_j)$ or $P_\theta(x) = P(x)$. After $P(x)$ changes into $P'(x)$, BI cannot obtain the posterior that is equal to $P'(x|y_j)$ in Eq. (1). Hence, the Bayesian prediction is not compatible with the classical Bayes' prediction. Therefore, we need an inference tool that is like the TPF $P(y_j|x)$ and is constructed with parameters.

(2) **About logical probability**

BI does not use logical probability because logical probability is not normalized; nevertheless, all probabilities BI uses are normalized. Consider labels "Non-rain", "Rain", "Light rain", "Moderate rain", "Light to moderate rain", ..., in a weather forecast. The sum of their logical probabilities is greater than 1. The conditional logical probability or truth value of a label with the maximum 1 is also not normalized. BI uses neither truth values nor truth functions and hence cannot solve the denotation (or semantic meaning) of a label. Fuzzy mathematics [18, 19] uses membership functions, which can also be used as truth functions. Therefore, we need an inference method that can derive truth functions or membership functions from sampling distributions.

(3) **About prior knowledge**

In BI, $P(\theta)$ is subjective. However, we often need objective prior knowledge. For example, to make probability prediction about a disease according to a medical test result "positive" or "negative", we need to know the prior distribution $P(x)$ [16]. To predict a car's real position according to a GPS indicator on a GPS map, we need to know the road conditions, which tell $P(x)$[1].

(4) **About optimization criterion**

According to Popper's theory [20], a hypothesis with less LP can convey more information. Shannon's information theory [9] contains a similar conclusion. The MAP is not well compatible with the information criterion.

The following example can further explain why we need LBI.

Example 1. Given the age population prior distribution $P(x)$ and posterior distribution $P(x|$ "adult" is true), which are continuous, please answer:

(1) How do we obtain the denotation (e.g. the truth function) of "adult" (see Fig. 1)?
(2) Can we make a new probability prediction or produce new likelihood function with the denotation when $P(x)$ is changed into $P'(x)$?
(3) If the set {Adult} is fuzzy, can we obtain its membership function?

It is difficult to answer these questions using either LI or BI. Nevertheless, using LBI, we can easily obtain the denotation and make the new probability prediction.

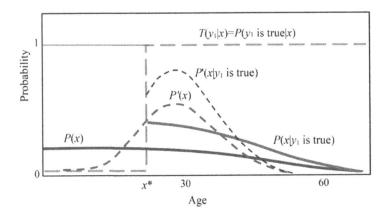

Fig. 1. Solving the denotation of y_1 = "x is adult" and probability prediction $P'(x|y_1$ is true).

3   Mathematical Basis: Three Kinds of Probabilities and Three Kinds of Bayes' Theorems

All probabilities [3] can be divided into three kinds:

(1) Statistical probability: relative frequency or its limit of an event;
(2) Logical Probability (LP): how frequently a hypothesis is judged true or how true a hypothesis is;
(3) Predicted (or subjective) probability: possibility or likelihood.

We may treat the predicted probability as the hybrid of the former two kinds. Hence, there are only two kinds of basic probabilities: the statistical and the logical.

A Hypothesis or Label Has Both Statistical (or Selected) Probability and LP. They are very different. Consider labels in a weather forecast: "Light rain" and "Light to heavy rain". The former has larger selected probability and less LP. The LP of a tautology, such as "He is old or not old", is 1 whereas its selected probability is close to 0.

Each of existing probability systems [3, 5–7] only contains one kind of probabilities. Now we define a probability system with both statistical probabilities and LPs.

Definition 2. A label y_j is also a predicate $y_j(X) = $ "$X \in A_j$." For y_j, U has a subset A_j, every x in which makes y_j true. Let $P(Y = y_j)$ denote the statistical probability of y_j, and $P(X \in A_j)$ denote the LP of y_j. For simplicity, let $P(y_j) = P(Y = y_j)$ and $T(A_j) = P(X \in A_j)$.

We call $P(X \in A_j)$ the LP because according to Tarski's theory of truth [21], $P(X \in A_j)$ = $P("X \in A_j"$ is true) = $P(y_j$ is true). Hence, the conditional LP $T(A_j|X)$ of y_j for given X is the feature function of A_j and the truth function of y_j. Hence the LP of y_j is

$$T(A_j) = \sum_i P(x_i)T(A_j|x_i) \qquad (9)$$

According to Davidson's truth-conditional semantics [22], $T(A_j|X)$ ascertains the semantic meaning of y_j. Note that statistical probability distributions, such as $P(Y)$, $P(Y|x_i)$, $P(X)$, and $P(X|y_j)$, are normalized; however, LP distributions are not normalized. In general, $T(A_1) + T(A_2) + \dots + T(A_n) > 1$; $T(A_1|x_i) + T(A_2|x_i) + \dots + T(A_n|x_i) > 1$.

If A_j is fuzzy, $T(A_j|X)$ becomes the membership function, and $T(A_j)$ becomes the fuzzy event probability defined by Zadeh [19]. For fuzzy sets, we use θ_j to replace A_j. Then $T(\theta_j|X)$ becomes the membership function of θ_j. That means

$$m_{\theta_j}(X) = T(\theta_j|X) = T(y_j|X) \tag{10}$$

We can also treat θ_j as a sub-model of a predictive model θ. In this paper, the likelihood function $P(X|\theta_j)$ is equal to $P(X|y_j; \theta)$ in popular likelihood method. $T(\theta_j|X)$ is different from $P(\theta|X)$ and is longitudinally normalized, e.g.,

$$\max\left(T(\theta_j|X)\right) = \max\left(T(\theta_j|x_1), T(\theta_j|x_2), \dots, T(\theta_j|x_m)\right) = 1 \tag{11}$$

There are three kinds of Bayes' Theorems, which are used by Bayes [23], Shannon [9], and the author respectively.

Bayes' Theorem I (used by Bayes): Let sets $A, B \in 2^U$, A^c be the complementary set of A, $T(A) = P(X \in A)$, and $T(B) = P(X \in B)$. Then

$$T(B|A) = T(A|B)T(B)/T(A), T(A) = T(A|B)T(B) + T(A|B^c)T(B^c) \tag{12}$$

There is also a symmetrical formula for $T(A|B)$. Note there are only one random variable X and two logical probabilities $T(A)$ and $T(B)$.

Bayes' Theorem II (used by Shannon):

$$P(X|y_j) = P(y_j|X)P(X)/P(y_j), \; P(y_j) = \sum_i P(x_i)P(y_j|x_i) \tag{13}$$

There is also a symmetrical formula for $P(y_j|X)$ or $P(Y|x_i)$. Note there are two random variables and two statistical probabilities.

Bayes' Theorem III:

$$P(X|\theta_j) = T(\theta_j|X)P(X)/T(\theta_j), \quad T(\theta_j) = \sum_i P(x_i)T(\theta_j|x_i) \tag{14}$$

$$T(\theta_j|X) = P(X|\theta_j)T(\theta_j)/P(X), \; T(\theta_j) = 1/\max(P(X|\theta_j)/P(X)) \tag{15}$$

The two formulas are asymmetrical because there is a statistical probability and a logical probability. $T(\theta_j)$ in Eq. (15) may be called longitudinally normalizing constant.

The Proof of Bayes' Theorem III: The joint probability $P(X, \theta_j) = P(X = x, X \in \theta_j)$, then $P(X|\theta_j)T(\theta_j) = P(X = x, X \in \theta_j) = T(\theta_j|X)P(X)$. Hence there is

$$P(X|\theta_j) = P(X)T(\theta_j|X)/T(\theta_j), \quad T(\theta_j|X) = T(\theta_j)P(X|\theta_j)/P(X)$$

Since $P(X|\theta_j)$ is horizontally normalized, $T(\theta_j) = \sum_i P(x_i)T(\theta_j|x_i)$. Since $T(\theta_j|X)$ is longitudinally normalized, there is

$$1 = \max[T(\theta_j)P(X|\theta_j)/P(X)] = T(\theta_j)\max[P(X|\theta_j)/P(X)]$$

Hence $T(\theta_j) = 1/\max[P(X|\theta_j)/P(X)]$. **QED.**

Equation (15) can also be expressed as

$$T(\theta_j|X) = [P(X|\theta_j)/P(X)]/\max[P(X|\theta_j)/P(X)] \tag{16}$$

Using this formula, we can answer questions of Example 1 in Sect. 2 to obtain the denotation of "Adult" and the posterior distribution $P'(x|y_1$ is true) as shown in Fig. 1.

4   Logical Bayesian Inference (LBI)

LBI has three tasks:

(1) To derive truth functions or a semantic channel from **D** or sampling distributions (e.g., multi-label learning [24, 25]);
(2) To select hypotheses or labels to convey information for given x or $P(X|j)$ according to the semantic channel (e.g., multi-label classification);
(3) To make logical Bayes' prediction $P(X|\theta_j)$ according to $T(\theta_j|X)$ and $P(X)$ or $P'(X)$. The third task is simple. We can use Eq. (14) for this task.

For the first task, we first consider continuous sampling distributions from which we can obtain the Shannon channel $P(Y|X)$. The TPF $P(y_j|X)$ has an important property: $P(y_j|X)$ by a constant k can make the same probability prediction because

$$\frac{P'(X)kP(y_j|X)}{\sum_i P'(x_i)kP(y_j|x_i)} = \frac{P'(X)P(y_j|X)}{\sum_i P'(x_i)P(y_j|x_i)} = P'(X|y_j) \tag{17}$$

A semantic channel $T(\theta|X)$ consists of a set of truth functions $T(\theta_j|X)$, $j = 1$, 2, ..., n. According to Eq. (17), if $T(\theta_j|X) \propto P(y_j|X)$, then $P(X|\theta_j) = P(X|y_j)$. Hence the optimized truth function is

$$T^*(\theta_j|X) = P(y_j|X)/\max(P(y_j|X)) \tag{18}$$

We can prove that the truth function derived from (18) is the same as that from Wang's random sets falling shadow theory [26]. According to the Bayes' Theorem II, from Eq. (18), we obtain

$$T^*(\theta_j|X) = [P(X|y_j)/P(X)]/\max[P(X|y_j)/P(X)] \tag{19}$$

Equation (19) is more useful in general because it is often hard to find $P(y_j|X)$ or $P(y_j)$ for Eq. (18). Equations (18) and (19) fit cases involving large samples. When samples are not large enough, we need to construct truth functions with parameters and to optimize them.

The semantic information conveyed by y_j about x_i is defined with log-normalized-likelihood [12, 14]:

$$I(x_i;\theta_j) = \log \frac{P(x_i|\theta_j)}{P(x_i)} = \log \frac{T(\theta_j|x_i)}{T(\theta_j)} \tag{20}$$

For an unbiased estimation y_j, its truth function may be expressed by a Gaussian distribution without the coefficient: $T(\theta_j|X) = \exp[-(X - x_j)^2/(2d^2)]$. Hence

$$I(x_i;\theta_j) = \log\left[1/T(\theta_j)\right] - (X - x_j)^2/(2d^2) \tag{21}$$

The $\log[1/T(\theta_j)]$ is the Bar-Hillel-Carnap semantic information measure [27]. Equation (21) shows that the larger the deviation is, the less information there is; the less the LP is, the more information there is; and, a wrong estimation may convey negative information. These conclusions accord with Popper's thought [20].

To average $I(x_i; \theta_j)$, we have

$$I(X;\theta_j) = \sum_i P(x_i|y_j) \log \frac{P(x_i|\theta_j)}{P(x_i)} = \sum_i P(x_i|y_j) \log \frac{T(\theta_j|x_i)}{T(\theta_j)} \tag{22}$$

where $P(x_i|y_j)$ $(i = 1, 2, ...)$ is the sampling distribution, which may be unsmooth or discontinuous. Hence, the optimized truth function is

$$T * (\theta_j|X) = \underset{T(\theta_j|X)}{\arg\max}\, I(X;\theta_j) = \underset{T(\theta_j|X)}{\arg\max} \sum_i P(x_i|y_j) \log \frac{T(\theta_j|x_i)}{T(\theta_j)} \tag{23}$$

It is easy to prove that when $P(X|\theta_j) = P(X|y_j)$ or $T(\theta_j|X) \propto P(y_j|X)$, $I(X;\theta_j)$ reaches its maximum and is equal to the KL information. If we only know $P(X|y_j)$ without knowing $P(X)$, we may assume that X is equiprobable to obtain the truth function.

To average $I(X; \theta_j)$ for different Y, we have [12, 14]

$$I(X;\theta) = H(\theta) - H(\theta|X)$$

$$H(\theta) = - \sum_j P(y_j) \log T(\theta_j), \quad H(\theta|X) = \sum_j \sum_i P(x_i, y_j)(x_i - x_j)^2/\left(2d_j^2\right) \tag{24}$$

Clearly, the MSI criterion is like the RLS criterion. $H(\theta|X)$ is like the mean squared error, and $H(\theta)$ is like the negative regularization term. The relationship between the log normalized likelihood and generalized KL information is

$$\log \prod_i \left[\frac{P(x_i|\theta_j)}{P(x_i)}\right]^{N_{ji}} = N_j \sum_i P(x_i|y_j) \log \frac{P(x_i|\theta_j)}{P(x_i)} = N_j I(X;\theta_j) \tag{25}$$

The MSI criterion is equivalent to the ML criterion because $P(X)$ does not change when we optimize θ_j.

For the second task of LBI, given x_i, we select a hypothesis or label by the classifier

$$y_j \ast= h(x) = \arg\max_{y_j} \log I(\theta_j; x) = \arg\max_{y_j} \log \frac{T(\theta_j|x)}{T(\theta_j)} \qquad (26)$$

This classifier produces a noiseless Shannon's channel. Using $T(\theta_j)$, we can overcome the class-imbalance problem [24]. If $T(\theta_j|x) \in \{0,1\}$, the classifier becomes

$$y_j \ast= h(x) = \arg\max_{y_j \text{ with } T(A_j|x)=1} \log[1/T(A_j)] = \arg\min_{y_j \text{ with } T(A_j|x)=1} T(A_j) \qquad (27)$$

It means that we should select a label with the least LP and hence with the richest connotation. The above method of multi-label learning and classification is like the Binary Relevance (BR) method [25]. However, the above method does not demand too much of samples and can fit cases where $P(X)$ changes (See [15] for details).

5   Logical Bayesian Inference for Machine Learning

In Sect. 4, we have introduced the main method of using LBI for multi-label learning and classification. From LBI, we can also obtain an iterative algorithm, the Channels' Matching (CM) algorithm, for the MLE [16] and mixture models [17].

For unseen instance classifications, we assume that observed condition is $Z \in C = \{z_1, z_2, \ldots\}$; the classifier is $Y = f(Z)$; a true class or true label is $X \in U = \{x_1, x_2, \ldots\}$; a sample is $D = \{(x(t); z(t)) | t = 1, 2, \ldots, N; x(t) \in U; z(t) \in C\}$. From D, we can obtain $P(X, Z)$. The iterative process is as follows.

Step I (the semantic channel matches the Shannon channel): For a given classifier $Y = f(Z)$, we obtain $P(Y|X)$, $T(\theta|X)$, and conditional information for given Z

$$I(X; \theta_j | Z) = \sum_i P(X_i|Z) \log \frac{T(\theta_j|X_i)}{T(\theta_j)}, \quad j = 1, 2, \ldots, n \qquad (28)$$

Step II (the Shannon channel matches the semantic channel): The classifier is

$$y_j = f(Z) = \arg\max_{y_j} I(X; \theta_j | Z), \quad j = 1, 2, \ldots, n \qquad (29)$$

Repeating the above two steps, we can achieve the MSI and ML classification. The convergence can be proved with the help of $R(G)$ function [16].

For mixture models, the aim of **Step II** is to minimize the Shannon mutual information R minus the semantic mutual information G [17]. The convergence of the CM for mixture models is more reliable than that of the EM algorithm[2].

---

[2] For the strict convergence proof, see http://survivor99.com/lcg/CM/CM4MM.html.

6   Confirmation Measure $b*$ for Induction

Early logical Bayesians [4–7] were also inductivists who used the conditional LP or truth function to indicate the degree of inductive support. However, contemporary inductivists use the confirmation measure or the degree of belief between -1 and 1 for induction [28, 29]. By LBI, we can derive a new confirmation measure $b* \in [-1, 1]$.

Now we use the medical test as an example to introduce the confirmation measure $b*$. Let x_1 be a person with a disease, x_0 be a person without the disease, y_1 be the test-positive, and y_0 be the test-negative. The y_1 also means a universal hypothesis "For all people, if one's testing result is positive, then he/she has the disease". According to Eq. (18), the truth value of proposition $y_1(x_0)$ (x_0 is the counterexample of y_1) is

$$b'^* = T^*\left(\theta_1|x_0\right) = P\left(y_1|x_0\right)/P\left(y_1|x_1\right) \tag{30}$$

We define the confirmation measure $b*$ of y_1 by $b'^* = 1 - |b*|$. The $b*$ can also be regarded as the optimized degree of belief of y_1. For this measure, having fewer counterexamples or $P(y_1|x_0)$ is more important than having more positive examples or $P(y_1|x_1)$. Therefore, this measure is compatible with Popper's falsification theory [30].

With the TPH $P(y_1|X)$, we can use the likelihood method to obtain the Confidence Level (CL) of y_1, which reflects the degree of inductive support of y_1. Using $T^*(\theta_1|X)$, we can obtain the same CL. And, the $b*$ is related to CL by

$$b* = \begin{cases} 1 - CL'/CL, & \text{if } CL > 0.5 \\ CL/CL' - 1, & \text{if } CL \leq 0.5 \end{cases} \tag{31}$$

where $CL' = 1 - CL$. If the evidence or sample fully supports a hypothesis, then $CL = 1$ and $b* = 1$. If the evidence is irrelevant to a hypothesis, $CL = 0.5$ and $b* = 0$. If the evidence fully supports the negative hypothesis, $CL = 0$ and $b* = -1$. The $b*$ can indicate the degree of inductive support better than CL because inductive support may be negative. For example, the confirmation measure of "All ravens are white" should be negative.

If $|U| > 2$ and $P(y_j|X)$ is a distribution over U, we may use the confidence interval to convert a predicate into a universal hypothesis, and then, to calculate its confidence level and the confirmation measure.

BI provides the credible level with given credible interval [3] for a parameter distribution instead of a hypothesis. The credible level or the Bayesian posterior does not well indicate the degree of inductive support of a hypothesis. In comparison with BI, LBI should be a better tool for induction.

7   Summary

This paper proposes the Logical Bayesian Inference (LBI), which uses the truth function as the inference tool like logical Bayesians and uses the likelihood method as frequentists. LBI also use frequencies to explain logical probabilities and truth functions, and hence is the combination of extreme frequentism and extreme Bayesianism. The truth function LBI uses can indicate the semantic meaning of a hypothesis or label and can be used for probability prediction that is compatible with the classical Baye's prediction. LBI is based on the third kind of Bayes' Theorem and the semantic information method. They all together form a new mathematical frame for semantic communication, machine learning, and induction. This new frame may support and improve many existing methods, such as likelihood method and fuzzy mathematics method, rather than replace them. As a new theory, it must be imperfect. The author welcomes researchers to criticize or improve it.

References

1. Fienberg, S.E.: When did Bayesian inference become "Bayesian"? Bayesian Anal. **1**(1), 1–40 (2006)
2. Anon: Bayesian inference, Wikipedia: the Free Encyclopedia. https://en.wikipedia.org/wiki/Bayesian_probability. Accessed 20 July 2018
3. Hájek, A.: Interpretations of probability. In: Zalta, E.N. (ed.) The Stanford Encyclopedia of Philosophy (Winter 2012 edn). https://plato.stanford.edu/archives/win2012/entries/probability-interpret/. Accessed 27 July 2018
4. Keynes, I.M.: A Treaties on Probability. Macmillan, London (1921)
5. Carnap, R.: Logical Fundations of Probability. The University of Chicago Press, Chicago (1962)
6. Jaynes, E.T.: Probability Theory: The Logic of Science, Edited by Larry Bretthorst. Cambridge University Press, New York (2003)
7. Reichenbach, H.: The Theory of Probability. University of California Press, Berkeley (1949)
8. Fisher, R.A.: On the mathematical foundations of theoretical statistics. Philos. Trans. R. Soc. **A222**, 309–368 (1922)
9. Shannon, C.E.: A mathematical theory of communication. Bell Syst. Tech. J. **27**(3), 379–429 and 623–656 (1948)
10. Goodfellow, I., Bengio, Y.: Deep Learning. The MIP Press, Cambridge (2016)
11. Lu, C.: B-fuzzy quasi-Boolean algebra and generalized mutual entropy formula. Fuzzy Syst. Math. (in Chinese) **5**(1), 76–80 (1991)
12. Lu, C.: A Generalized Information Theory (in Chinese). China Science and Technology University Press, Hefei (1993)
13. Lu, C.: Meanings of generalized entropy and generalized mutual information for coding (in Chinese). J. China Inst. Commun. **15**(6), 37–44 (1994)
14. Lu, C.: A generalization of Shannon's information theory. Int. J. Gener. Syst. **28**(6), 453–490 (1999)
15. Lu, C.: Semantic channel and Shannon's channel mutually match for multi-label classifications. In: ICIS2018, Beijing, China (2018)
16. Lu, C.: Semantic channel and Shannon channel mutually match and iterate for tests and estimations with maximum mutual information and maximum likelihood. In: 2018 IEEE International Conference on Big Data and Smart Computing, pp. 227–234, IEEE Conference Publishing Services, Piscataway (2018)
17. Lu, C.: Channels' matching algorithm for mixture models. In: Shi, Z., Goertzel, B., Feng, J. (eds.) ICIS 2017. IAICT, vol. 510, pp. 321–332. Springer, Cham (2017). https://doi.org/10.1007/978-3-319-68121-4_35
18. Zadeh, L.A.: Fuzzy sets. Inf. Control **8**(3), 338–353 (1965)

19. Zadeh, L.A.: Probability measures of fuzzy events. J. Math. Anal. Appl. **23**(2), 421–427 (1986)
20. Popper, K.: Conjectures and Refutations. Repr. Routledge, London and New York (1963/2005)
21. Tarski, A.: The semantic conception of truth: and the foundations of semantics. Philos. Phenomenol. Res. **4**(3), 341–376 (1944)
22. Davidson, D.: Truth and meaning. Synthese **17**(1), 304–323 (1967)
23. Bayes, T., Price, R.: An essay towards solving a problem in the doctrine of chance. Philos. Trans. R. Soc. Lond. **53**, 370–418 (1763)
24. Zhang, M.L., Zhou, Z.H.: A review on multi-label learning algorithm. IEEE Trans. Knowl. Data Eng. **26**(8), 1819–1837 (2014)
25. Zhang, M.L., Li, Y.K., Liu, X.Y., Geng, X.: Binary relevance for multi-label learning: an overview. Front. Comput. Sci. **12**(2), 191–202 (2018)
26. Wang, P.Z.: From the fuzzy statistics to the falling fandom subsets. In: Wang, P.P. (ed.) Advances in Fuzzy Sets, Possibility Theory and Applications, pp. 81–96. Plenum Press, New York (1983)
27. Bar-Hillel, Y., Carnap, R.: An outline of a theory of semantic information. Tech. Rep. No. 247, Research Lab. of Electronics, MIT (1952)
28. Hawthorne, J.: Inductive logic. In: Zalta, E.N. (ed.) The Stanford Encyclopedia of Philosophy. https://plato.stanford.edu/entries/logic-inductive/. Accessed 22 July 2018
29. Tentori, K., Crupi, V., Bonini, N., Osherson, D.: Comparison of confirmation measures. Cognition **103**(1), 107–119 (2017)
30. Lu, C.: Semantic information measure with two types of probability for falsification and confirmation. https://arxiv.org/abs/1609.07827. Accessed 27 July 2018

Solution of Brain Contradiction
by Extension Theory

Germano Resconi[1(✉)] and Chunyan Yang[2]

[1] Mathematical and Physical Department,
Catholic University, Via Trieste 17, Brescia, Italy
resconi42@gmail.com
[2] Research Institute of Extenics and Innovation Methods,
Guangdong University of Technology,
Guangzhou 510006, People's Republic of China

Abstract. In fuzzy theory any degree to belong to a set can be considered as a positive distance from complementary set. So the distance moves from zero to one when the object belongs to the set. The extension theory considers a negative value of the distance. This is in conflict with the classical definition of the distance is a positive scalar. So we have a classical contradiction. To solve this conflict we define the distance as a vector with two different directions one positive and the other negative. The distances are vectors with positive norm. In this way we have positive norm for the two directions. In extension theory we define the dependent function and suitable transformations in a way to build a nonlinear neuron that can solve a very old conflicting problem in brain linear neural computation.

Keywords: Dependent function · Vector distance · Nonlinear neuron
Solution of conflicts · Solution of Boolean function by neuron

1 Introduction

A new approach to implementation of Boolean function by nonlinear neuron is introduce by the implementation of characteristic nonlinear function introduce for the first time by Cai Wen in the extension theory [1–4]. We show that with the new neuron we can easily compress the tradition AND OR NOT Boolean functions network into one step system to have a more efficient system for brain implementation of complex functions.

2 From Vector Distance to the Dependent Functions

In Extenics, one of the dependent functions is given by the expression

$$k(x) = \frac{\left|x - \frac{a+b}{2}\right| - \frac{b-a}{2}}{\left(\left|x - \frac{a+b}{2}\right| - \frac{b-a}{2}\right) - \left(\left|x - \frac{a_0+b_0}{2}\right| - \frac{b_0-a_0}{2}\right)} = \frac{\rho(x,a,b)}{\rho(x,a,b) - \rho(x,a_0,b_0)} \tag{1}$$

Z. Shi et al. (Eds.): ICIS 2018, IFIP AICT 539, pp. 24–29, 2018.
https://doi.org/10.1007/978-3-030-01313-4_3

For continuous function we transform (1) into (2)

$$k'(x) = \frac{\left(x - \frac{a+b}{2}\right)^2 - \left(\frac{b-a}{2}\right)^2}{\left(\left(x - \frac{a+b}{2}\right)^2 - \left(\frac{b-a}{2}\right)^2\right) - \left(\left(x - \frac{a_0+b_0}{2}\right)^2 - \left(\frac{b_0-a_0}{2}\right)^2\right)}$$

$$= \frac{(x-a)(x-b)}{(x-a)(x-b) - (x-a_0)(x-b_0)} \tag{2}$$

So we have the form

$$V = \begin{bmatrix} k(15) \\ k(20) \\ k(50) \\ k(85) \end{bmatrix} = \begin{bmatrix} 0 \\ 1 \\ 1 \\ 0 \end{bmatrix} \text{ and } a = 20, b = 50, a_0 = 15, b_0 = 85$$

We have the Fig. 1.

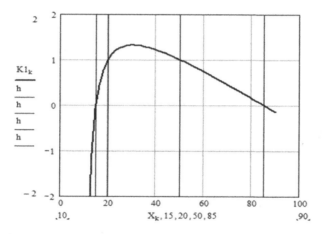

Fig. 1. Graph of the expression (2) with a = 20, b = 50, a_0 = 15, b_0 = 85.

3 Neural Solution of Boolean Function by Dependent Function $k'(x)$

For Linear neuron when y = 1 if $c_1x_1 + c_2x_2 - \theta > 0$ and y = 0 if Now given the Boolean function $(\neg x_1 \wedge x_2) \vee (x_1 \wedge \neg x_2) = x_1 XOR x_2 = y$ We have a contradiction so we cannot obtain the Boolean function XOR by the classical neuron. In fact we have $c_1 0 + c_2 0 - \theta \leq 0, c_1 0 + c_2 1 - \theta > 0, c_1 1 + c_2 0 - \theta > 0, c_1 1 + c_2 1 - \theta \leq 0$ and $-\theta \leq 0$ True, $c_2 - \theta > 0$ True if $c_2 > 0$, $c_1 - \theta > 0$ True if $c_1 > 0$, $c_1 + c_2 - \theta \leq 0$ True if $c_1 + c_2 \leq 0$.

Boolean function XOR solved by a nonlinear neuron

$$
\begin{bmatrix}
X & Y & (\neg X \wedge Y) \vee (X \wedge \neg Y) \\
0 & 0 & 0 \\
1 & 0 & 1 \\
0 & 1 & 1 \\
1 & 1 & 0
\end{bmatrix}
\Rightarrow
\begin{bmatrix}
x & k'(x) \\
0 & 0 \\
1 & 1 \\
2 & 1 \\
3 & 0
\end{bmatrix}, k'(x)
$$

$$
= \frac{(x-0)(x-3)}{(x-0)(x-3)-(x-1)(x-2)} \tag{3}
$$

$$
\begin{bmatrix}
X & Y & X \rightarrow Y \\
0 & 0 & 1 \\
1 & 0 & 0 \\
0 & 1 & 1 \\
1 & 1 & 1
\end{bmatrix}
\Rightarrow
\begin{bmatrix}
x & k'(x) \\
0 & 1 \\
1 & 0 \\
2 & 1 \\
3 & 1
\end{bmatrix}, k'(x) = \frac{(x-1)}{(x-1)-(x-0)(x-2)(x-3)} \tag{4}
$$

We remark that for this Boolean function $k'(x)$ has a singular point. Now because for dependent function, we have the scheme (Figs. 2 and 3). We can change the form of $k'(x)$ with the same original properties but without the negative value of $k'(x)$ and also without the singularity. So we have

$$
k''(x) = \frac{A(x)^2}{A(x)^2 + B(x)^2} = \frac{(x-2)^2}{(x-2)^2 + [(x-0)(x-1)(x-3)]^2} \tag{5}
$$

For the Boolean function and dependent function we can create the scheme for a nonlinear neuron by which we solve in one step and without linear hidden neurons the old problem of the learning process in Brain by AND, OR, NOT logic functions (Fig. 4).

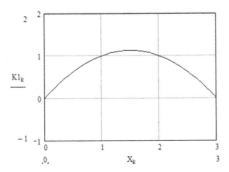

Fig. 2. Graph of the dependent function (3)

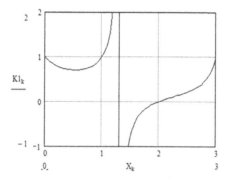

Fig. 3. Graph of the dependent function (4)

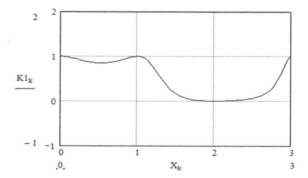

Fig. 4. Graph of the dependent function (5)

4 Machine and Systems by Nonlinear Neuron

Given the machine with x the input, q the states and y the output the system can be represented by the Boolean equations (Figs. 5 and 6). Given the machine with x the input, q the states and y the output

$$
\begin{bmatrix}
q\backslash x & 0 & 1 & y \\
1 & 3 & 6 & 0 \\
2 & 3 & 4 & 1 \\
3 & 2 & 5 & 0 \\
4 & 5 & 2 & 0 \\
5 & 6 & 3 & 0 \\
6 & 5 & 1 & 0
\end{bmatrix}
\Rightarrow
\begin{bmatrix}
q\backslash x & 0 & 1 & y \\
000 & 100 & 110 & 0 \\
010 & 100 & 001 & 1 \\
100 & 010 & 111 & 0 \\
001 & 111 & 010 & 0 \\
111 & 110 & 100 & 0 \\
110 & 111 & 000 & 0
\end{bmatrix}
, \text{by code}
\begin{bmatrix}
1 & \rightarrow & 000 \\
2 & \rightarrow & 010 \\
3 & \rightarrow & 100 \\
4 & \rightarrow & 001 \\
5 & \rightarrow & 111 \\
6 & \rightarrow & 110
\end{bmatrix}
,
$$

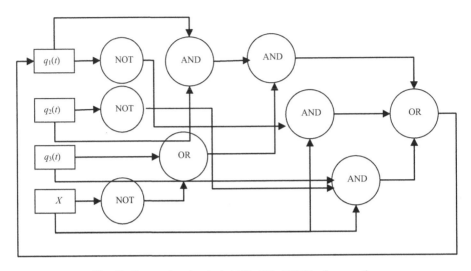

Fig. 5. System by classical AND, OR, NOT logic operation

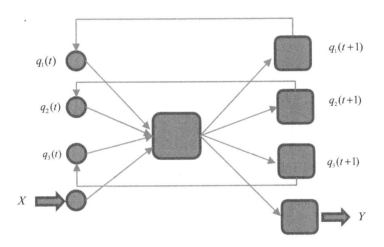

Fig. 6. System by non linear neurons

Boolean system

$$\begin{cases} q_1(t+1) = \overline{q_1(t)}\overline{X} + \overline{q_2(t)q_3(t)}X + q_1(t)q_2(t)(q_3(t)+\overline{X}) \\ q_2(t+1) = q_1(t)\overline{X} + Xq_2(t)\overline{q_3(t)} + \overline{q_1(t)q_2(t)}) \\ q_3(t+1) = q_1(t)q_2(t)\overline{q_3(t)}\overline{X} + \overline{X}q_1(t)\overline{q_2(t)} + q_2(t)\overline{q_1(t)}) + q_3(t)\overline{q_1(t)}\overline{X} \end{cases} \quad (6)$$

$$q_1(t+1) = k_1''(x) = \frac{[(x-1)(x-10)(x-12)(x-11)]^2}{[(x-1)(x-10)(x-12)(x-11)]^2 + [(x-0)(x-2)(x-4)(x-7)(x-3)(x-8)(x-9)(x-15)]^2}$$

$$q_2(t+1) = k_2''(x) = \frac{[(x-0)(x-2)(x-10)(x-15)(x-11)]^2}{[(x-0)(x-2)(x-10)(x-15)(x-11)]^2 + [(x-1)(x-4)(x-7)(x-3)(x-8)(x-15)(x-11)]^2}$$

$$q_3(t+1) = k_3''(x) = \frac{[(x-0)(x-2)(x-1)(x-7)(x-8)(x-12)(x-15)(x-11)]^2}{[(x-0)(x-2)(x-1)(x-7)(x-8)(x-12)(x-15)(x-11)]^2 + [(x-4)(x-3)(x-10)(x-9)]^2}$$

$$Y = k_4''(x) = \frac{[(x-0)(x-1)(x-4)(x-7)(x-3)(x-8)(x-9)(x-12)(x-15)(x-11)]^2}{[(x-0)(x-1)(x-4)(x-7)(x-3)(x-8)(x-9)(x-12)(x-15)(x-11)]^2 + [(x-2)(x-10)]^2}$$

$$(7)$$

5 Conclusions

In this paper, we show how is possible by extension theory [1] to solve a very important problem in realization of Boolean function by one step nonlinear neurons.

Acknowledgments. This paper is sponsored by National Natural Science Foundation Project (61503085) and Science and Technology Planning Project of Guangdong Province (2016A040404015).

References

1. Cai, W., Yang, C.Y., Lin, W.: Extension Engineering Methods. Science Press, Beijing (2003)
2. Yang, C.Y., Cai, W.: Extenics: Theory, Method and Application. Science Press & The Educational Publisher, Beijing (2013)
3. Cai, W.: Extension theory and its application. Chin. Sci. Bull. **38**(8), 1538–1548 (1999)
4. Li, W.H., Yang, C.Y.: Extension information-knowledge-strategy system for semantic interoperability. J. Comput. **3**(8), 32–39 (2008)
5. Resconi, G., Xu, X., Xu, G.: Introduction to Morphogenetic Computing. SCI, vol. 703. Springer, Cham (2017). https://doi.org/10.1007/978-3-319-57615-2

Cognitive Features of Students Who Are Tired of Learning Geometry

Yan Wang[1,2] and Xiaohui Zou[1,2(✉)] (iD)

[1] Shanghai Shenyue Software Technology Co., Ltd.,
Rm. B10. North D., Bld. 8, No. 619 Longchang Rd., Shanghai 200090, China
yuki@geomking.com
[2] Sino-American Searle Research Center,
Tiangongyuan, Paolichuntianpai Building 2, Room 1235, Daxing, Beijing 102629, China
949309225@qq.com

Abstract. The purpose of this paper is to analyse the cognitive features of students who are tired of learning geometry, explore ways to help them overcome the disgust. The method is: analyse the cognitive features and interests of students who are tired of learning geometry, then as result, through the above analysis, we found that if we want to find a way to transform geometry learning from the perspective of interest into a way that students are interested in, we can use the methods, tools and means of simplifying, to take positive actions against their fears and boredom, especially to eliminate cognitive problems that lead to student misunderstandings. The significance lies in: let the students know if the misunderstanding are eliminated and the features of the cognitive knowledge of the geometric language are clearly defined, they are always interested in geometry learning, never feel tired. This could be realized through creating interesting connections among graphic language, natural language and symbolic language, finding out the cognitive ways in which students can be interested in geometry learning and improve their interest in learning.

Keywords: Disgusted emotions · Geometric language · Cognitive features
Interest cultivation

1 Introduction

Plane geometry is one of the most basic and important subjects in developing junior high school students' thinking ability (including intuitive thinking, logical thinking and innovation). In the specific age of junior middle school students' growth, no other subject could replace it inland and abroad (Fangqu and Wen 2017).

However, plane geometry is precisely a subject that is difficult for teachers to teach and students to learn for a long time. In this paper, we will try to explore and analyse the reasons why students are bored with mathematics' weariness, and then find some feasible solutions to conquer it.

© IFIP International Federation for Information Processing 2018
Published by Springer Nature Switzerland AG 2018. All Rights Reserved
Z. Shi et al. (Eds.): ICIS 2018, IFIP AICT 539, pp. 30–34, 2018.
https://doi.org/10.1007/978-3-030-01313-4_4

2 Cognitive Features and Interests of Students Who Are Tired of Geometry Learning

Beginning with some examples between the teacher and the student in the reality.

1. Near the senior middle school entrance examination, one student went to the game shop because he hated mathematics classes, the persuasion and education from the parents would not help. It is thoughtful that he was even an "outstanding" student of the school.
2. In one classrooms, there was a loud noise like "how dare to swear me" and "Who told you to throw my book" "Whose exercise is not complete?" Afterwards, it was a farce that a student was criticized by his head teacher for his incomplete math homework.
3. In a classroom, teachers told the students that the circumference is a certain point to the equidistant point on the same plane. The students copied in the notebook, but did not understand what the circle was, after the teacher took chalk to draw a circle on the blackboard. The students cheered immediately: Aha, that's the circle, understand!

So, if we analyse the phenomenon coming from the above mentioned examples and reality, we could come the conclusion that the students who are tired of geometry learning are normally:

1. Will not learn
2. Can not learn
3. feel that mathematics is boring
4. don't establish a good foundation since the beginning, so that all the following series of problems can't be understood and solved

But, if we talk with these students, we can find that they are interested in many things except geometry learning, like play games, watch TV, play basketball/football, dancing, chasing stars, etc. All these things bring them interest and fun, they never feel tired or bored by doing these things.

3 Way to Transform Geometry Learning from the Perspective of Interest for Students

Against above mentioned problems, after research and praxis in several schools, we come to the conclusion that following methods, tools and means of simplifying could have effect, to take positive actions against students' fears and boredom, especially to eliminate cognitive problems that lead to student misunderstandings.

1. Choose the teaching mode and build a good classroom atmosphere

The first thing that should be done is that teachers should choose the teaching method scientifically from the cognitive basis of the students according to the characteristics of the teaching materials, and choose the teaching method suitable for the students' appetite. For example, the teaching method like "teacher-student interaction" and "small step walking" provides teachers way to cut a relatively difficult problem into smaller questions so that the problem would be easier for students to understand. All students could follow teacher's thought, they are possible to raise their hand to ask, use their brains, and learn what they want to learn in a relaxed classroom atmosphere. Of course, we should also pay attention to the students' actual knowledge accepting ability to avoid simple copy. In recent years, China has advocated quality education and innovative education, several new mathematics teaching methods are invented. We could have a wide range of teaching methods to choose, such as "exploratory mathematics teaching", "mathematics questioning teaching", "activity-based mathematics teaching", "opened mathematical teaching", "entire and example teaching", "mathematical modelling teaching", etc., Anyway we should avoid the adopting of a monotonous teaching method and having lessons in a dogmatic way.

2. Second, meticulously design teaching links to stimulate students' curiosity

After choosing the appropriate teaching method, the second factor that we should consider is how to design every teaching step meticulously, cultivate students' positive attitude by mathematics, and consciously strengthen the connection between teaching content and real life so that each student can feel that the things to learn are practical or have learning value. For example, to find similar items, teachers send algebraic signs to students. One student in each corner of the classroom holds a sign. Other students look for "similar items" in the four corners. Although in classroom it's a little chaotic, this simple activity motivate the students' interest in learning mathematics in a pleasant atmosphere. Another example is the concept of variance learning, teacher takes a weight to the class, picks three students with similar size to weigh, write down the numbers, calculate the average and variance according to the formula. Then picks the weight of the three fattest/thinnest/average classmates and calculate the average and variance. The results shows that the average number of students in the two groups is similar, but the variance is very different. This activity enables students to feel the significance of the variance and will never forget it. Like these wonderful scenes, it can both attract students and connect them with new knowledge, allow them to experience the establishment of knowledge points themselves. The understanding of the whole story of the knowledge point could inspire students' curiosity and receive a multiplier effect.

3. Third, establish a good teacher-student relationship

In the teacher-student relationship, the teacher plays a leading role in the adjustment. The teacher's facial expressions should reflect equality and democracy. First of all, a teacher's single action and smile can infect students and make students feel emotional and excited. Especially in class, the teacher stands on the podium, with a kind face and a smiling face to relieve the student's nervousness. Secondly, mathematics teaching is often carried out through problem solving, through problem solving to develop students' ability to analyse and solve problems. When students make mistakes, teachers' eyes should be strict and sincere; when students make progress, they should be praise and trust. In particular, some backward students should not be treated with contempt, disgust, or disdain. This will damage their self-esteem and cause them to develop rebellious attitudes. Instead, they should be looked with hope and trust to let them see hope and increase their strength. Students in the learning process will inevitably be influenced or tempted by social family or classmates, interfere with normal learning emotions, and having some malignant consequences such as learning weariness and playing truant. At this time, we should not shout loudly, and make simple judgement. Instead, we should get to know the origin of the problems, and help students overcome difficulties and regain the joy of learning. Only by understanding students and paying attention to students in an equal, democratic manner and emotion, students can then respect and like teachers from the bottom of their hearts. Then students can be converted from like mathematics teachers to like mathematics. Let the students know that the teacher treats them sincerely and truly. This will greatly enhance the emotional communication between teachers and students.

4   Case Study

This is an ordinary middle school in Qingpu District of Shanghai. It has a strong team of teachers who love to study assiduously. However, in terms of teaching concepts, teaching abilities, and the quality of students, etc., there is still a certain gap compared with many advanced regions and schools. Particularly, it is always felt that the teaching of plane geometry is a difficult point in teaching and it is a "bottleneck" that restricts the improvement of teaching quality. For a long time, plane geometry has been a difficult task for teachers to teach and students to learn.

While the school was thinking hard and looking for a "breakthrough" in the study of geometric teaching, in September 2011, the Municipal Education Commission of Shanghai identified this school as an experimental application of "Geomking – junior middle school plane geometry learning software" (abbreviated as "Geomking" software)in the Shanghai Rural School Education Information Technology Promotion Project. From October 2011, this school began to learn and use the "Geomking" software.

For this reason, according to the requirements of teaching and research, after the teaching content and teaching goal of each lesson are determined, we abandon the practice of finding the exercise in the sea of books, but select the exercise from the "Geomking" software as the teaching content and guide students to gradually achieve the learning goals. The "smart search" function by "Geomking" software is the most intelligent function of the software. By simply input very simple information, it's possible to search for all the required learning content or topics in a very short time, so that teachers and students feel very convenient when using them.

By example explanation, the solution is generally presented to the students directly. Instead, they are asked to discuss the matter first, to understand how to think after receiving the problem, and it is normal when they meet problems during the thinking. The conversation of thinking and hypothesis produces a motivational effect by students' thinking, the teacher gives appropriate guidance, and displays or let students read the corresponding content in the software in a timely manner to achieve students' initiative learning and independence. Independent thinking is an important foundation for knowledge construction; teamwork, discussion and communication are effective links to promote internalization, deepening, and improvement of knowledge. The learning outcomes produced by different learning methods and the effects on student development are different.

Through nearly a year of teaching, research, and practice, we have achieved certain results and experience, which has made our mathematics teachers' teaching ability and teaching level significantly improved, so that they can adapt to the requirements of cultivating students' geometric analysis ability in teaching.

5   Conclusion

Through the above-mentioned case, it can be concluded that teachers' changing concepts and updating teaching methods are the key to cultivating and inspiring students to learn mathematics. The key to the improvement of cognitive features of students who are tired of learning geometry is for teachers to leave the dominant power to themselves, subject power to the students, realize the emotional interaction between teachers and students. They should try to cultivate and inspire students' interest in mathematic. Raise students' interest for mathematics can not achieve obvious results overnight. It needs to persist in teaching for a long time, and teachers should constantly sum up experience in teaching and constantly learn from each other. Only in this way the expected results could be achieved.

Reference

Fangqu, X., Wen, X.: Transparent Geometry - New Practice of Internet + Planar Geometry. Shanghai Education Publishing House, Shanghai (2017)

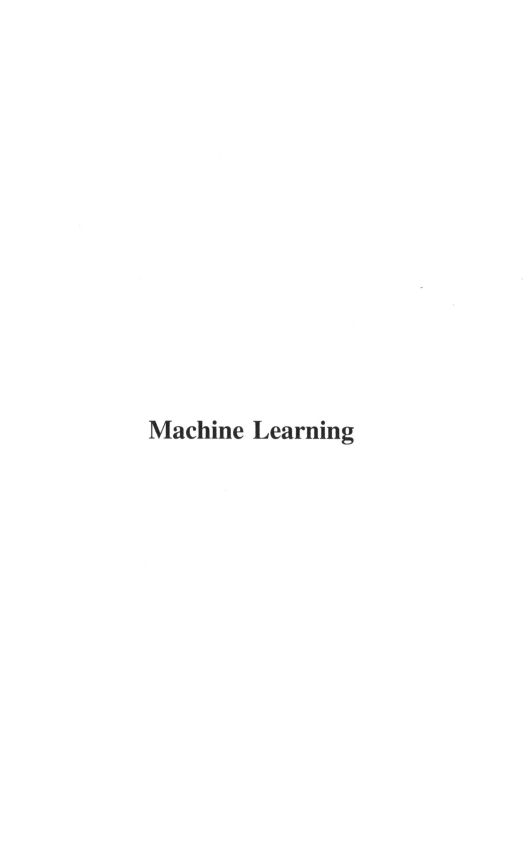

Machine Learning

Semantic Channel and Shannon's Channel Mutually Match for Multi-label Classification

Chenguang Lu[⊠]

College of Intelligence Engineering and Mathematics,
Liaoning Engineering and Technology University,
Fuxin 123000, Liaoning, China
lcguang@foxmail.com

Abstract. A semantic channel consists of a set of membership functions or truth functions which indicate the denotations of a set of labels. In the multi-label learning, we obtain a semantic channel from a sampling distribution or Shannon's channel. If samples are huge, we can directly convert a Shannon's channel into a semantic channel by the third kind of Bayes' theorem; otherwise, we can optimize the membership functions by a generalized Kullback–Leibler formula. In the multi-label classification, we partition an instance space with the maximum semantic information criterion, which is a special Regularized Least Squares (RLS) criterion and is equivalent to the maximum likelihood criterion. To simplify the learning, we may only obtain the truth functions of some atomic labels to construct the truth functions of compound labels. In a label's learning, instances are divided into three kinds (positive, negative, and unclear) instead of two kinds as in the One-vs-Rest or Binary Relevance (BR) method. Every label's learning is independent as in the BR method. However, it is allowed to train a label without negative examples and a number of binary classifications are not used. In the label selection, for an instance, the classifier selects a compound label with the most semantic information. This classifier has taken into the consideration the correlation between labels already. As a predictive model, the semantic channel does not change with the prior probability distribution (source) of instances. It still works when the source is changed. The classifier will vary with the source and hence can overcome the class-imbalance problem. It is shown that the old population's increase will change the classifier for label "Old person" and has been impelling the evolution of the semantic meaning of "Old". The CM iteration algorithm for unseen instance classification is introduced.

Keywords: Shannon's channel · Bayes' theorem
Natural language processing · Semantic information · Multi-label classification
Membership function · Semi-supervised learning

1 Introduction

Multi-label classification generally includes two steps: multi-label learning and multi-label selection. In multi-label learning, for every label y_j, we need to train its posterior $P(y_j|X; \theta)$ by a sample, where X is a random variable denoting an instance, and θ is a

Z. Shi et al. (Eds.): ICIS 2018, IFIP AICT 539, pp. 37–48, 2018.
https://doi.org/10.1007/978-3-030-01313-4_5

predictive model with parameters. In multi-label selection, we need to partition the instance space into different classes with specific criteria and to label every class.

There have been many valuable studies about multi-label classifications [1, 2]. Information, cross-entropy, and uncertainty criteria also have been used [3–5]. In label learning, if there are more than two labels to be learned, it is hard to obtain the posterior probability function $P(y_j|X; \theta)$ ($j = 1, 2, \ldots n$) because of the need of normalization ($\sum_j P(y_j|X; \theta) = 1$). Therefore, most researchers convert multi-label learning into multiple single label learnings [1]. The One-vs-Rest is a famous method [1]. However, in some cases, the conversion is improper. For instance, a sample has two examples (age 25, "Youth") and (age 24, "Adult"); it is unreasonable to regard (age 24, "Adult") as a negative example of "Youth". Therefore, the Binary Relevance (BR) method [2] requires that every instance is related to n labels with n "Yes" or "No". However, it demands too much of samples. It has another problem because of labels' correlation. In natural language, many negative labels, such as "Non-youth" and "Non-old person", are rarely used; it is unnecessary to add label "Adult" or "Non-youth" to an instance with label "Old person". So, the canonical BR relevance method needs improvements.

For the above reasons, the author develops a new method. It is similar to the BR method but has the following distinct features:

- It uses membership functions instead of posterior probability functions so that the normalization is unnecessary; the learned membership functions can be used to classify other samples with different distribution $P(X)$.
- In the label learning, we directly obtain membership functions from sampling distribution $P(X, Y)$ (Y is a label), by a new Bayes' formula or a generalized Kullback–Leibler (KL) formula, without binary classifications. This method allows us to train a label with only positive examples. It is not necessary to prepare n data sets for n labels.
- In label selection or classification, we use the Maximum Semantic Information (MSI) criterion to partition the instance space and to label classes. The classifier changes with $P(X)$.

This paper is based on the author's studies on the semantic information theory [6–8], maximum likelihood estimations (semi-supervised learning) [9], and mixture models (unsupervised learning) [10]. In the recent two decades, the cross-entropy method has become popular [11]. The author's above studies and this paper use not only cross-entropy but also mutual cross-entropy.

The main contributions of this paper are:

- Providing a new Bayes' formula, which can directly derive labels' membership functions from continuous sampling distributions $P(X, Y)$.
- Proving that the Maximum Semantic Information (MSI) criterion is a special Regularized Least Squares (RLS) criterion.
- Simplifying multi-label classification by the mutual matching of the semantic channel and Shannon's channel, without a number of binary classifications and labels' correlation problem.
- Overcoming the class-imbalance problem and explaining the classification of "Old people" changes with the age population distribution in natural language.

The rest of this paper is organized as follows. Section 2 provides mathematical methods. Section 3 discusses multi-label classifications. Section 4 introduces an iterative algorithm for unseen instance classifications. Section 5 is the summary.

2 Mathematical Methods

2.1 Distinguishing Statistical Probability and Logical Probability

Definition 1. Let U denote the instance set, and X denote a discrete random variable taking value from $U = \{x_1, x_2, ...\}$. For the convenience of theoretical analyses, we assume that U is one-dimensional. Let L denote the set of selectable labels, including some atomic labels and compound labels and let $Y \in L = \{y_1, y_2, ...\}$. Similarly, let L_a denote the set of some atomic labels and let $a \in L_a = \{a_1, a_2, ...\}$.

Definition 2. A label y_j is also a predicate $y_j(X) = $ "$X \in A_j$." For each y_j, U has a subset of A_j, every instance of which makes y_j true. Let $P(Y = y_j)$ denote the statistical probability of y_j, and $P(X \in A_j)$ denote the Logical Probability (LP) of y_j. For simplicity, let $P(y_j) = P(Y = y_j)$ and $T(y_j) = T(A_j) = P(X \in A_j)$.
We call $P(X \in A_j)$ the logical probability because according to Tarski's theory of truth [12], $P(X \in A_j) = P(\text{"}X \in A_j\text{"}$ is true$) = P(y_j$ is true$)$. Hence the conditional LP of y_j for given X is the feature function of A_j and the truth function of y_j. We denote it with $T(A_j|X)$. There is

$$T(A_j) = \sum_i P(x_i)T(A_j|x_i) \tag{2.1}$$

According to Davidson's truth-conditional semantics [13], $T(A_j|X)$ ascertains the semantic meaning of y_j. Note that statistical probability distribution, such as $P(Y)$, $P(Y|x_i)$, $P(X)$, and $P(X|y_j)$, are normalized, whereas the LP distribution is not normalized. For example, in general, $T(A_1|x_i) + T(A_2|x_i) + ... + T(A_n|x_i) > 1$.

For fuzzy sets [14], we use θ_j as a fuzzy set to replace A_j. Then $T(\theta_j|X)$ becomes the membership function of θ_j. We can also treat θ_j as a sub-model of a predictive model θ. In this paper, likelihood function $P(X|\theta_j)$ is equal to $P(X|y_j; \theta)$ in the popular method.

2.2 Three Kinds of Bayes' Theorems

There are three kinds of Bayes' theorem, which are used by Bayes [15], Shannon [16], and the author respectively.

Bayes' Theorem I. (used by Bayes): Assume that sets A, $B \in 2^U$, A^c is the complementary set of A, $T(A) = P(X \in A)$, and $T(B) = P(X \in B)$. Then

$$T(B|A) = T(A|B)T(B)/T(A), T(A) = T(A|B)T(B) + T(A|B^c)T(B^c) \tag{2.2}$$

There is also a symmetrical formula for $T(A|B)$. Note there are only one random variable X and two logical probabilities.

Bayes' Theorem II (used by Shannon): Assume that $X \in U$, $Y \in L$, $P(x_i) = P(X = x_i)$, and $P(y_j) = P(Y = y_j)$. Then

$$P(x_i|y_j) = P(y_j|x_i)P(x_i)/P(y_j), \ P(y_j) = \sum_i P(x_i)P(y_j|x_i) \qquad (2.3)$$

There is also a symmetrical formula for $P(y_j|x_i)$. Note there are two random variables and two statistical probabilities.

Bayes' Theorem III. Assume $T(\theta_j) = P(X \in \theta_j)$. Then

$$P(X|\theta_j) = T(\theta_j|X)P(X)/T(\theta_j), \quad T(\theta_j) = \sum_i P(x_i)T(\theta_j|x_i) \qquad (2.4)$$

$$T(\theta_j|X) = P(X|\theta_j)T(\theta_j)/P(X), \ T(\theta_j) = 1/\max(P(X|\theta_j)/P(X)) \qquad (2.5)$$

The two formulas are asymmetrical because there is a statistical probability and a logical probability. $T(\theta_j)$ in (2.5) may be called longitudinally normalizing constant.

The Proof of Bayes' Theorem III. Assume the joint probability $P(X, \theta_j) = P(X = \text{any}, X \in \theta_j)$, then $P(X|\theta_j)T(\theta_j) = P(X = \text{any}, X \in \theta_j) = T(\theta_j|X)P(X)$. Hence there is

$$P(X|\theta_j) = P(X)T(\theta_j|X)/T(\theta_j), \quad T(\theta_j|X) = T(\theta_j)P(X|\theta_j)/P(X)$$

Since $P(X|\theta_j)$ is horizontally normalized, $T(\theta_j) = \sum_i P(x_i) T(\theta_j|x_i)$. Since $T(\theta_j|X)$ is longitudinally normalized and has the maximum 1, we have

$$1 = \max[T(\theta_j)P(X|\theta_j)/P(X)] = T(\theta_j)\max[P(X|\theta_j)/P(X)]$$

Hence $T(\theta_j) = 1/\max[P(X|\theta_j)/P(X)]$. **QED.**

2.3   From Shannon's Channel to Semantic Channel

In Shannon's information theory [16], $P(X)$ is called the source, $P(Y)$ is called the destination, and the transition probability matrix $P(Y|X)$ is called the channel. So, a channel is formed by a set of transition probability functions: $P(y_j|X)$, $j = 1, 2, \ldots, n$.

Note that $P(y_j|X)$ (y_j is constant and X is variable) is different from $P(Y|x_i)$ and also not normalized. It can be used for Bayes' prediction to get $P(X|y_j)$. When $P(X)$ becomes $P'(X)$, $P(y_j|X)$ still works. $P(y_j|X)$ by a constant k can make the same prediction because

$$\frac{P'(X)kP(y_j|X)}{\sum\limits_i P'(x_i)kP(y_j|x_i)} = \frac{P'(X)P(y_j|X)}{\sum\limits_i P'(x_i)P(y_j|x_i)} = P'(X|y_j) \qquad (2.6)$$

Similarly, a set of truth functions forms a semantic channel: $T(\theta|X)$ or $T(\theta_j|X)$, $j = 1, 2, \ldots, n$. According to (2.6), if $T(\theta_j|X) \propto P(y_j|X)$, there is $P(X|\theta_j) = P(X|y_j)$. Conversely, let $P(X|\theta_j) = P(X|y_j)$, then we have $T(\theta_j|X) = P(y_j|X)/\max(P(y_j|X))$.

2.4   To Define Semantic Information with Log (Normalized Likelihood)

The (amount of) semantic information conveyed by y_j about x_i is defined with log-normalized-likelihood [8, 9]:

$$I(x_i; \theta_j) = \log \frac{P(x_i|\theta_j)}{P(x_i)} = \log \frac{T(\theta_j|x_i)}{T(\theta_j)} \tag{2.7}$$

For an unbiased estimation y_j, its truth function may be a Gaussian function:

$$T(\theta_j|X) = \exp\left[-(X - x_j)^2/(2d^2)\right] \tag{2.8}$$

Then $I(x_i; \theta_j) = \log[1/T(\theta_j)] - (X - x_j)^2/(2d^2)$. It shows that this information criterion reflects Popper's thought [17]. It tells that the larger the deviation is, the less information there is; the less the logical probability is, the more information there is; and, a wrong estimation may convey negative information. To average $I(x_i; \theta_j)$, we have

$$I(X; \theta_j) = \sum_i P(x_i|y_j) \log \frac{P(x_i|\theta_j)}{P(x_i)} = \sum_i P(x_i|y_j) \log \frac{T(\theta_j|x_i)}{T(\theta_j)} \tag{2.9}$$

$$I(X; \theta) = \sum_j P(y_j) \sum_i P(x_i|y_j) \log \frac{P(x_i|\theta_j)}{P(x_i)}$$

$$= \sum_j \sum_i P(x_i, y_j) \log \frac{T(\theta_j|x_i)}{T(\theta_j)} = H(\theta) - H(\theta|X) \tag{2.10}$$

$$H(\theta) = -\sum_j P(y_j) \log T(\theta_j), \; H(\theta|X) = -\sum_j \sum_i P(x_i, y_j) \log T(\theta_j|x_i)$$

where $I(X; \theta_j)$ is the generalized Kullback–Leibler (KL) information, and $I(X; \theta)$ is the semantic mutual information (a mutual cross-entropy). When $P(x_i|\theta_j) = P(x_i|y_j)$ for all i, j, $I(X; \theta)$ reaches its upper limit: Shannon mutual information $I(X; Y)$. To bring (2.8) into (2.10), we have

$$I(X; \theta) = H(\theta) - H(\theta|X)$$

$$= -\sum_j P(y_j) \log T(\theta_j) - \sum_j \sum_i P(x_i, y_j)(x_i - x_j)^2 / \left(2d_j^2\right) \tag{2.11}$$

It is easy to find that the maximum semantic mutual information criterion is a special Regularized Least Squares (RLS) criterion [18]. $H(\theta|X)$ is similar to mean squared error and $H(\theta)$ is similar to negative regularization term.

Assume that a sample is $D = \{(x(t); y(t))|t = 1, 2, \ldots, N; x(t) \in U; y(t) \in L\}$, a conditional sample is $D_j = \{x(1), x(2), \ldots, x(N_j)\}$ for given y_j, and the sample points come from independent and identically distributed random variables. If N_j is big enough, then $P(x_i|y_j) = N_{ij}/N_j$, where N_{ij} is the number of x_i in D_j. Then we have the log normalized likelihood:

$$\log \prod_i \left[\frac{P(x_i|\theta_j)}{P(x_i)}\right]^{N_{ji}} = N_j \sum_i P(x_i|y_j) \log \frac{P(x_i|\theta_j)}{P(x_i)} = N_j I(X; \theta_j) \qquad (2.12)$$

3 Multi-label Classification for Visibal Instances

3.1 Multi-label Learning (the Receiver's Logical Classification) for Truth Functions Without Parameters

From the viewpoint of semantic communication, the sender's classification and the receiver's logical classification are different. The receiver learns from a sample to obtain labels' denotations, e.g., truth functions or membership functions, whereas the sender needs, for a given instance, to select a label with most information. We may say that the learning is letting a semantic channel match a Shannon's channel and the sender's classification is letting a Shannon's channel match a semantic channel.

We use an example to show the two kinds of classifications. Assume that U is a set of different ages. There are subsets of U: $A_1 = \{young\ people\} = \{X|15 \leq X \leq 35\}$, $A_2 = \{adults\} = \{X|X \geq 18\}$, $A_3 = \{juveniles\} = \{X|X < 18\} = A_2^c$ (^{c} means complementary set), which form a cover of U. Three truth functions $T(A_1|X)$, $T(A_2|X)$, and $T(A_3|X)$ represent the denotations of y_1, y_2, and y_3 respectively, as shown in Fig. 1.

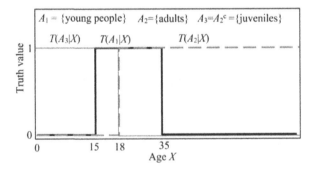

Fig. 1. Three sets form a cover of U, indicating the semantic meanings of y_1, y_2, and y_3.

In this example, $T(A_2) + T(A_3) = 1$. If $T(A_1) = 0.3$, then the sum of the three logical probabilities is $1.3 > 1$. However, the sum of three statistical probabilities $P(y_1) + P(y_2) + P(y_3)$ must be 1. $P(y_1)$ may change from 0 to 0.3.

Theorem 1. If $P(X)$ and $P(X|y_j)$ come from the same sample D that is big enough so that every possible example appears at least one time, then we can directly obtain the numerical solution of feature function of A_j (as shown in Fig. 2(a)) according to Bayes' Theorem III and II:

$$T^*(A_j|X) = \frac{P(X|y_j)}{P(X)} \Big/ \max\left(\frac{P(X|y_j)}{P(X)}\right) = P(y_j|X)/\max(P(y_j|X)) \qquad (3.1)$$

It is easy to prove that changing $P(X)$ and $P(Y)$ does not affect $T*(A_j|X)$ because $T*(A_j|X)$ ($j = 1, 2, \ldots$) reflect the property of Shannon's channel or the semantic channel. This formula is also tenable to a fuzzy set θ_j and compatible with Wang's random set falling shadow theory about fuzzy sets [20]. The compatibility will be discussed elsewhere. If $P(X|y_j)$ is from another sample instead of the sample with $P(X)$, then $T*(A_j|X)$ will not be smooth as shown in Fig. 2(a) and (b). The larger the size of D is, the smoother the membership function is.

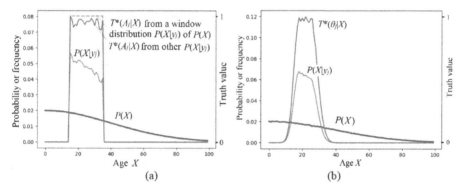

Fig. 2. The numerical solution of the membership function according to (3.1); (a) for a set and (b) for a fuzzy set.

3.2   Selecting Examples for Atomic Labels' Learning

According to mathematical logic, k atomic propositions may produce 2^k independent clauses. The logical add of some of them has 2^{2**k} results. So, there are 2^{2**k} possible compound labels. To simplify the learning, we may filter examples in a multi-label sample to form a new sample D_a with k atomic labels and k corresponding negative labels. We may use First-Order-Strategy [1] to split examples in D with multi-labels or multi-instances into simple examples, such as, to split $(x_1; a_1, a_2)$ into $(x_1; a_1)$ and $(x_1; a_2)$, and to split $(x_1, x_2; a_1)$ into $(x_1; a_1)$ and $(x_2; a_1)$. Let Y_a denote one of the $2k$ labels, i.e. $Y_a \in \{a_1, a_1', a_2, a_2', \ldots, a_k, a_k'\}$. Consider that some a_j' does not appear in D_a, $|D_a|$ may be less than $2k$. From D_a, we can obtain $P(X, Y_a)$ and corresponding semantic channel $T*(\theta_a|X)$ or $T*(\theta_{aj}|X)$ ($j = 1, 2, \ldots, k + k'$).

3.3   Multi-label Learning for Truth Functions with Parameters

If $P(Y, X)$ is obtained from a not large enough sample, we can optimize the truth function with parameters of every compound label by

$$T^*(\theta_j|X) = \arg\max_{T(\theta_j|X)} I(X; \theta_j) = \arg\max_{T(\theta_j|X)} \sum_i P(x_i|y_j) \log \frac{T(\theta_j|x_i)}{T(\theta_j)} \qquad (3.2)$$

It is easy to prove that when $P(X|\theta_j) = P(X|y_j)$, $I(X; \theta_j)$ reaches the maximum and is equal to the KL information $I(X; y_j)$. So, the above formula is compatible with (3.1). Comparing two truth functions, we can find logical implication between two labels. If $T(\theta_j|X) \le T(\theta_k|X)$ for every X, then y_j implies y_k, and θ_j is the subset of θ_k.

We may learn from the BR method [2] to optimize the truth function of an atomic label with both positive and negative instances by

$$T^*(\theta_{aj}|X) = \arg\max_{T(\theta_j|X)} [I(X; \theta_{aj}) + I(X; \theta_{aj}^c)]$$

$$= \arg\max_{T(\theta_{aj}|X)} \sum_i \left[P(x_i|a_j) \log \frac{T(\theta_{aj}|x_i)}{T(\theta_{aj})} + P(x_i|a_j') \log \frac{1 - T(\theta_{aj}|x_i)}{1 - T(\theta_{aj})} \right] \qquad (3.3)$$

$T*(\theta_{aj}|x_i)$ is only affected by $P(a_j|X)$ and $P(a_{j'}|X)$. For a given label, this method divides all examples into three kinds (the positive, the negative, and the unclear) instead two kinds (the positive and the negative) as in One-vs-Rest and BR methods. $T*(\theta_{aj}|x_i)$ is not affected by unclear instances or $P(X)$. The second part may be 0 because the new method allows that a negative label a_j' does not appear in D or D_a.

In many cases where we use three or more labels rather than two to tag some dimension of instance spaces, the formula (3.2) is still suitable. For example, the truth functions of "Child", "Youth", and "Adult" may be separately optimized by three conditional sampling distributions; "Non-youth" will not be used in general. A number of binary classifications [1, 2] are not necessary.

3.4   Multi-label Selection (the Sender's Selective Classification)

For the visible instance X, the label sender selects y_j^* by the classifier

$$y_j^* = h(x_i) = \arg\max_{y_j} \log I(\theta_j; x_i) = \arg\max_{y_j} \log[T(\theta_j|x_i)/T(\theta_j)] \qquad (3.4)$$

Using $T(\theta_j)$ can overcome the class-imbalance problem. If $T(\theta_j|X) \in \{0, 1\}$, the information measure becomes Bar-Hillel and Carnap's information measure [19]; the classifier becomes

$$y_j^* = h(x_i) = \arg\max_{y_j \text{ with } T(A_j|x_i)=1} \log[1/T(A_j)] = \arg\min_{y_j \text{ with } T(A_j|x_i)=1} T(A_j) \qquad (3.5)$$

For $X = x_i$, if several labels are correct or approximatively correct, y_j^* will be one of 2^k independent clauses. When $k = 2$, these clauses are $a_1 \wedge a_2, a_1 \wedge a_2', a_1' \wedge a_2$, and $a_1' \wedge a_2'$. Therefore, this result is similar to what the BR method provides. When sets are fuzzy, we may use a slightly different fuzzy logic [6] from what Zadeh provides [14] so that a compound label is a Boolean function of some atomic labels. We use

$$T(\theta_1 \cap \theta_2^c | X) = \max(0, T(\theta_1 | X) - T(\theta_2 | X)) \tag{3.6}$$

so that $T(\theta_1 \cap \theta_1^c | X) = 0$ and $T(\theta_1 \cup \theta_1^c | X) = 1$. Figure 3 shows the truth functions of 2^2 independent clauses, which form a partition of plan $U * [0, 1]$.

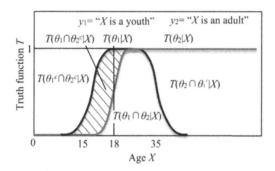

Fig. 3. The truth functions of 2^2 independent clauses

3.5 Classifier $h(X)$ Changes with $P(X)$ to Overcome Class-Imbalance Problem

Although optimized truth function $T * (\theta_j | X)$ does not change with $P(X)$, the classifier $h(X)$ changes with $P(X)$. Assume that $y_4 =$ "Old person", $T * (\theta_4 | X) = 1/[1 + \exp(-0.2(X - 75))]$, $P(X) = 1 - 1/[1 + \exp(-0.15(X - c))]$. The $h(X)$ changes with c as shown in Table 1.

Table 1. The classifier $h(X)$ for $y_4 =$ "Old person" changes with $P(X)$

c	Population density decreasing ages	x^* for classifier ($y_1 = f(X|X \geq x^*)$)
50	40–60	49
60	50–70	55
70	60–80	58

The dividing point x^* of $h(X)$ increases when old population increases because the semantic information criterion encourages us to reduce the failure of reporting small probability events. Macrobian population's increase will change $h(X)$ and Shannon's channel. Then, the new semantic channel will match new Shannon's channel, and so on. Therefore, the semantic meaning of "Old" should have been evolving with human lifetimes in this way. Meanwhile, the class-imbalance problem is overcome.

4 The CM Iteration Algorithm for the Multi-label Classification of Unseen Instances

For unseen instances, assume that observed condition is $Z \in C = \{z_1, z_2, \ldots\}$; the classifier is $Y = f(Z)$; a true class or true label is $X \in U = \{x_1, x_2, \ldots\}$; a sample is $D = \{(x(t); z(t)) | t = 1, 2, \ldots, N; X(t) \in U; z(t) \in C\}$. From D, we can obtain $P(X, Z)$. If D is not big enough, we may use the likelihood method to obtain $P(X, Z)$ with parameters. The problem is that Shannon's channel is not fixed and also needs optimization. Hence, we treat the unseen instance learning as semi-supersized learning. We can use the Channels' Matching (CM) iteration algorithm [9, 10].

Let C_j be a subset of C and $y_j = f(Z | Z \in C_j)$. Hence $S = \{C_1, C_2, \ldots\}$ is a partition of C. Our aim is, for given $P(X, Z)$ from D, to find optimized S, which is

$$S^* = \arg \max_S I(X; \theta | S) = \arg \max_S \sum_j \sum_i P(C_j) P(x_i | C_j) \log \frac{T(\theta_j | x_i)}{T(\theta_j)} \tag{4.1}$$

First, we obtain the Shannon channel for given S:

$$P(y_j | X) = \sum_{z_k \in C_j} P(z_k | X), \quad j = 1, 2, \ldots, n \tag{4.2}$$

From this Shannon's channel, we can obtain the semantic channel $T(\theta | X)$ in numbers or with parameters. For given Z, we have the conditional semantic information

$$I(X_i; \theta_j | Z) = \sum_i P(X_i | Z) \log \frac{T(\theta_j | X_i)}{T(\theta_j)} \tag{4.3}$$

Then let the Shannon channel match the semantic channel by

$$y_j = f(Z) = \arg \max_{y_j} I(X; \theta_j | Z), \quad j = 1, 2, \ldots, n \tag{4.4}$$

Repeat (4.2)–(4.4) until S does not change. The convergent S is the S^* we seek. Some iterative examples show that the above algorithm is fast and reliable [10].

5 Summary

This paper provides a ne multi-label learning method: using the third kind of Bayes' theorem (for larger samples) or the generalized Kullback–Leibler formula (for not big enough samples) to obtain the membership functions from sampling distributions, without the special requirement for samples. The multi-label classification is to partition instance space with the maximum semantic information criterion, which is a special regularized least squares criterion and is equivalent to the maximum likelihood criterion. To simplify multi-label learning, we discuss how to use some atomic labels' membership functions to form a compound label's membership function. We also discuss how the classifier changes with the prior distribution of instances and how the class-imbalance problem is overcome for better generalization performance. We treat unseen instance classification as semi-supervised learning and solve it by the Channel Matching (CM) iteration algorithm, which is fast and reliable [9].

References

1. Zhang, M.L., Zhou, Z.H.: A review on multi-label learning algorithm. IEEE Trans. Knowl. Data Eng. **26**(8), 1819–1837 (2014)
2. Zhang, M.L., Li, Y.K., Liu, X.Y., et al.: Binary relevance for multi-label learning: an overview. Front. Comput. Sci. **12**(2), 191–202 (2018)
3. Gold, K., Petrosino, A.: Using information gain to build meaningful decision forests for multilabel classification. In: Proceedings of the 9th IEEE International Conference on Development and Learning, pp. 58–63. Ann Arbor, MI (2010)
4. Doquire, G., Verleysen, M.: Feature selection for multi-label classification problems. In: Cabestany, J., Rojas, I., Joya, G. (eds.) IWANN 2011. LNCS, vol. 6691, pp. 9–16. Springer, Heidelberg (2011). https://doi.org/10.1007/978-3-642-21501-8_2
5. Reyes, O., Morell, C., Ventura, S.: Effective active learning strategy for multi-label learning. Neurocomputing **273**(17), 494–508 (2018)
6. Lu, C.: B-fuzzy quasi-Boolean algebra and a generalize mutual entropy formula. Fuzzy Syst. Math. (in Chinese) **5**(1), 76–80 (1991)
7. Lu, C.: A Generalized Information Theory (in Chinese). China Science and Technology University Press, Hefei (1993)
8. Lu, C.: A generalization of Shannon's information theory. Int. J. Gener. Syst. **28**(6), 453–490 (1999)
9. Lu, C.: Semantic channel and Shannon channel mutually match and iterate for tests and estimations with maximum mutual information and maximum likelihood. In: 2018 IEEE International Conference on Big Data and Smart Computing, pp. 227–234. IEEE Conference Publishing Services, Piscataway (2018)
10. Lu, C.: Channels' matching algorithm for mixture models. In: Shi, Z., Goertzel, B., Feng, J. (eds.) ICIS 2017. IAICT, vol. 510, pp. 321–332. Springer, Cham (2017). https://doi.org/10.1007/978-3-319-68121-4_35
11. Anon, Cross entropy, Wikipedia: the Free Encyclopedia. https://en.wikipedia.org/wiki/Cross_entropy. Edited on 13 Jan 2018
12. Tarski, A.: The semantic conception of truth: and the foundations of semantics. Philos. Phenomenol. Res. **4**(3), 341–376 (1944)
13. Davidson, D.: Truth and meaning. Synthese **17**(1), 304–323 (1967)
14. Zadeh, L.A.: Fuzzy sets. Inf. Control **8**(3), 338–353 (1965)
15. Bayes, T., Price, R.: An essay towards solving a problem in the doctrine of chance. Philos. Trans. R. Soc. Lond. **53**, 370–418 (1763)
16. Shannon, C.E.: A mathematical theory of communication. Bell Syst. Tech. J. **27**, 379–429 and 623–656 (1948)
17. Popper, K.: Conjectures and refutations. Repr. Routledge, London and New York (1963/2005)

18. Goodfellow, I., et al.: Generative Adversarial Networks (2014). arXiv:1406.2661[cs.LG]
19. Bar-Hillel, Y., Carnap, R.: An outline of a theory of semantic information. Technical report No. 247, Research Lab. of Electronics, MIT (1952)
20. Wang, P.Z.: Fuzzy Sets and Random Sets Shadow. Beijing Normal University Press, Beijing (1985). (in Chinese)

Exploiting the Similarity of Top 100 Beauties for Hairstyle Recommendation via Perceptual Hash

Chentong Zhang and Jiajia Jiao[✉]

College of Information Engineering,
Shanghai Maritime University, Shanghai, China
zhangchentong@stu.shmtu.edu.cn,
jiaojiajia@shmtu.edu.cn

Abstract. In recent years, with the fast development of the fashion industry, the definition of beauty is constantly changing and the diversity of women's hairstyles always gives us a dazzling feeling. Most young female would like to pursue the varying fashion trend and change a hairstyle to meet the fashion requirements. How to choose an appropriate fashion hairstyle has become a critical issue for modern fashion women. In this paper, we design and implement a C++ based female hairstyle recommendation system. At first, the selected top 100 beauties per year in the world are collected for building the beauty standard. Then, considering the changing female status (e.g., age, weight), perceptual hash algorithm (pHash) is used to calculate the global similarity between the user and 100 beauties for selecting the appropriate hairstyle. Finally, the comprehensive rank results from more than 350 demonstrate the recommended hairstyle is preferred.

Keywords: Hairstyle recommendation · Top 100 beauties
Perceptual hash algorithm · Image similarity

1 Introduction

Modern women pay more and more attention to fashion experience. Fashion expression is not only on clothing, but also on hairstyle. The common recommendation modules include Amazon's "guessing you like" block, and so on. It uses an object based collaborative filtering algorithm [1]. The key is that the system will record the user's previous search records to understand the user's preferences. At present, there is no recommendation system software for users to recommend hairstyles, most of them are similar to Meitu and other beauty software included in the wearing hairstyle model, users to speculate whether the hairstyle is suitable for itself. Obviously, there is a problem that the accuracy of hairdressing after face beauty treatment.

This paper discusses the techniques of recommendation based on machine learning, image data processing and image similarity calculation, which are recommended by the data set with the similarity of the user's face, which is more pertinent, and can solve the current problem which is difficult to personalized recommendation for the user's hair.

Z. Shi et al. (Eds.): ICIS 2018, IFIP AICT 539, pp. 49–59, 2018.
https://doi.org/10.1007/978-3-030-01313-4_6

The implementation of the hairstyle recommendation system can provide a quick hairstyle choice for female users.

This paper is organized as follow: Sect. 1 introduces the brief information and Sect. 2 presents the related work. Section 3 provides the detailed hairstyle recommendation design. Section 4 describes the implementation for hairstyle recommendation and Sect. 5 is the results analysis, and Sect. 6 concludes the entire work.

2   Related Work

A perfect hairstyle could enhance anyone's self-confidence, especially women. However, in order to choose a good hairstyle, one was limited to rely on knowledge of a beauty expert. Wisuwat Sunhem, et al. presented a hairstyle recommendation system for women based on hairstyle experts' knowledge and a face shape classification scheme [2]. The system classified the user's face shape into five categories which were suitable for hairstyle recommendation using support vector machine algorithms. The hairstyle e rules were based on beauty experts' suggestions. The system is based on the fine grained similarity of face shape. However, human face shape change over time. For example, getting fat due to higher work pressure can change the shape of a person's face. Therefore, our focus is to design and implement a coarse-grain similarity of face shape for hairstyle recommendation.

Everyone has different understanding of the concept of beauty. It is difficult to define most of the things that most people think of beauty. However, at this stage, there are many media widely selected beauty. The result of the selection makes people have a certain grasp of the fashion trend. Because of the social effect, women will choose the hair, makeup, dress and so on, which makes the hairstyle recommendation have the standard of beauty. Therefore, how to define and verify the beauty is also our important work in the hairstyle recommendation.

3   Proposed Hairstyle Recommendation System

3.1   Overall Framework

The three parts of the design of the hairstyle recommendation system based on machine learning are functional requirement analysis, module analysis and algorithm design analysis. Functional requirement analysis mainly explains the modules needed by the system and the functions that each module needs to achieve. Module analysis is also the structural division of the system. At the present stage, data acquisition, data updating, face detection, recommendation processing and recommendation results display are divided into three parts. Algorithm analysis is mainly about the content of image similarity.

The overall design idea of the system is shown in Fig. 1. User function modules include registration, login, password modification, uploading and saving pictures, face detection, hairstyle data set update, user view of recommended hairstyle results and data collection material display.

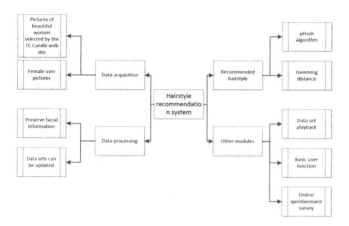

Fig. 1. Overall framework map

3.2   Data Collection and Pre-handling

Although people can feel beauty and recognize beauty, there are many different answers to the question of "what is beauty". Beauty is a subjective feeling. If something is in line with people's aesthetic sense, then it is said that this thing is beautiful. Similarly, the beauty of the world elected annually is most commonly recognized as "beauty". Beauty is the effect produced by all things. A person is beautiful, and we can think that the hairstyle that she owns is beautiful. Therefore, the 2013–2017 years of the TC Candle website [7] recommended by the hairstyle recommender system is selected as a dataset. The acquisition of data sets adopts the way of intercepting and saving video frames in a fixed time interval. Female users recommend different data sets according to their favorite trends. For example, some people like retro wind and some like new trends. The advantage of the system data set design is to provide users with more varied choices of hairstyles.

The data sets are used for face detection, and the face part is detected and preserved. This step completes the image preprocessing operation, and the final recommendation results are based on the similarity ratio between the user's face and the beautiful face of the data set. Figure 2 shows the process of image processing.

(a)User picture   (b)Detection picture   (c)Face picture

Fig. 2. pHash algorithm flow chart

3.3   Perceptual Hash Based Recommendation

When most pictures are converted to frequency domain, their coefficients are basically 0, which is determined by the redundancy and relevance of pictures. The key technique for the search of similar pictures in Google or Baidu's map function is called "perceptual Hash algorithm" [3]. The function is to generate a "fingerprint" string for each picture, and to compare the similarity of the picture by comparing the fingerprints produced by different pictures. The common perceptual hashing algorithm is divided into low frequency perceptual hashing algorithm, dHash algorithm and pHash algorithm. The pHash algorithm is used in this system.

(1) **pHash algorithm.**

Although the mean hash algorithm based on low frequency is simple, there are many factors that can affect the mean value, and the reliability is reduced. In each step of perceptual hash algorithm, histogram equalization is needed, which will affect the mean size and affect the hash value. Then, the enhanced version of pHash algorithm can solve this problem well, and get low-frequency components with DCT transform. The working process of the pHash algorithm is shown in Fig. 3.

Fig. 3. pHash algorithm flow chart

- Reduce the size: use small pictures, but at least more than 8*8, where 32*32 is the best size. Reducing the size of the picture simplifies the calculation of DCT.
- Simplify the color: transform the image into grayscale image, discard the color factors of the image, and further simplify the computation.
- Calculate DCT: DCT transform and get the DCT coefficient matrix of 32*32.
- Reduce DCT: just keep the 8*8 matrix in the upper left corner, because this part presents the lowest frequency in the picture.

- Calculate the mean value: the mean of DCT is calculated.
- Calculate the hash value: if the DCT value of the pixel is greater than or equal to the DCT mean, then it is set to "1", and the less than DCT mean is set to "0". In this way, a 64 bit binary number, the hash value of a picture, is formed by comparison.

When most pictures are converted to frequency domain, their coefficients are basically 0, which is determined by the redundancy and relevance of pictures. The more similar the pictures' hash value is, the more similar the picture is. The system uses the perceptual Hash algorithm to find the largest image of the person's face in a number of data sets and recommends it. In the perceptual Hash algorithm, the pHash algorithm is improved on the original low frequency perception Hash algorithm. In summary, the system uses the pHash algorithm to convert the data to a hash value.

(2) Similarity algorithm

The core of recommendation algorithm is to measure the similarity of faces between users and data sets. Here we use the pHash algorithm in perceptual hashing algorithm to get hash value. Next, we need to compare the similarity of hash values produced by different images. The following is the image Hash similarity measure method. Let S_1, S_2 be a hash sequence of length N, where $S_1(i)$ and $S_2(i)$ are the i elements in S_1 and S_2 sequences, respectively, $D(S_1, S_2)$ is a similar measure of hash S_1 and S_2.

(1) *Hamming distance*

Hamming distance, also called Hing distance, requires hash sequence as a 0/1 form. The method of measuring the difference is by comparing the number of different values of the corresponding position 0/1 of the two sequences. The size of the Hamming distance is proportional to the size of the corresponding image difference, that is, the larger the value is, the larger the image difference is, the more novel the value is, the more novel a map is. The smaller the difference in content is [12]. The formula of Hamming distance described in Eq. (1):

$$D(s_1, s_2) = \sum |s_1(i) - s_2(i)| \tag{1}$$

(2) *Correlation coefficient*

Correlation coefficient is used to distinguish the linear degree between the hash sequences obtained by the image. The calculation formula of correlation coefficient is Eq. (2). The larger value represents the higher similarity.

$$D(s_1, s_2) = \frac{C(s_1, s_2)}{\sqrt{V(s_1)} \sqrt{V(s_2)}} \tag{2}$$

(3) *Euclidean distance*

Euclidean distance, also known as Euclidean distance, measures the absolute distance between two hash sequences. The Euclidean distance of the two images of hash S1 and S2 is as follows in Eq. (3):

$$D(s_1, s_2) = \sqrt{\sum_{i=1}^{N} [s_1(i) - s_2(i)]^2} \tag{3}$$

Generally, the smaller the value, the more similar between the two images; on the contrary, the greater the value, the greater the difference in the image, but the Euclidean distance measurement has a very obvious disadvantage, which is influenced by the scale of different units of the index.

(4) *Norm*

The norm is also the Minkowski distance. The norm between the two image hash sequences is defined as follows in Eq. (4), where P is a parameter.

$$D(s_1, s_2) = \left(\sum_{i=1}^{N} [s_1(i) - s_2(i)]^P \right)^{\frac{1}{P}} \tag{4}$$

(5) *Angle cosine*

The angle cosine method compares the degree of difference in the two vector directions, takes the Hash sequence as a vector, and uses the angle cosine method to make the qualitative measure. The definition of the angle cosine is shown in Eq. (5):

$$D(s_1, s_2) = \cos \theta = \sqrt{\frac{(s_1^T s_2)^2}{s_1^T s_1 s_2^T s_2}} \tag{5}$$

The angle of the angle is from 0 to 90°. The bigger the angle cosine value is, the greater the difference is.

In the process of similarity measurement, pHash chooses Hamming distance [4]. The advantage is that the algorithm runs fast and ignores image scaling. Therefore, Hairstyle recommendation system adopts Hamming distance to measure similarity due to its simple idea and high efficiency.

3.4   Haar Feature Classified

The program uses the haar feature classifier [5] in OpenCV to realize the face detection function. OpenCV provides a set of object detection function. After training, it can detect any object you need. The library has many test parameters that can be directly used for many scenes, such as face, eyes, mouth, body, upper body, lower body and smiling face. The detection engine consists of a very simple detector cascade. These detectors are called Haar feature detectors, each of which has different scales and weights. In the training phase, the decision tree is optimized by known correct and wrong pictures. After initializing the face detector with the required parameters, it can be used for face detection. In the detection process, the trained classifier traverses each pixel of the input image with different scales to detect faces of different sizes.

4   Implementation

4.1   Programming Language

The system uses the C++ language [6] to develop, the hair recommendation system mainly uses the image processing technology. Because of the large amount of data, the time consumption of image processing is more than that of the general algorithm. Then, it is very important to improve the processing efficiency for improving the time consumption. The C++ language has the following advantages: (1) high flexibility. For example, C++ language uses pointer and pointer to realize dynamic memory allocation. (2) The rate of memory utilization is improved. There is a limit to the memory of the computer. Large quantities of images take up a lot of memory resources. However, the C++ language can achieve direct allocation and release of memory, thus improving the memory utilization. (3) It has rapid execution efficiency. The compiled language generated by the C++ language is the executable code. It can run directly on the processor.

4.2   Critical Persuade Code for Hairstyle Recommendation

As Fig. 4 shows, the hairstyle recommendation module recommends the corresponding hairstyle pictures based on the name of the input user picture, and the fifth line pseudo code calls the CalHashValue function to transform the picture into a string by using the pHash algorithm. In line 3 to line 7, pseudo code realizes the Hamming distance value of the string corresponding to the user picture and the string corresponding to each picture in the dataset. The operation of pseudo code in line 8 to line 13 is to find out the smallest picture with Hamming distance. The corresponding picture is finally displayed.

5   Results and Analysis

The system is composed of six main parts. The system login interface, as shown in Fig. 5(a), is successfully entered into the main interface of the system through the login system after the login account. In the main interface shown in Fig. 5(b), each function module and the interface to add and save the picture are included. The face detection interface realizes the recognition and preservation of the face parts in the picture, and lists the number of the detected faces at the bottom of the interface and the detection

	Input: User picture name & Output: Recommend hairstyle picture
1	**Begin:**
2	Call function **CalHashValue**(Mat &src)
3	Str1:=Hash value of user picture
4	**for** ImageNum=1 to ImageSumNumber:
5	Str2:=Hash value of a data set picture
6	Call Function **CalHanmingDistance**(string &str1,string&str2)
7	StoreHamingdistance[ImageNum]:=**CalHanmingDistance**(str1,str2)
8	**for** ImageNum=1 to ImageSumNumber:
9	if storeHamingdistance[ImageNum]<min && storeHamingdistance[ImageNum]!=0
10	Then min:=storeHamingdistance[ImageNum]
11	storemostsimilarImage:=ImageNum
12	Search the corresponding picture of the data set the name of the storemostsimilarImage
13	Display a corresponding hairstyle picture
14	**End**

Fig. 4. Critical persuade code for hairstyle recommendation

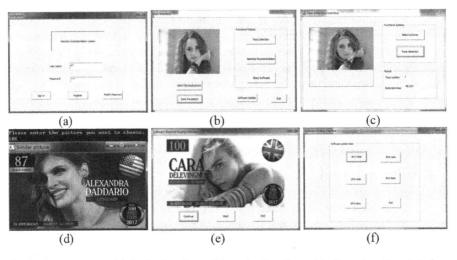

(a)   (b)   (c)

(d)   (e)   (f)

Fig. 5. System page (a) login interface; (b) main interface; (c) face detection interface; (d) hairstyle recommendation interface; (e) software material display interface; (f) software update interface

time, as shown in Fig. 5(c). Figure 5(d) shows the hairstyle recommendation interface. The user enters the picture name, and the system will show the recommended hairstyle pictures. The content of the dataset is displayed in the software material display interface, which is similar to the video player interface, as shown in Fig. 5(e). The software update interface, as shown in Fig. 5(f), implements the updated hairstyle data set, which includes 2013–2017 years of data.

After the completion of the system design and the test work, in order to understand the satisfaction of the user to the results of the recommendation, an online questionnaire is set up. The content of the questionnaire is to give a user picture as a topic, and let the investigators anonymous to the user's hairstyle recommended by the hairstyle recommendation system (in this case, the recommended hairstyle 1) and the other two

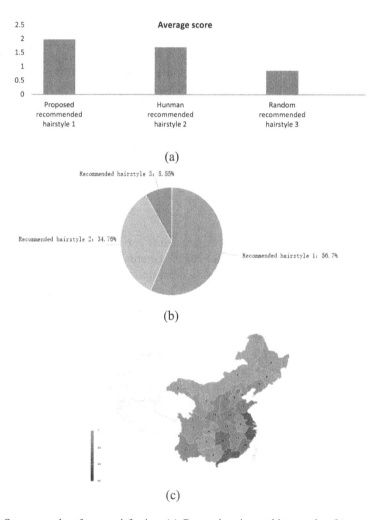

(a)

(b)

(c)

Fig. 6. Survey results of user satisfaction. (a) Comprehensive ranking results; (b) recommended hairstyle for the first place; (c) geographic distribution map of the respondents.

hairstyle pictures (recommended hairstyle 2 and recommended hairstyle 3) are sorted. After the online questionnaire was released, a total of 352 questionnaires were collected. The wide range of respondents has enhanced the credibility of the survey results. After recommending the system to test the user's picture, the recommended hairstyle is recommended hairstyle 1. The average comprehensive score of the options is calculated automatically according to the sorting of all the candidates. It reflects the comprehensive ranking of the options, and the higher the score is, the better the comprehensive ranking is. The calculation method is as follows in Eq. (6):

$$\text{Averagescore} = \frac{\sum (\text{frequency} * \text{weight})}{\text{Number}} \tag{6}$$

The variable Number refers to the number of people filling in the questionnaire and the weight is determined by the position of the option being arranged. For example, there are 3 options for sorting, the weight in the first place is 3, the second position is 2, and the third position is 1. The results of the statistical analysis of the questionnaire are shown in Fig. 6(a) and the ratio of the recommended hairstyle 1 to the first is shown in Fig. 6(b). From the graph, we can see that the user satisfaction of hairstyle recommendation system recommended is high, which proves that the hairstyle recommendation system has great application prospects. Considering the disagreement between people in different regions, the survey of online questionnaires is widely distributed, accounting for 88.6% of China's region, as shown in Fig. 6(c).

6   Conclusion

The hairstyle recommendation system recommends is based on the face coarse-grain global similarity, the benefit lies in ignoring the external influence factors of the face change, and can provide the hair recommendation reference for the fashionable female users quickly and effectively. Based on a lot of survey work and analysis, this paper has stipulated the standard of hairstyle beauty. The dataset is updatable and scalable, so that users can choose different data sets according to their preferences. The user uploads the picture to be recommended. Then, face detection module keeps only the valid face information of the user's picture. Finally, the system compares the face part of the user picture with the data set, and draws the most similar face, and recommends its corresponding hairstyle. In addition, user satisfaction surveys were conducted on the results of the system hairstyle recommendation, and the user satisfaction rate of the recommended hairstyle was 23.4% higher than the average satisfaction.

Acknowledgement. The work is granted by National Natural Science Foundation of China, numbered 61502298 and Shanghai Maritime University Innovative Program Grant.

References

1. Wang, X., Wang, C.: Recommendation System of e-Commerce Based on Improved Collaborative Filtering Algorithm, pp. 332–335. IEEE, Piscataway (2017)
2. Sunhem, W., Pasupa, K., Jansiripitikul, P.: Hairstyle Recommendation System for Women, pp. 166–169. IEEE, Piscataway (2016)
3. Liu, X., Zhang, Q., Luan, R., Yu, F.: Applications of Perceptual Hash Algorithm in Agriculture Images, pp. 698–702. IEEE, Piscataway (2013)
4. Haifeng, H., Zhang, L., Wu, J.: Hamming Distance Based Approximate Similarity Text Search Algorithm, pp. 1–6. IEEE, Piscataway (2015)
5. Sharifara, A., Rahim, M.S.M., Anisi, Y.: A General Review of Human Face Detection Including a Study of Neural Networks and Haar Feature-Based Cascade Classifier in Face Detection, pp. 73–78. IEEE, Piscataway (2014)
6. Bhattacharya, P., Neamtiu, I.: Assessing Programming Language Impact on Development and Maintenance: A Study on C and C++, pp. 171–180. IEEE, Piscataway (2011)
7. http://independentcritics.com/

Attribute Coordinate Comprehensive Evaluation Model Combining Principal Component Analysis

Xiaolin Xu[1(✉)], Yan Liu[1(✉)], and Jiali Feng[2]

[1] Shanghai Polytechnic University, Shanghai, China
xlxu2001@163.com, liuyan@sspu.edu.cn
[2] Shanghai Maritime University, Shanghai, China
jlfeng@shmtu.edu.cn

Abstract. Attribute coordinate comprehensive evaluation method features subjective weighting in which the weights of indicators are determined by evaluators, which possibly leads to the arbitrariness in setting the weights. When there are many indicators, it is difficult to accurately judge if the sample is better or worse than others. To address the problem, this paper applies principal component analysis on the attribute coordinate comprehensive evaluation method. When there are many indicators, they can be reduced to new indicators with related meanings given through the method of principal component analysis. With the simplification, it will greatly facilitate experts to rate samples, which is the paramount basis that provides the preference of experts for the attribute coordinate comprehensive evaluation method to further calculate all the satisfaction degrees of objects to be evaluated. Experimental results show the advantages of the improved algorithm over the original algorithm.

Keywords: Attribute coordinate comprehensive evaluation
Principal component analysis · Barycentric coordinates
Global satisfaction degree

1 Introduction

In regards to comprehensive evaluation, the most important problem needed to address is how to set the weight of each evaluated indicator. The setting of weights can fall into two categories. One is the subjective weighting, such as AHP [1, 2], and the other is the objective weighting, such as the least square method and the principal component analysis [3, 4]. The two types have their own advantages and disadvantages. Subjective weighting is that the weights are given by experts and could be arbitrary in some cases, while objective weighting is not able to reflect the experiences or preferences of experts. Attribute coordinate comprehensive evaluation, belonging to the former, whose characteristic is that it can construct the corresponding psychological preference curve through evaluators rating the sample data in light of their own experiences or

The work was supported by the Key Disciplines of Computer Science and Technology of Shanghai Polytechnic University (No. XXKZD1604).

Z. Shi et al. (Eds.): ICIS 2018, IFIP AICT 539, pp. 60–69, 2018.
https://doi.org/10.1007/978-3-030-01313-4_7

preferences, has made certain progress both in theory and practice [5–13]. However, when indicators are many, it is difficult for experts to accurately distinguish satisfactory samples from unsatisfactory samples, which might result in arbitrary ratings on some samples. To address the obstacle, the principal component analysis method is used to reduce the number of indicators and give the related meanings of new indicators, so it is easier for experts to rate on samples with new indicators.

This paper first introduces the steps of simplification of indicators by means of the principal component analysis, then explores the core idea of the attribute coordinate comprehensive evaluation method, and next elaborates the process of combining the two methods through the simulation and the comparison of results before and after the model is improved.

2 Reduction of Indicators by Principal Component Analysis

Principal component analysis is a method of dimensionality reduction in mathematics. The basic idea is to try to make the original indicators $X_1, X_2, \dots X_t$ (for example, there are t indicators) recombined into a set of relatively unrelated comprehensive indicators F_m with fewer numbers than the number of original indicators. The specific steps of the principal component analysis are as follows:

(1) **Calculate the covariance matrix**

Calculate the covariance matrix $s = (s_{ij})_{p \times p}$ of sample data

$$s_{ij} = \frac{1}{n-1} \sum_{k=1}^{n} (x_{ki} - \bar{x}_i)(x_{kj} - \bar{x}_j) \quad i,j = 1, 2, \dots, p \tag{1}$$

Among them, s_{ij} $(i, j = 1, 2, \dots, p)$ is the correlation coefficient between the original variable x_i and x_j. p is the number of indicators. n is the number of samples. $\bar{x}_i$ and $\bar{x}_j$ is respectively the mean of values of indicator i and j. x_{ki} is the value of indicator i of a certain sample, and x_{kj} is the value of indicator j of a certain sample.

(2) **Calculate the eigenvalues λ_i of S and orthogonal unit eigenvectors a_i.**

The first m larger eigenvalues of S, $\lambda 1 \geq \lambda 2 \geq \dots \lambda m > 0$, is the variance of the first m principal components, and the unit eigenvector a_i corresponding to λ_i is the coefficient of the principal component F_i, and then the ith principal component Fi is:

$$F_i = a_i X \tag{2}$$

The variance (information) contribution rate of principal components reflects the information magnitude, γ_i is:

$$\gamma_i = \lambda_i / \sum_{i=1}^{m} \lambda_i \tag{3}$$

(3) **Determine the principal components**

The final principal components to be selected are $F1$, $F2$, ... Fm, and m is determined by the cumulative contribution rate of variance $G(m)$.

$$G(m) = \sum_{i=1}^{m} \lambda_i / \sum_{k=1}^{p} \lambda_k \tag{4}$$

When the cumulative contribution rate is greater than 85%, it will be considered enough to reflect the information of the original variables, and m is the extracted first m principal components.

(4) **Calculate the load of the principal components**

The principal component load reflects the degree of correlation between the principal component F_i and the original variable X_j, and the load $l_{ij}(i = 1, 2, ..., m; j = 1,2, ..., p)$ of the original variable X_j ($j = 1,2, ... p$) on the principal component F_i ($i = 1, 2, ..., m$) is:

$$l(F_i, X_j) = \sqrt{\lambda_i} a_{ij} (i = 1, 2, ..., m; j = 1, 2, ..., p) \tag{5}$$

(5) **Calculate the scores of the principal components**

The scores on the m principal components of the sample:

$$F_i = a_{1i}X_1 + a_{2i}X_2 + \cdots + a_{pi}X_p \qquad i = 1, 2, ..., m \tag{6}$$

(6) **Select the principal components and give the new meanings**

Provide the new meaning of the new evaluation indicator F_i ($i = 1, 2, ..., m$) for experts to rate on the new samples.

3 Attribute Coordinate Comprehensive Evaluation Model

3.1 Explore Barycentric Coordinates Reflecting Evaluators' Preference Weight

Attribute coordinate comprehensive evaluation method combines machine learning with experts' ratings on sample data. Set T_0 to be the critical total score, T_{max} the largest total score, we evenly select several total scores: T_1, T_2, ... T_{n-1} from (T_0, T_{max}) regarding the curve fitting requirements, and then select some samples on each total score $T_i(i = 1, 2, 3 ... n - 1)$ and rate them according to experts' preference or experiences, which is taken as the process of the learning of samples, so as to get the barycentric coordinate for T_i ($i = 1, 2, 3 ... n - 1$) according to (7).

$$b\big(\{f^h(z)\}\big) = \left(\frac{\sum\limits_{h=1}^{t} v_1^h f_1^h}{\sum\limits_{h=1}^{t} v_1^h}, \ldots, \frac{\sum\limits_{h=1}^{t} v_m^h f_m^h}{\sum\limits_{h=1}^{t} v_m^h}\right) \tag{7}$$

Where, $\{f_k, k = 1, \ldots s\} \subseteq S_T \cap F$ is the set for sample fi with the total score equal to T. In Formula (7), $b(\{v^h(z)\})$ is the barycentric coordinate of $\{v^h(z)\}$, $\{f^h, h = 1, \ldots t\}$ is the values of indicators of t sets of samples the evaluator Z selects from $\{f_k\}$, $\{v^h(z)\}$ is the ratings (or taken as weight) the evaluator gives on the samples.

3.2 Calculate the Most Satisfactory Solution

Use the interpolation formula $G_j(T) = a_{0j} + a_{1j} T + a_{2j} T_2 + \ldots + a_{n+1j} T_{n+1}$ and barycentric coordinates obtained above to do curve fitting and construct the psychological barycentric line (or most satisfactory local solution line) $L(b(\{f^h(z)\}))$; and then calculate the global satisfaction degree according to (8), and sort them in descending order to obtain the most satisfactory solution.

$$sat(f, Z) = \left(\frac{\sum\limits_{i=1}^{m} f_{ij}}{\sum\limits_{j=1}^{m} F_j}\right)^{\left(\frac{\sum\limits_{i=1}^{m} f_j}{3\left(\sum\limits_{j=1}^{m} f_{ij}\right)}\right)} * \exp\left(-\frac{\sum\limits_{j=1}^{m} w_j |f_j - b(f^h(z_j))|}{\sum\limits_{j=1}^{m} w_j \delta_j}\right) \tag{8}$$

Where, $sat(f, Z)$ is the satisfaction of evaluated object f from evaluator Z, whose value is expected to be between 0 and 1. f_j is the value of each indicator. $|f_j - b(f^h(z_j))|$ is to measure the difference between each attribute value and the corresponding barycentric value. w_j and δ_j are used as the factor which can be adjusted to make the satisfaction comparable value in the case where the original results are not desirable. $\sum\limits_{j=1}^{m} F_j$ is the sum of F_j with each indicator value full score. $\sum\limits_{ij=1}^{m} f_{ij}$ is the sum of the values of all the indicators F_{ij} of F_i.

4   Simulation Experiment

To verify the effectiveness of the improved method, we chose the grades of nine courses from 2008 students in the final exam in a high school as the experimental data, nine courses being taken as nine indicators including Chinese, mathematics, English, physics, chemistry, politics, history, geography and biology. The sample data is shown in Table 1.

First of all, we use the attribute coordinate comprehensive evaluation method to respectively construct the psychological barycentric curves of several courses without applying the principal component analysis. And then we improve the method in the

Table 1. Sample data of nine courses

Item	Chinese	Math	English	Physics	Chemistry	Politics	History	Geography	Biology
1	91	68	82	27	55	72	78	71.5	75.5
2	91	77	50	53	65	47	75	70.5	80.5
3	91	88	15	78	48	65	63	66.5	79

way that the principal component analysis is used to simplify the indicators, further the attribute coordinate comprehensive evaluation method is applied to construct the psychological barycentric lines of the new indicators.

We also compare the global satisfaction degrees between two students before and after the improved method is applied.

4.1 Attribute Coordinate Comprehensive Evaluation Without Using Principal Component Analysis

Respectively, we choose the total score of 1000, 701 and 620 as the three evaluation planes, and select some samples for the experts to rate. The last column (Rating) of Tables 2 and 3 are respectively the rating data for total score 701 and 620.

Table 2. The samples and ratings for total score 701

Item	Chinese	Math	English	Physics	Chemistry	Politics	History	Geography	Biology	Rating
1004	97	96	71	75	68	55	78.5	71.5	90.5	8
1005	89	86	66	80	81	69	74.5	69	87	9
1476	94	102	48	74	76	60	77.5	82.5	87	9
398	69	84	80	82	74	70	75	81	85	6

Table 3. The samples and ratings for total score 620

Item	Chinese	Math	English	Physics	Chemistry	Politics	History	Geography	Biology	Rating
1345	85	95	61.5	71	54	47	68	69.5	70	7
1	91	68	82	27	55	72	78	71.5	75.5	9
947	95	90	72.5	79	35	55	65.5	61	67	10
1023	86	80	72.5	52	63	51	76	59	80.5	9
1548	85	89	67.5	69	41	41	71.5	79	77	8

According to (7), the barycentric coordinates of total score 701 and 620 with (Chinese, math, geography) are respectively (88.65625, 92.625, 76.4375) and (88.79069767, 83.93023256, 67.51162791).

Next, according to the interpolation theorem, we calculate the barycentric curves of Chinese, mathematics, geography (respectively shown in Figs. 1, 2, 3). It can be seen that the barycenter curve of Chinese is very unreasonable, as the curve should be monotonically increasing, while in this curve, the curve for total score of 650 is even lower than that of the total score of 600. From Figs. 2 and 3, we can see that barycentric curves of mathematics and geography are almost the same, which is not obvious to see the expert put more weight on arts or science.

The most likely reason for the result is that so many indicators make it difficult for experts to accurately distinguish good samples from bad samples among nine indicators, which could result in arbitrary ratings.

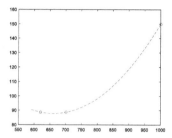

Fig. 1. The barycenter curve of Chinese

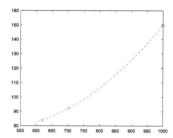

Fig. 2. The barycenter curve of Math

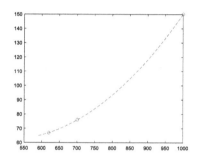

Fig. 3. The barycenter curve of Geography

4.2 Attribute Coordinate Comprehensive Evaluation with Principal Component Analysis

We apply the improved algorithm, first carrying out principal component analysis to reduce the quantity of indicators.

(1) Calculate the covariance matrix S (correlation coefficient matrix) between indicators.

	x1	x2	x3	x4	x5	x6	x7	x8	x9
x1	0.1725	0.3276	−0.2918	0.5982	0.0915	−0.3520	−0.3481	0.2933	−0.2836
x2	0.5151	−0.3978	0.5263	0.4750	0.0890	0.2451	−0.0012	−0.0697	−0.0318
x3	0.2543	0.7953	0.4751	−0.1837	−0.0851	0.1512	0.0638	0.0963	0.0034
x4	0.4274	−0.1859	0.0203	−0.2104	−0.6264	−0.5471	0.0743	0.1566	0.1282
x5	0.3474	−0.0728	−0.0053	−0.3652	0.7509	−0.3622	0.1624	0.1384	−0.0019
x6	0.1596	0.1882	−0.2401	0.2653	0.0964	−0.0190	0.0561	−0.2374	0.8614
x7	0.2310	0.1357	−0.3541	0.1628	−0.0799	0.0730	0.7306	−0.3300	−0.3492
x8	0.3152	−0.0823	−0.4071	−0.1534	−0.0618	0.5785	−0.0036	0.5999	0.0786
x9	0.3983	0.0389	−0.2510	−0.2912	−0.0266	0.1494	−0.5532	−0.5754	−0.1788

(2) **Calculate the eigenvalue vector of the correlation coefficient matrix**

(1.5315, 0.2945, 0.2291, 0.1658, 0.1331, 0.1170, 0.1006, 0.0881, 0.0778)

(3) **Calculate the principal component contribution rate vector λ and cumulative contribution rate G(M).**

The contribution rate vector λ = (55.9456, 10.7591, 8.3684, 6.0553, 4.8631, 4.2733, 3.6763, 3.2172, 2.8418)

The contribution rate of the first three principal components is G(M) = 75.0731%, although there will be some information loss, it is not so great to affect the overall situation.

According to the coefficient matrix S, the expressions of the first three principal components (f1, f2, f3) are respectively as follows.

f1 = 0.1725x1 + 0.5151x2 + 0.2543x3 + 0.4274x4 + 0.3474x5 + 0.1596x6 + 0.231x7 + 0.3152x8 + 0.3983x9

f2 = 0.3276x1 − 0.3978x2 + 0.7953x3 − 0.1859x4 − 0.0728x5 + 0.1882x6 + 0.1357x7 − 0.0823x8 + 0.0389x9

f3 = −0.2918x1 + 0.5263x2 + 0.4751x3 + 0.0203x4 − 0.0053x5 − 0.2401x6 − 0.3541x7 − 0.4071x8 − 0.2510x9

Respectively, x1, x2 … x9 represents Chinese, math…biological.

From the expression of the first principal component f1, it has the positive load on each variable, indicating that the first principal component represents the comprehensive components.

From the expression of the second principal component f2, the value of f2 decreases with the increase of x2(Math), x4(physics) and x5(chemistry), whereas increases with the increase of x3(English), x6(politics), x7(history) and x9(biology), which indicates f2 reflects a student's level of liberal arts.

From the expression of the third principal component f3, the value of f3 increases with the increase of x2(Math), x3(English) and x4(physics), whereas decreases with the increase of x1(Chinese), x6(politics), x7(history), x8(geography) and x9(biology), which indicates f3 reflects a student's level of science.

In this way we can simplify the nine indicators into three ones: f1, f2 and f3. Now we can calculate students' scores with the new indicator system. Table 4 is new sample data with the new indicator system.

Table 4. Sample data with the new indicators

Item	f1 (comprehensive)	f2 (liberal arts)	f3 (science)
1	184.3496	80.13484	55.48093
2	192.4282	40.70861	51.70827
3	192.2151	7.118305	43.40139

(4) **Attribute coordinate comprehensive evaluation**

Respectively we provide three total score planes 460, 345 and 311 for the expert to rate. The scores of the last two total samples are shown in Tables 5 and 6 respectively. The expert's preference can be seen directly from the ratings (the last column). When the total score is higher, the expert pays more attention to the comprehensive level of students. When the total score is relatively lower, the expert values students' science scores more. This evaluation is easier than that without principal component analysis.

Table 5. The samples and ratings for total score around 345

Item	f1 (comprehension)	f2 (liberal arts)	f3 (science)	Total score	Ratings
263	238.4894	44.51444	62.95037	345.9542	7
999	228.5096	62.76003	54.17796	345.4476	6
1007	217.0385	52.21342	75.05868	344.3106	10
1066	245.756	37.56391	60.98198	344.3019	9

Table 6. The samples and ratings for total score around 311

Item	f1 (comprehension)	f2 (liberal arts)	f3 (science)	Total score	Ratings
1074	224.2676	29.82561	57.03609	311.1293	6
1699	166.3697	68.02551	76.22716	310.6224	10
798	205.9612	43.66174	60.83696	310.4599	8
735	199.5349	43.14419	67.52119	310.2003	9

We can obtain the barycentric coordinates of 460, 345 and 311 respectively (268.2157, 92.1146, 100), (223.98, 51.46642, 6922313) and (200.2936, 44.12158, 66.05498). We draw the barycentric curves of indicator f1, f2 and f3 respectively (shown in Figs. 4, 5, 6). It can be seen that the three curves are all monotonically increasing, which are more reasonable than those drawn with the old model.

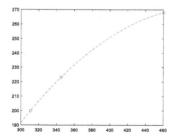
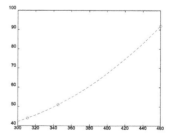

Fig. 4. The barycentric curve of indicator f1   **Fig. 5.** The barycentric curve of indicator f2

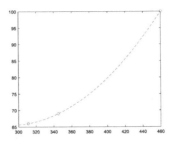

Fig. 6. The barycentric curve of indicator f3

(5) **the Comparison of Satisfaction Degree Before and After Improvement**

Finally, we examine the satisfaction degree obtained respectively using the two models. The followings are the scores of two students No. 466 and No. 196. They almost have the same total score, however, it is obvious that No. 196 is better at science than No. 466. So normally the satisfaction degree of No. 196 should be greater than that of No. 466 under the condition that the evaluator values the science scores more. However the result is opposite in the case of the unimproved method, which is unreasonable (shown in Table 7). Comparatively, the improved algorithm fixes the flaw and obtains the reasonable result, better reflecting the preference of the evaluator (shown in Table 8).

Table 7. The comparison of satisfaction degrees using the unimproved method

Item	Chinese	Math	English	Physics	Chemistry	Politics	History	Geography	Biology	Satisfaction
466	98	65	80	77	53	64	75	55	79.5	0.7896
196	80	72	31	81	71	62	82	80.5	86	0.7745

Table 8. The comparison of satisfaction degrees using the improved method

Item	f1 (comprehension)	f2 (liberal arts)	f3 (science)	Satisfaction
196	255.5199	93.5100	126.7342	0.5708
466	220.5927	118.4874	110.3656	0.5524

5   Conclusion

The improved method integrates principal component analysis into the original method to reduce the number of indicators so as to make the experts' rating process more simple and effective. The simulation examines the comparison of the results before and after using the principal component analysis and shows that the barycentric curves look more favorable, and the satisfaction degrees of the evaluated objects more accurately reflect the preferences and experiences of experts.

References

1. Ji, Z.: Research on personal information security evaluation based on analytic hierarchy process. Value Eng. **5**, 57–60 (2018)
2. Gong, X., Chang, X.: Comprehensive prejudgment of coal mine fire rescue path optimization based on fault tree-analytic hierarchy process. Mod. Electron. Tech. **8**, 151–154 (2018)
3. Pan, X., Liu, F.: Evaluation mode of risk investment based on principal component analysis. Sci. Technol. Prog. Policy **3**, 65–67 (2004)
4. Zhou, B., Wang, M.: A method of cloud manufacturing service QoS evaluation based on PCA. Manuf. Autom. **14**, 28–33 (2013)
5. Xu, X., Xu, G., Feng, J.: Study on updating algorithm of attribute coordinate evaluation model. In: Huang, D.-S., Hussain, A., Han, K., Gromiha, M.M. (eds.) ICIC 2017. LNCS (LNAI), vol. 10363, pp. 653–662. Springer, Cham (2017). https://doi.org/10.1007/978-3-319-63315-2_57
6. Xu, G., Xu, X.: Study on evaluation model of attribute barycentric coordinates. Int. J. Grid Distrib. Comput. **9**(9), 115–128 (2016)
7. Xu, X., Feng, J.: A quantification method of qualitative indices based on inverse conversion degree functions. In: Enterprise Systems Conference, pp. 261–264 (2014)
8. Xu, G., Min, S.: Research on multi-agent comprehensive evaluation model based on attribute coordinate. In: IEEE International Conference on Granular Computing (GrC), pp. 556–562 (2012)
9. Xu, X., Xu, G., Feng, J.: A kind of synthetic evaluation method based on the attribute computing network. In: IEEE International Conference on Granular Computing (GrC), pp. 644–647 (2009)
10. Xu, X., Xu, G.: Research on ranking model based on multi-user attribute comprehensive evaluation method. In: Applied Mechanics and Materials, pp. 644–650 (2014)
11. Xu, X., Xu, G.: A recommendation ranking model based on credit. In: IEEE International Conference on Granular Computing (GrC), pp. 569–572 (2012)
12. Xu, X., Feng, J.: Research and implementation of image encryption algorithm based on zigzag transformation and inner product polarization vector. In: IEEE International Conference on Granular Computing, vol. 95, no. 1, pp. 556–561 (2010)
13. Xu, G., Wang, L.: Evaluation of aberrant methylation gene forecasting tumor risk value in attribute theory. J. Basic Sci. Eng. **16**(2), 234–238 (2008)

A Specialized Probability Density Function for the Input of Mixture of Gaussian Processes

Longbo Zhao and Jinwen Ma[(⊠)]

Department of Information Science, School of Mathematical Sciences and LMAM,
Peking University, Beijing 100871, People's Republic of China
jwma@math.pku.edu.cn

Abstract. Mixture of Gaussian Processes (MGP) is a generative model being powerful and widely used in the fields of machine learning and data mining. However, when we learn this generative model on a given dataset, we should set the probability density function (pdf) of the input in advance. In general, it can be set as a Gaussian distribution. But, for some actual data like time series, this setting or assumption is not reasonable and effective. In this paper, we propose a specialized pdf for the input of MGP model which is a piecewise-defined continuous function with three parts such that the middle part takes the form of a uniform distribution, while the two side parts take the form of Gaussian distribution. This specialized pdf is more consistent with the uniform distribution of the input than the Gaussian pdf. The two tails of the pdf with the form of a Gaussian distribution ensure the effectiveness of the iteration of the hard-cut EM algorithm for MGPs. It demonstrated by the experiments on the simulation and stock datasets that the MGP model with these specialized pdfs can lead to a better result on time series prediction in comparison with the general MGP models as well as the other classical regression methods.

Keywords: Gaussian distribution · Mixture of Gaussian processes
Hard-cut EM algorithm · Probability density function
Time series prediction

1 Introduction

Gaussian process (GP) is a powerful model and widely used in machine learning and data mining [1–3]. However, there are two main limitations. Firstly, it cannot fit the multi-modal dataset well because GP model employs a global scale parameter [4]. Secondly, its parameter learning consumes $O(N^3)$ computational time [5,6], where N is the number of training samples. In order to overcome those difficulties, Tresp [4] proposed mixture of Gaussian processes (MGP) in 2000, which was developed from the mixture of experts. Since then, many kinds

Z. Shi et al. (Eds.): ICIS 2018, IFIP AICT 539, pp. 70–80, 2018.
https://doi.org/10.1007/978-3-030-01313-4_8

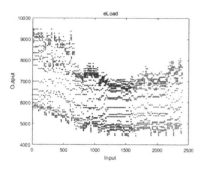

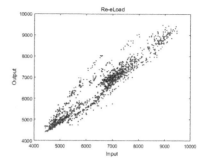

Fig. 1. The sketch of the eLoad data.

Fig. 2. The sketch of the transformed eLoad data.

of MGP model have been proposed and can be classified into two main forms: the generative model [7–10] and conditional model [4,6,11–13]. In comparison with the conditional model, the generative model has two main advantages: (1) The missing features can be easily inferred from the outputs; (2) The influence of the inputs on the outputs is more clear [8]. Therefore, many scholars have studied the generative model [14–20].

However, when we learn the generative model on a given dataset, we should set the probability density function (pdf) of the input in advance. In general, it can be set as a Gaussian distribution [14–20]. But, for some actual data like time series, this setting or assumption is not so reasonable and effective. When we learn MGP model on these actual data, we usually need to utilize the ARMA model [14–21] to transform the data, and then use the transformed data on the MGP model. However, this transformation can destroy the correlation of samples, which is very important for MGP model. Figure 1 shows the eLoad data [14] from which we can see that samples in three different colors (blue, black, and red) represent three temporally sequential samples, respectively. Figure 2 shows the transformed eLoad data from which we can find that three temporally sequential samples are mixed together and cannot be classified effectively. In this paper, we propose a specialized pdf for the input of the MGP model to solve this problem. As shown in Fig. 3, this pdf consists of three components. The left and right side parts are Gaussian distributions, while the middle is a uniform distribution. For the training of the MGP model, we use the hard-cut EM algorithm [17] as the basic learning framework for parameter estimation. Actually, the hard-cut EM algorithm can get better result than some popular learning algorithms.

The rest of the paper is organized as follows. Section 2 introduces the GP and MGP models. We describe the specialized probability density function in Sect. 3. We further propose the learning algorithm for the MGP model of the specialized pdfs in Sect. 4. The experimental results are contained in Sect. 5. Finally, we make a brief conclusion in Sect. 6.

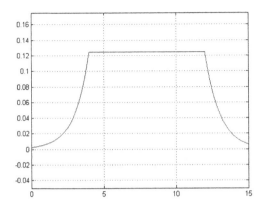

Fig. 3. The sketch of the specialized input distribution.

2   GP and MGP Models

2.1   GP Model

We mathematically define the GP model as follows:

$$Y \sim N(m(X), K(X, X)) \tag{1}$$

where $\mathbf{D} = \{X, Y\} = \{(x_i, y_i): i = 1,2,...,N\}$, x_i denotes a d-dimensional input vector, and y_i is the corresponding output. $m(X)$ and $K(X,X)$ denote the mean vector and covariance matrix, respectively. Without loss of generality, we assume $m(X) = 0$. There are many choices for covariance function, such as linear, Gaussian noise, squared exponential function and so on. Here, we adopt the squared exponential (SE) covariance function [10]:

$$K(x_i, x_j; \theta) = \sigma_f^2 exp(-\frac{\sigma_l^2}{2}\|x_i - x_j\|^2) + \sigma_n^2 I_{(i=j)} \tag{2}$$

where $\theta = \{\sigma_f^2, \sigma_l^2, \sigma_n^2\}$ denote the vector. On the given sample dataset $\mathbf{D}$, the log-likelihood function can be expressed as follows:

$$\log p(Y|X, \theta) = \log N(Y|0, K(\mathbf{X}, \mathbf{X})) \tag{3}$$

In order to obtain the estimation of parameters θ, we perform the maximum likelihood estimation (MLE) procedure [10], that is, we get

$$\hat{\theta} = argmax_\theta \log N(Y|0, K(X, X)) \tag{4}$$

2.2   MGP Model

Denote C and N as the number of GP components and training samples in the MGP model, respectively. On the basis of the GP model, we define MGP model by the following steps:

Step 1. Partition samples into each GP components by the Multinomial distribution:

$$p(z_n = c) = \pi_c \tag{5}$$

where $c = 1,...,C$ and $n = 1,...,N$.

Step 2. Accordingly, each input $\boldsymbol{x}_i$ fulfills the following distribution:

$$p(\boldsymbol{x}_i|z_n = c) \sim p(\boldsymbol{x}|\boldsymbol{\psi}_c) \tag{6}$$

where $\{\boldsymbol{\psi}_c : c = 1, ..., C\}$ is the parameter set. In general, $p(\boldsymbol{x}|\boldsymbol{\psi}_c)$ is a Gaussian distribution.

Step 3. Denote $\boldsymbol{I}_c = \{n|z_n = c\}$, $\boldsymbol{X}_c = \{\boldsymbol{x}_n|z_n = c\}$, $\boldsymbol{Y}_c = \{y_n|z_n = c\}$ $(c=1,...,C, n=1,...,N)$ as the sample indexes, inputs and outputs of the training samples in the c-th component, respectively. Given $\boldsymbol{X}_c$, the corresponding c-th GP component can be mathematically defined as follows:

$$\boldsymbol{Y}_c \sim N(\boldsymbol{0}, K(\boldsymbol{X}_c, \boldsymbol{X}_c)) \tag{7}$$

where $K(\boldsymbol{X}_c, \boldsymbol{X}_c)$ is given by Eq.(2) with the hyper-parameter $\boldsymbol{\theta}_c = \{\sigma_{fc}^2, \sigma_{lc}^2, \sigma_{nc}^2\}$.

Based on Eqs. (5), (6) and (7), we mathematically define the MGP model. The log-likelihood function is derived as follows:

$$
\begin{aligned}
\log(p(\boldsymbol{Y}_c|\boldsymbol{X}_c, \boldsymbol{\Theta}, \boldsymbol{\Psi})) = \sum_{c=1}^{C} (& \sum_{n \in \boldsymbol{I}_c} (\log(\pi_c p(\boldsymbol{x}_n|\boldsymbol{\mu}_c, \boldsymbol{S}_c))) \\
& + \log(p(\boldsymbol{Y}_c|\boldsymbol{X}_c, \boldsymbol{\theta}_c)))
\end{aligned}
\tag{8}
$$

where $\boldsymbol{\Theta} = \{\boldsymbol{\theta}_c : c = 1, ..., C\}$ and $\boldsymbol{\Psi} = \{\boldsymbol{\psi}_c, \pi_c : c = 1, ..., C\}$ denote the hyper-parameters and parameters of the MGP model, respectively.

3   Specialized Input Distribution and Its Learning Algorithm

For many real world datasets, such as UCI machine learning repository, Gaussian distribution is not appropriate for the input. In order to solve this problem, we propose a specialized distribution for this situation.

3.1   Specialized PDF

This specialized distribution is a piecewise-defined continuous function, which consists of three parts, the middle part is a uniform distribution density, both sides are Gaussian distribution densities, shown in Fig. 3. We mathematically defined the specialized distribution as follows:

$$
P(\boldsymbol{x}; \boldsymbol{\psi}) = \begin{cases}
\frac{\lambda_1}{(\sqrt{2\pi}\tau_1)} \exp^{-\frac{(x-a)^2}{2\tau_1^2}} & x < a \\
\lambda & a \leq x \leq b \\
\frac{\lambda_2}{(\sqrt{2\pi}\tau_2)} \exp^{-\frac{(x-b)^2}{2\tau_2^2}} & x > b
\end{cases}
\tag{9}
$$

where we redefine $\boldsymbol{\psi} = \{\lambda, \lambda_1, \lambda_2, \tau_1, \tau_2, a, b\}$ as the parameter vector.

3.2 Learning Algorithm for the Specialized PDF

In order to learn ψ, we set that the input interval (a,b) contains the number of the samples with probability p_0. Denote X and N as the training sample set and the number of training sample, respectively. We summarize the algorithm framework as following steps:

Step 1. Learn a, b, and λ:

$$a = X_{\frac{N(1-p_0)}{2}}; b = X_{\frac{N(1+p_0)}{2}}; \lambda = \frac{p_0}{(b-a)} \tag{10}$$

where $p(x < X_{\frac{N(1-p_0)}{2}} | x \in X) = \frac{(1-p_0)}{2}$. In order to reduce the effect of the misclassified (or outlier) point on the middle part, we estimate a and b as Eq.(10) do.

Step 2. Estimate λ_1, λ_2, τ_1 and τ_2.

Denote p_1 and p_2 as the sample ratio at both left side and right side, respectively. The probability density function is continuously integrable, and the integral of the probability density function is equal to 1. In other word:

$$\int P(x;\psi)dx = \begin{cases} p_1 & x < a \\ p_0 & a \leq x \leq b \; ; \\ p_2 & x > b \end{cases} \qquad p_0 + p_1 + p_2 = 1 \tag{11}$$

According to the continuity of the probability density function, we only need do same simple calculations to get $\{\lambda_1, \lambda_2, \tau_1, \tau_2\}$:

$$\lambda_1 = 2p_1; \lambda_2 = 2p_2; \tau_1 = \frac{\lambda_1}{\sqrt{2\pi\lambda}}; \tau_2 = \frac{\lambda_2}{\sqrt{2\pi\lambda}} \tag{12}$$

4 The MGP Model of the Specialized PDFs and Its Learning Algorithm

We now consider the MGP model with these specialized pdfs. For the parameter learning of the MGP model, there are main three kinds of learning algorithms: MCMC methods [22,23], variational Bayesian inference [24,25], and EM algorithm [5,9,11]. However, the MCMC methods and variational Bayesian inference methods have their own limitations: the time complexity of the MCMC method is very high, and variational Bayesian inference may lead to a rather deviation from the true objective function. EM algorithm is an important and effective iterative algorithm to do maximum likelihood or maximum a posterior(MAP) estimates of parameters for mixture model. However, for such a complex MGP model, the posteriors of latent variables and Q function are rather complicated. In order to overcome this difficulty, we implement the hard-cut EM algorithm [17] to learn parameter, which makes certain approximations in E-step.

Denote z_{nc} be the latent variables, where z_{nc} is a Kronecker delta function, $z_{nc} = 1$, if the sample (x_n, y_n) belongs to the c-th GP component. Therefore, we can obtain the log likelihood function of the complete data from Eq. (8) as follows:

$$\log(p(\boldsymbol{Y}, \boldsymbol{Z} | \boldsymbol{X}, \boldsymbol{\Theta}, \boldsymbol{\Psi})) = \sum_{c=1}^{C}(\sum_{n=1}^{N}(\boldsymbol{z}_{nc} \log(\pi_c p(\boldsymbol{x}_n | \boldsymbol{\psi}_c)))$$
$$+ \log(p(\boldsymbol{Y}_c | \boldsymbol{X}_c, \boldsymbol{\theta}_c))) \tag{13}$$

The main idea of hard-cut EM algorithm can be expressed as the following steps:

E-step. Assign the samples to the corresponding GP component according to the maximum a posterior (MAP) criterion:

$$\widehat{k}_n = argmax_{1 \leq c \leq C}\{\pi_c p(\boldsymbol{x}_n | \boldsymbol{\psi}_c) p(y_n | \boldsymbol{\theta}_c)\} \tag{14}$$

that is, latent variable $z_{\widehat{k}_n n} = 1$.

M-step. With the known partition, we can estimate the parameters $\boldsymbol{\Psi}$ and hyper-parameters $\boldsymbol{\Theta}$ via the MLE procedure:

(1) For learning the parameters $\{\boldsymbol{\psi}_c\}_c$, we perform the learning algorithm in the last section.
(2) For estimating the hyper-parameter $\boldsymbol{\Theta}$, we perform the MLE procedure on each c-th component to estimate $\boldsymbol{\theta}_c$ as shown in Eq. (4).

5   Experimental Results

In order to test the accuracy and effectiveness of the specialized pdf for MGP model, we carry out several experiments on the simulation and stock datasets. We employ the root mean squared error (RMSE) to measure the prediction accuracy, which is defined as follows:

$$RMSE = \sqrt{\frac{\sum_{n=1}^{N}(\boldsymbol{y}_n - \widehat{\boldsymbol{y}}_n)^2}{N}} \tag{15}$$

where $\widehat{\boldsymbol{y}}_n$ and $\boldsymbol{y}_n$ denote the predicted value and true value, respectively. We also compare our algorithm with some classical machine learning algorithms: kernel, RBF, SVM, and denote 'OURS' as our proposed model with the hard-cut EM algorithm.

5.1   Simulation Experiments

In the simulation experiments, we generate three groups of synthetic datasets from MGP model. those three MGP models contain 4, 6, 8 GP components, respectively. The number of samples in each group is 2600, 3900, 5000, respectively. In each group, there are three datasets, which are the same except the degree of overlap. Figure 4 shows the dataset with the smallest degree of overlap

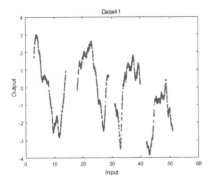

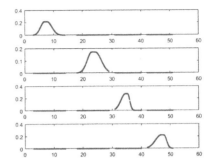

Fig. 4. The dataset with the least degree of overlap from MGP with 4 components.

Fig. 5. The distribution of the probability density function of the input on each Gaussian processes component.

with 4 GP components. On each group dataset, we run each algorithm 100 times, and randomly extract training samples and test samples, where 1/3 are training samples and other 2/3 are test samples. The RMSE of each algorithm is listed in Table 1. From the Table 1, We can see that our proposed algorithm obtains the better results. Figure 5 shows the specialized pdfs on the first group dataset with the smallest overlapping degree. We can obtain that the specialized pdf at both ends of the data in the form of a Gaussian distribution of attenuation, the specialized pdf in the middle of the data is a uniform distribution. This shape of the specialized pdf is more consistent with the uniform distribution than Gaussian distribution. The attenuation of both ends of the specialized pdf in the form of a Gaussian distribution ensures the effectiveness of the iteration of the hard-cut EM algorithm. Then, the class label of the samples can be updated according to the MAP criteria in the iteration of hard-cut EM algorithm. If we apply uniform distribution only, the iterative steps of hard-cut EM algorithm is invalid.

5.2 Prediction on Stock Data

In this section, we obtain the closing price data of three stocks from Shanghai Stock Exchange, and the IDs are 300015, 002643, and 601058, respectively.

From Eq. (10), we can know that the specialized pdf is closely related to the interval length of the middle data. In order to check the effect of different input lengths on the prediction accuracy of the algorithm, we do some transformations on the input. Since the range of output changes is too large, we use a linear function to narrow the output down to the same range as the synthetic data. In summary, we transform the datasets as follows:

(i) Transform the input as following equation:

$$X_n = \frac{n}{\delta} \tag{16}$$

where $i = 1,...,N$, N is the sample number, $\delta = \{101, 51, 23, 11, 7, 3, 1\}$.

Table 1. The RMSEs of the four algorithms on the three groups.

C = 4	Data41	Data42	Data43
Kernel	0.2143 ± 0.0000	0.6871 ± 0.0000	0.6537 ± 0.0000
RBF	0.1965 ± 0.0000	0.7065 ± 0.0000	0.6594 ± 0.0000
SVM	0.2548 ± 0.0002	0.3739 ± 0.0060	0.3704 ± 0.0038
OURS	0.0604 ± 0.0003	0.2991 ± 0.0319	0.3326 ± 0.0174
C = 6	Data61	Data62	Data63
Kernel	0.5212 ± 0.0000	0.5211 ± 0.0000	0.5392 ± 0.0000
RBF	0.5411 ± 0.0000	0.5498 ± 0.0000	0.5554 ± 0.0000
SVM	0.2678 ± 0.0013	0.3378 ± 0.0033	0.3942 ± 0.0080
OURS	0.2558 ± 0.0366	0.3653 ± 0.0384	0.3392 ± 0.0210
C = 8	Data81	Data82	Data83
Kernel	0.4247 ± 0.0000	0.4245 ± 0.0000	0.4993 ± 0.0000
RBF	0.4530 ± 0.0000	0.4265 ± 0.0000	0.4934 ± 0.0000
SVM	0.3642 ± 0.003	0.4238 ± 0.0023	0.4831 ± 0.0017
OURS	0.3220 ± 0.0335	0.3639 ± 0.0898	0.4668 ± 0.3434

Table 2. The RMSEs of the four algorithms on the three groups of the transformed stock datasets.

300015	X1	X2	X3	X4	X5	X6	X7
Kernel	0.6347	0.5439	0.4121	0.2753	0.2241	0.1846	0.2149
RBF	0.5218	0.3717	0.2918	0.2671	0.3435	0.6930	0.9934
SVM	0.3634	0.2532	0.1971	0.1707	0.1869	0.1707	0.1842
OURS	0.3531	0.2325	0.1784	0.1676	0.1601	0.1467	0.1583
002643	X1	X2	X3	X4	X5	X6	X7
Kernel	0.6487	0.4722	0.3116	0.2323	0.1972	0.1533	0.1542
RBF	0.5357	0.4462	0.2744	0.2280	0.2116	0.2110	0.8882
SVM	0.2815	0.2487	0.2364	0.2175	0.1998	0.1902	0.1792
OURS	0.2681	0.2024	0.1835	0.1799	0.1930	0.1805	0.1544
601058	X1	X2	X3	X4	X5	X6	X7
Kernel	0.7200	0.5267	0.3939	0.2738	0.2150	0.1503	0.1256
RBF	0.5296	0.4547	0.3636	0.2644	0.2279	0.2785	0.9705
SVM	0.4079	0.2325	0.1748	0.1502	0.1439	0.1439	0.1458
OURS	0.3271	0.1966	0.1318	0.1507	0.1545	0.1315	0.1472

(ii) Transform the output by a linearly compressed, and the compressed interval is $[-4.5, 4.5]$.

$$\tilde{y} = \frac{9y}{M - m} + \frac{4.5}{M - m} \tag{17}$$

where M and m denote the maximum value and minimum value of the stock, respectively.

Through the above transformations, each stock can produce 7 datasets. In each 7 datasets of three stock datasets, we repeat each regression algorithm 100 times, and randomly extracted 1/3 as training samples and the other 2/3 as test samples. The RMSE of each algorithm on those three transformed stock datasets is listed in Table 2. From Table 2, we can obtain that our proposed algorithm can get a better predict accuracy than other classical regression algorithms, and our algorithm obtain the better result with the smaller δ, but this is not absolute.

6 Conclusion

We have designed a specialized pdf for the input of MGP model which consists of three parts: the right and left side parts still take the form of Gaussian distributions, while the middle part takes the form of a uniform distribution. This specialized pdf has the advantages of both the Gaussian distribution and the uniform distribution. That is, the tail Gaussian distributions in the left and right side parts ensure that the hard-cut EM algorithm can perform more efficiently during each iteration, and the uniform distribution in the middle part is more reasonable for the time series data. The experiments are conducted on three groups of synthetic datasets and stock datasets. It is demonstrated by the experimental results that the hard-cut EM algorithm for the MGPs with the specialized pdfs can obtain a better prediction accuracy than the other classical regression algorithms. This specialized input pdf is more effective for the time series data.

Acknowledgment. This work was supported by the National Science Foundation of China under Grant 61171138.

References

1. Rasmussen, C.E.: Evaluation of Gaussian Processes and Other Methods for Non-linear Regression. University of Toronto (1999)
2. Williams, C.K.I., Barber, D.: Bayesian classification with Gaussian processes. IEEE Trans. Pattern Anal. Mach. Intell. **20**(12), 1342–1351 (1998)
3. Rasmussen, C.E., Kuss, M.: Gaussian processes in reinforcement learning. In: NIPS, vol. 4, p. 1 (2003)
4. Tresp, V.: Mixtures of Gaussian processes. In: Advances in Neural Information Processing Systems, pp. 654–660 (2001)
5. Yuan, C., Neubauer, C.: Variational mixture of Gaussian process experts. In: Advances in Neural Information Processing Systems, pp. 1897–1904 (2009)

6. Stachniss, C., Plagemann, C., Lilienthal, A.J., et al.: Gas Distribution Modeling using Sparse Gaussian Process Mixture Models. In: Robotics: Science and Systems, vol. 3 (2008)
7. Yang, Y., Ma, J.: An efficient EM approach to parameter learning of the mixture of Gaussian processes. In: Liu, D., Zhang, H., Polycarpou, M., Alippi, C., He, H. (eds.) ISNN 2011. LNCS, vol. 6676, pp. 165–174. Springer, Heidelberg (2011). https://doi.org/10.1007/978-3-642-21090-7_20
8. Meeds, E., Osindero, S.: An alternative infinite mixture of Gaussian process experts. In: Advances in Neural Information Processing Systems, pp. 883–890 (2006)
9. Sun, S., Xu, X.: Variational inference for infinite mixtures of Gaussian processes with applications to traffic flow prediction. IEEE Trans. Intell. Transp. Syst. 12(2), 466–475 (2011)
10. Williams, C.K.I., Rasmussen, C.E.: Gaussian processes for machine learning, MIT Press 2(3), 4 (2006)
11. Nguyen, T., Bonilla, E.: Fast allocation of Gaussian process experts. In: International Conference on Machine Learning, pp. 145–153 (2014)
12. Lázaro-Gredilla, M., Van Vaerenbergh, S., Lawrence, N.D.: Overlapping mixtures of Gaussian processes for the data association problem. Pattern Recogn. 45(4), 1386–1395 (2012)
13. Ross, J., Dy, J.: Nonparametric mixture of Gaussian processes with constraints. In: International Conference on Machine Learning, 1346–1354 (2013)
14. Wu, D., Ma, J.: A two-layer mixture model of Gaussian process functional regressions and its MCMC EM algorithm. IEEE Trans. Neural Netw. Learn. Syst. (2018)
15. Wu, D., Chen, Z., Ma, J.: An MCMC based EM algorithm for mixtures of Gaussian processes. In: Hu, X., Xia, Y., Zhang, Y., Zhao, D. (eds.) ISNN 2015. LNCS, vol. 9377, pp. 327–334. Springer, Cham (2015). https://doi.org/10.1007/978-3-319-25393-0_36
16. Wu, D., Ma, J.: A DAEM algorithm for mixtures of Gaussian process functional regressions. In: Huang, D.-S., Han, K., Hussain, A. (eds.) ICIC 2016. LNCS (LNAI), vol. 9773, pp. 294–303. Springer, Cham (2016). https://doi.org/10.1007/978-3-319-42297-8_28
17. Chen, Z., Ma, J., Zhou, Y.: A precise hard-cut EM algorithm for mixtures of Gaussian processes. In: Huang, D.-S., Jo, K.-H., Wang, L. (eds.) ICIC 2014. LNCS (LNAI), vol. 8589, pp. 68–75. Springer, Cham (2014). https://doi.org/10.1007/978-3-319-09339-0_7
18. Chen, Z., Ma, J.: The hard-cut EM algorithm for mixture of sparse Gaussian processes. In: Huang, D.-S., Han, K. (eds.) ICIC 2015. LNCS (LNAI), vol. 9227, pp. 13–24. Springer, Cham (2015). https://doi.org/10.1007/978-3-319-22053-6_2
19. Zhao, L., Chen, Z., Ma, J.: An effective model selection criterion for mixtures of Gaussian processes. In: Hu, X., Xia, Y., Zhang, Y., Zhao, D. (eds.) ISNN 2015. LNCS, vol. 9377, pp. 345–354. Springer, Cham (2015). https://doi.org/10.1007/978-3-319-25393-0_38
20. Zhao, L., Ma, J.: A dynamic model selection algorithm for mixtures of Gaussian processes. In: 2016 IEEE 13th International Conference on Signal Processing (ICSP), pp. 1095–1099. IEEE (2016)
21. Liu, S., Ma, J.: Stock price prediction through the mixture of Gaussian processes via the precise hard-cut EM algorithm. In: Huang, D.-S., Han, K., Hussain, A. (eds.) ICIC 2016. LNCS (LNAI), vol. 9773, pp. 282–293. Springer, Cham (2016). https://doi.org/10.1007/978-3-319-42297-8_27

22. Shi, J.Q., Murray-Smith, R., Titterington, D.M.: Bayesian regression and classification using mixtures of Gaussian processes. Int. J. Adapt. Control. Signal Process. **17**(2), 149–161 (2003)
23. Tayal, A., Poupart, P., Li, Y.: Hierarchical double Dirichlet process mixture of Gaussian processes. In: AAAI (2012)
24. Chatzis, S.P., Demiris, Y.: Nonparametric mixtures of Gaussian processes with power-law behavior. IEEE Trans. Neural Netw. Learn. Syst. **23**(12), 1862–1871 (2012)
25. Kapoor, A., Ahn, H., Picard, R.W.: Mixture of Gaussian processes for combining multiple modalities. In: Oza, N.C., Polikar, R., Kittler, J., Roli, F. (eds.) MCS 2005. LNCS, vol. 3541, pp. 86–96. Springer, Heidelberg (2005). https://doi.org/10.1007/11494683_9

Research of Port Competitiveness Evaluation Based on Attribute Evaluation Method

Xueyan Duan[✉] and JieQiong Liu

Department of Economic and Management, Shanghai Polytechnic University,
Shanghai 201209, China
{xyduan, jqliu}@sspu.edu.cn

Abstract. Attribute Evaluation Method can simulate the psychological preferences of decision makers and give the evaluation results in line with the psychology of decision makers. The effectiveness of this method is verified through the ranking of 11 ports' competitiveness, which also provides a new idea for port competitiveness evaluation.

Keywords: Port · Competitiveness · Evaluation · Attribute Evaluation Method
Preferences

1 Introduction and Literature Review

Attribute Theory was put forward by Professor Jiali Feng. Attribute Coordinate Comprehensive Evaluation Method (ACCEM) is a comprehensive evaluation method based on Attribute Theory. ACCEM can simulate normal thinking pattern of people which can reflect the preferences or preference curve of evaluators and can be used in the empirical decision-making and uncertainty mathematical analysis. Some scholars applied ACCEM and obtained some results. ACCEM has already be used in College entrance examination evaluation system (Xie 2002), Enterprise's Productive Forces Evaluation (Xu 2002), 3PL's Core Competence Evaluation (Duan et al. 2006), Software Enterprises' Competence (Xu et al. 2006), Optimized Selection of Supplier in SCM (Li 2007), whose results turned out to be rather satisfactory. ACCEM are proved to be validity from these applications.

A port is the hub of a transportation system. Correct assessment of port competitiveness will directly affect the future development space of port. Some methods have been applied to port evaluation, such as AHP, FCE (Fuzzy Comprehensive Evaluation), DEA, TOPSIS. These studies mainly start from objective analysis and do not integrate into the psychological preferences of decision makers, which needs further study.

ACCEM provides a new idea for studying the evaluation of port competitiveness. It can not only objectively analyze the characteristics of port competitiveness, but also simulate the psychological preference curve of decision makers, which can provide a more reasonable evaluation result.

© IFIP International Federation for Information Processing 2018
Published by Springer Nature Switzerland AG 2018. All Rights Reserved
Z. Shi et al. (Eds.): ICIS 2018, IFIP AICT 539, pp. 81–87, 2018.
https://doi.org/10.1007/978-3-030-01313-4_9

2   Attribute Coordinate Comprehensive Evaluation Method

ACCEM establishes on the theory of Attribute Theory. Attribute theory method simulates people's normal thinking mode. Attribute theory reflects decision maker's psychological preference and psychological preference curve based on qualitative mapping theory.

In ACCEM, we assume that $X = \left\{ x_i = (x_{i1}, x_{i2}, \cdots, x_{im}) \big| x_{ij} (0 \leq x_{ij} \leq 100) \right\}$ is the set of all the solutions. $S_T = \left\{ x_i = (x_{i1}, x_{i2}, \cdots, x_{im}) \Big| \sum_{j=1}^{m} x_{ij} = T \right\}$ is the set of all the solutions that the total score is equal to T. Their intersection $S_T \cap X$ forms a hyperplane of equal total $T (T_0 \leq T \leq 100)$ or $m-1$ dimensional simplex. $\{x_k, k = 1, \cdots, s\} \subseteq S_T \cap X$ is the set of sample scheme x whose total score is equal to T.

Suppose decision maker z picked t schemes that he was fairly satisfied, $\{x^h, h = 1, \cdots, t\}$, and he gave each scheme a score $v^h(x^h)$. Then the center of gravity can be get with $v^h(x^h)$ as the weight. We can use machine learning to get the locally satisfactory solution. When the training sample set $\{x_k\}$ is large enough, the training times are enough, and the scheme set $\{x^h(z)\}$ that decision maker z picked is enough, the center of gravity $b(\{x^h(z)\})$ will gradually approach the local optimal solution. Thus decision maker z can use the local satisfaction function (1) to evaluate the satisfaction of all schemes $\{x_i = (x_{i1}, x_{i2}, \cdots, x_{im})\}$ in $S_T \cap X$.

$$sat(x_i, z) = \exp\left(-\frac{\sum_{j=1}^{m} w_j \left|x_{ij} - b_{ij}\left(\{x^h(z)\}\right)\right|}{\sum_{j=1}^{m} w_j \left(b_{ij}(\{x^h(z)\}) - \delta_j\right)}\right) \tag{1}$$

Considering the continuous change of decision make z's psychological standard, the set of locally satisfactory solutions $\left\{ b(\{x^h(z)\}) \big|_{T \in [100, 100 \times n]} \right\}$ can be viewed as a line $L(b(\{x^h(z)\}))$, which is called the local optimal line or mental standard line of decision maker z. The result of $L(b(\{x^h(z)\}))$ can be calculated by interpolation.

Being proved mathematically and debugged many times, global consistency coefficient $\lambda(x, z)$ is introduced in order to get a global satisfaction solution from the local optimal line $L(b(\{x^h(z)\}))$.

$$\lambda(x, z) = \left(\frac{\sum_{i=1}^{m} x_{ij}}{\sum_{j=1}^{m} X_j}\right)^{\frac{\sum_{j=1}^{m} x_j}{\sum_{i=1}^{m} x_{ij}}} \tag{2}$$

Global satisfactory solution is:

$$sat(x,z) = \lambda(x,z) * sat(x_i,z) = \left(\frac{\sum\limits_{i=1}^{m} x_{ij}}{\sum\limits_{j=1}^{m} X_j}\right)^{\frac{\sum\limits_{j=1}^{m} x_j}{m}} \sum\limits_{i=1}^{m} x_{ij} * \exp\left(-\frac{\sum\limits_{j=1}^{m} w_j \left|x_{ij} - b_{ij}(\{x^h(z)\})\right|}{\sum\limits_{j=1}^{m} w_j(b_{ij}(\{x^h(z)\}) - \delta_j)}\right)$$

(3)

Using the formula (3), we can evaluate the global satisfaction of the objects in the whole decision space. The specific steps of ACCEM are as described below:

(1) Determine the feasible schemes that influence the decision making, analyze the attribute characteristics of each feasible scheme, and evaluate and quantify the properties of each feasible scheme
(2) Qualitative mapping function is used to normalize attribute utility value nonlinearly.
(3) T_0 is set as the critical total score. In $(T_0, 100m)$. A number of points $(T_1, T_2, \cdots, T_{n-1})$ are uniformly selected according to the requirements of curve fitting. At each point with a total score of $T_i = (i = 1, 2, \cdots, n - 1)$, several sample schemes are selected for study to find the center of gravity coordinates with a total score of $T_i = (i = 1, 2, \cdots, n - 1)$, which is locally satisfactory solution.
(4) Using the interpolation formula, the curve fitting was carried out to find the psychological standard line which is also called local most satisfactory solution line.
(5) According to (3), the global satisfaction of each scheme is calculated, and the ranking is conducted from large to small to obtain the most satisfactory solution.

3 Attribute Characteristics of Port Competitiveness

Competitiveness is a strong comprehensive ability that participants of both or all parties expressed by comparing. Port competitive ability can be defined as port's competitive advantage in production capacity, value creation, sustainable development, etc. So we choose Port Infrastructure, Port Scale and Port Greening as one class attribute. Each one class attribute has three secondary attributes.

(1) Port Infrastructure: Port infrastructure refer to the necessary facilities for completing the most basic functions of port logistics. Generally including port channel, breakwater, anchorage, dock, berth, port traffic, etc.
(2) Port Scale: Port scale includes the following main contents: Parliamentary terminals, berths, warehouses and other facilities have reached the scale; Ownership and technical level of major equipment such as handling machinery; The number and quality of employees; Strive for economic benefits, etc.

(3) Port Greening: Green port refers to the development mode of environmental protection, ecology, low energy consumption and low emission. Energy consumption, pollutant discharge and damage to the ecological environment will be minimized to obtain the maximum socio-economic and ecological benefits.

4 Port Competitiveness Index System

We have drawn lessons from studies on port evaluation indicators at home and abroad, carried out theoretical analysis and expert consultation, and finally formed a port competitiveness evaluation system with three levels. Each Criterion layer indicator layer has three sub-indicators. Altogether there are 11 indicators in the third indicator layer. The structure of the whole index system is shown in the Table 1.

Table 1. Port competitiveness evaluation index system

Target layer	Criterion layer	Indicator layer
Port competitiveness (**U**)	Port infrastructure (**u1**)	Level channel proportion (**u11**)
		Berth quantity (**u12**)
		Shoreline length (**u13**)
	Port scale (**u2**)	Proportion of water transport freight turnover (**u21**)
		Port throughput (**u22**)
		GDP of port city (**u23**)
	Port greening (**u3**)	Environmental protection investment index (**u31**)
		Energy consumption per unit of GDP (**u32**)
		Afforestation coverage in the port (**u33**)

5 Empirical Study

We choose 11 ports to carry out empirical research. Index original values of 11 ports are standardized using Z-score method.

Then we process the data with formula as $x'_{ij} = x_{ij} * 10 + 50$ so that the data is between 0 and 100. Datas after standardization are shown in Table 2.

Then we can evaluate 11 ports using ACCEM after indicators standardization. Because the index system is divided into three layers, we need to analyze layer by layer.

For example, One class index *Port Infrastructure* has three indicators. Take the three attributes as axes and three score $x(a_j)(j = 1, 2, 3)$ of three attributes. Each port x_i

responses a $3D$ coordinate points $x_{ij} = (x_{i1}, x_{i2}, x_{i3})$ in the $3D$ decision-making coordinate. So we can establish 3 Lagrange interpolation equations as (4).

$$G(T) = G(g_1(T), g_2(T), g_3(T),) \tag{4}$$

Among them:
$$\begin{cases} g_1(T) = a_{01} + a_{11}T + a_{21}T^2 & (4-1) \\ g_2(T) = a_{02} + a_{12}T + a_{22}T^2 & (4-2) \\ g_3(T) = a_{03} + a_{13}T + a_{23}T^2 & (4-3) \end{cases}$$

Choose 3 total score which is $T_{\min}$, $T_{\max}$ and T_i through the study of the sample we can get evaluation criteria points. Plug three evaluation criteria points in formula (5).

$$g_i(T) = \frac{(T - x_1^*)(T - x_2^*)}{(x_0^* - x_1^*)(x_0^* - x_2^*)} a_{i0} + \frac{(T - x_0^*)(T - x_2^*)}{(x_1^* - x_0^*)(x_1^* - x_2^*)} a_{i1} + \frac{(T - x_0^*)(T - x_1^*)}{(x_2^* - x_0^*)(x_2^* - x_1^*)} a_{i2} \tag{5}$$

We can get nine coefficients of equation set. Plug nine coefficients in formula (4) to get the interpolation curve $G(T) = G(g_1(T), g_2(T), g_3(T),)$ of evaluation standard curve $L(b(\{x^h(z)\}))$. Then we can get any evaluation criteria in any total score plane. Using local satisfaction function and global satisfaction function, we can get the global satisfaction score of one class index *Port Infrastructure*.

In the same way we calculate global satisfaction of the other two indicators of one class index.

Table 2. Datas after standardization of 11 ports

	u11	u12	u13	u21	u22	u23	u31	u32	u33
Port 1	47.27	45.12	52.99	39.87	74.94	58.06	63.22	50.07	61.46
Port 2	33.17	46.91	46.78	60.94	53.61	50.63	59.93	55.08	37.22
Port 3	41.98	45.18	49.24	59.21	51.81	47.28	48.95	44.33	49.34
Port 4	59.62	56.95	52.77	57.77	48.83	43.12	48.01	56.52	46.92
Port 5	46.39	44.54	45.45	41.89	47.48	44.41	52.70	51.86	42.07
Port 6	62.26	47.34	42.04	57.19	46.89	44.45	51.77	34.29	59.03
Port 7	55.21	44.04	47.56	58.06	50.12	44.59	44.26	49.71	39.64
Port 8	50.80	44.63	41.52	51.99	45.41	56.91	29.52	53.29	44.49
Port 9	60.50	71.74	67.00	42.76	44.38	44.99	48.95	39.67	49.34
Port 10	42.87	44.21	39.86	40.16	42.54	43.74	46.14	67.99	59.03
Port 11	49.92	59.36	64.79	40.16	43.98	71.83	56.55	47.20	61.46

Then take the three indexes of one class as three attributes to calculate global satisfaction score of target layer and sort. The whole calculation process can be realized through the computer software programming.

The final scores and ranking list are shown in Fig. 1.

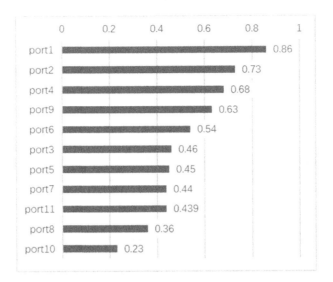

Fig. 1. Final evaluation scores and ranking of port competitiveness

6   Conclusion

In conclusion, ACCEM can simulate the psychological preference and change process of the evaluator. In addition, the port competitiveness can be evaluated and ranked according to the local psychological evaluation standard and the global psychological evaluation standard of the evaluator. From Fig. 1, we can see that the total score of port 3 and port 5 is similar. By analyzing the original data, we can find that the throughput of port 3 is higher. This reflects that the psychological preference of the evaluator is more inclined to the throughput index. This shows that the evaluation principle of this method is in accordance with the preference of the evaluator. Different evaluators may have different evaluation results. This is the characteristic of ACCEM. ACCEM can give the evaluation results that meet the requirements of decision makers by constantly learning their psychological preferences. The research of this paper proves the validity of ACCEM, and broadens the thinking for the evaluation of port competitiveness, which plays the role of throwing bricks and introducing jade.

References

Feng, J., Dong, Z.: A mathematical model of sensation neuron detection based on attribute abstraction and integration. Comput. Res. Dev. **34**(7), 481–486 (1997)

Feng, J.: The Research on Decision Support System of Nuclear Accident Emergency and It's Computer Realization. Institute of Atomic Energy of China, Beijing (2001)

Feng, J.: The qualitative mapping model of judgment and the classification of non-linear patterns. J. Guangxi Norm. Univ. Nat. Sci. Ed. **22**(1), 27–32 (2004a)

Feng, J.: The recognition methods based on qualitative mapping model. J. Guangxi Norm. Univ. Nat. Sci. Ed. **22**(2), 14–18 (2004b)

Wu, Q., Feng, J., Dong, Z., Zhang, Y.: A kind of evaluation and decision model based on analysis and learning of attribute coordinate. J. Nanjing Univ. (Nat. Sci.) **39**(2), 182–188 (2003). (in Chinese)

Xie, X., Liu, J., Lu, Q.: Admissions decisions in the university entrance exam system. J. Guilin Eng. **21**(4), 402–406 (2001). (in Chinese)

Duan, X., Liu, Y., Xu, G.: Evaluation on 3PL's core competence based on method of attribute theory. J. Shanghai Marit. Univ. **27**(1), 41–43 (2006). (in Chinese)

Xu, G., Liu, Y., Feng, J.: Evaluation of Chinese software enterprises' core competence in attribute theory. J. Guangxi Norm. Univ. (Nat. Sci. Ed.) **24**(4), 34–37 (2006)

Li, J., Liu, Y., Feng, J.: Comprehensive assessment & optimized selection of supplier in SCM based on attribute theory. Logist. Technol. **26**(5), 75–79 (2007). (in Chinese)

Universal Learning Machine – Principle, Method, and Engineering Model Contributions to ICIS 2018

Chuyu Xiong[✉]

New York, USA

Abstract. Universal learning machine is a computing system that can automatically learn any computational task with sufficient data, and no need for manual presetting and intervention. Universal learning machine and universal learning theory are very important research topic. Many disciplines (AI, AGI, machine epistemology, neuroscience, computational theory, mathematics, etc.) cross here. In this article, we discuss the principles, methods, and engineering models of universal learning machine. X-form is the central concept and tool, which is introduced by examining objective and subjective patterns in details. We also discuss conceiving space and governing space, data sufficiency, learning strategies and methods, and engineering model.

Keywords: Universal learning machine · Subjective pattern
X-form · Conceiving space · Governing space · Primary consciousness
Learning dynamics · Data sufficiency · Squeeze to higher abstraction
Embed to parameter space

1 Introduction

Universal learning machine is a computing system that could automatically learn any computational task without special presetting or manual intervention, if we provide it with sufficient data to drive the learning. But, why do we need universal learning machine?

A computer is a machine that processes information, and an intelligent computer is a machine that can process information more flexibly. Thus, it is natural to ask the question: Where does a computer's ability to process information come from? Currently, the processing ability mostly comes from human programming. This is surely effective. However, as the well-known AlphaGo demonstrated, we can not manually program a Go software to win over human players, but we can build a system that can acquire a Go program that wins over all human players

---

C. Xiong—Independent Researcher.
Great thanks for whole heart support of my wife. Thanks for Internet and research contents contributers to Internet.

Z. Shi et al. (Eds.): ICIS 2018, IFIP AICT 539, pp. 88–101, 2018.
https://doi.org/10.1007/978-3-030-01313-4_10

by learning from inputing data. This fact and many other similar facts tell us that we are coming to the end of manual programming and we need to make computing systems that can learn from its input data and acquire information processing ability automatically.

However, not all computing systems that can learn from its input are universal learning machine. For example, a Go program, which can learn playing Go game, is an effective learning machine, but not a universal learning machine. It can only learn playing Go, nothing else. Also, it needs a lot of manual presetting and manual intervention. Clearly, it is better if such system is a universal learning machine.

Machine learning is a hot research field at present, but there is no established principle and method for universal learning machine and universal learning theory. The current theory of machine learning, such as deep learning, focusing on certain special fields, is not good enough to explain the results it achieved, let alone guide universal learning. In short, we need to have a theoretical framework for universal learning. Thus, we started studies on universal learning [2]. We have found that, in order to achieve universal learning, computing system must have a specific structure. Based on this, we introduced concept of X-form, and tried to build universal learning around X-form. As a consequence of our work, we can see better what deep learning is doing, from the view of X-form [3].

In this article we discuss X-form, conceiving space, governing space, primary consciousness, learning dynamics, data sufficiency, learning strategies. We show that, in theory, universal learning machine can be made. We also briefly discuss engineering models to realize universal learning machine.

Remark, this article is the English and short version of our article in Chinese [5]. Due to restriction to size, many details are skipped, please check [2,5]

2   Universal Learning Machine

We start with the definition of the learning machine.

Definition 2.1 (Learning Machine). An N-M information processing unit (IPU) having an input space (N bits), an output space (M bits), and a processor. An IPU processes the input information through the processor and outputs it. If an IPU adjusts and changes its ability of processing information (i.e. modifying its own processor) during its information processing, according to a set of simple and fixed rules, this IPU is called a learning machine.

Note the phrase "a set of simple and fixed rules" in the definition. This is a very strong restriction. We use it to exclude human intervention.

Without loss of generality, we can make discussion easier by only considering N-1 learning machine. Most of the time, we only need to consider N-1 learning machine. It is worth mentioning that an N-1 learning machine is equivalent to an N-dim Boolean function from the perspective of input and output, but it is an N-dim Boolean function that is self-adjusting and adapting.

Learning machine is driven by series of data inputs, which is called as data sequence.

Definition 2.2 (Universal Learning Machine). For the learning machine M, assume that its current processor is P_0, and P_1 is another processor. If we have a data sequence T, and when we applied T to M, eventually, the processor changes from P_0 to P_1, then we say that M can learn P_1 starting from P_0. If for any given pair of processors P_0 and P_1, M can learn P_1 starting from P_0, we say M is a universal learning machine.

Simply say, universal learning machine is a learning machine that can learn any information processing starting from any existing information processing.

Inside the learning machine, obviously, there is some kind of learning mechanism. More specifically, this learning mechanism captures the information embedded in the data sequence and uses that information to modify its own information processing. However, we need to distinguish between two situations: (1) the learning mechanism can only modify the information processing, and the learning mechanism itself is not being modified; (2) the learning mechanism itself is also being modified. But how can you exactly describe these two situations? Actually, we can do so by using data sequence.

Definition 2.3 (Level 1 Learning Machine). Suppose M is a universal learning machine. For any given pair of processors P_0 and P_1, by definition, there is at least one data sequence T, so that the learning machine learns P_1 starting from P_0. If the data sequence T depends only on P_0 and P_1, and does not depend on any historical state of M, we call M as a level 1 universal learning machine.

Note that according to this idea, we can also define a level 0 learning machine, which is an IPU that its information processing cannot be changed. Similarly, we can also define a level 2 learning machine, which is a learning machine whose information processing can be changed and the learning mechanism can be changed as well, but the learning mechanism of the learning mechanism cannot be changed. We can define a J-level learning machine $J = 0, 1, 2 \ldots$ according to this rule.

The learning machine distinguishes input by patterns and not by bits. The learning machine should generate different outputs for different patterns, but should generate the same output for 2 inputs different in terms of bits if the 2 inputs are with same pattern. In other words, the goal of a learning machine is to learn to distinguish between patterns and how to handle patterns. Therefore, we need to understand patterns. But let's first explain what are spatial pattern and temporal pattern. We can briefly say that spatial pattern is independent of the time factor of the input (such as the order, the previous input, etc.), only related to the input itself, and temporal pattern is also related to the time factor of the input. The spatial pattern is relatively simpler. In this article, we will focus on the spatial patterns, and not discuss temporal pattern. Of course, we are very clear that temporal pattern is crucial. We will discuss it in later articles.

3   Subjective Pattern and X-form

Learning machines differ from most computing systems, instead of treating data by each bit, they treat the data by pattern. Pattern is very important for the learning machine, and everything about the learning machine is related to pattern. However, pattern is also very confusing. We can look at pattern from different angles and get completely different results. We can look at pattern objectively, that is, completely independent of the learning machine and learning. We can also look at pattern subjectively, that is, rely on the learning machine. In this section, we discuss pattern and tools to handle pattern.

3.1   Objective Pattern

Objective pattern is the starting point of us.

Definition 3.1 (Base Pattern Space). N-dim base pattern space, written as PS_N^0 (PS stands for pattern space), is an N-dim binary vector space:

$$PS_N^0 = \{b = (b_1, b_2, \ldots, b_N)| \ b_k \in \{0, 1\}, k = 1, \ldots, N\}$$

Each element in PS_N^0 is a base pattern. PS_N^0 has a total 2^N base patterns. When N is not very small, this is very huge. In fact, this hugeness is the root of the richness and difficulty of universal learning machine.

We skip details of objective pattern. We just state here: base patterns are starting point, most patterns are not base patterns; an objective pattern is a set of base patterns; operations on objective patterns are identical to set operations.

3.2   Subjective Pattern

The key is that pattern is not only related to the bits presented in the input data, but also how the learning machine perceives the input. For example, consider a simple 4-1 learning machine, which input space is 4-dim. When the input is $(1, 1, 1, 0)$, the learning machine can perceive that it is just a base pattern, but it can also perceive the input as that two base patterns $(1, 1, 0, 0)$ and $(1, 0, 1, 0)$ appear together, or other more complicated situations [2].

This simple fact and many more similar facts strongly tell us: Objective pattern alone is not good enough to understand a learning machine. Subjective pattern is more important. We must understand how learning machine subjectively perceive pattern.

Definition 3.2 (Perception Bit). For the learning machine M, if there is a bit pb inside it, which behaves in this way: there is a set $B \subset PB_N^0$, so for any $b \in B$, $pb = 1$, and for any $b \notin B$, $pb = 0$, then we call this bit pb a processing bit of M. If M has more than one such bit, i.e. there is a set: $\{pb_j| \ j = 1, \ldots, L\}$, where each pb_j is a processing bit, we call this set as a processing bit set of M, or just processing bits, or perception bits.

Since in a learning machine, processing is strongly related to its perception to input, we also use name perception bit. The perception bits reflect how a learning machine processes information and are its important structural elements. The perception bits of a learning machine can be quite complicated and not easy to see. However, we know that it does exist.

Theorem 3.1 (Perception Bits Exist). *For any learning machine M, its perception bits $\{pb_j|\ j = 1, \ldots, L\}$ exist and are not empty.*

We skip proof here. For a learning machine M, if b and b' are two different base patterns, i.e. two different N-dim binary vectors, then M may have the same or different perception bits for b and b'. If the perception bits are same, then b and b' are perceived same by M, even b and b' might be very different bit-wise. If M perceives b and b' different, then M must have different perception bits (at least one perception bit has a different behavior) for them. In this way, we have the following definition.

Definition 3.3 (Base Pattern Perceived). Suppose M is a learning machine, $\{pb_j|\ j = 1, \ldots, L\}$ are its perception bits, if for 2 base patterns b_1 and b_2, the values of their perception bits are $(pv_1^1, pv_2^1, \ldots, pv_L^1)$ and $(pv_1^2, pv_2^2, \ldots, pv_L^2)$, respectively, and for at least one k, $1 \le k \le L$, we have $pv_k^1 = pv_k^2 = 1$, we say that M subjectively perceive b_1 and b_2 the same, at the perception bit pb_k.

That is to say, for the two base patterns, if they are different at all perception bits, the learning machine cannot perceive them as same. However, if at least one perception bit, the value is the same, M may subjectively perceive them to be the same. Of course, M can also be subjectively perceive them to be different.

Definition 3.4 (Pattern Perceived). Suppose M is a learning machine, $\{pb_j|\ j = 1, \ldots, L\}$ are its perception bits. Then if p is a set of base patterns, and at one perception bit pb_k, $1 \le k \le L$, any base pattern in p will take the same value at pb_k, we say that M subjectively perceives p as one pattern at pb_k. That is to say, p (a set of base patterns) is one subjective pattern for M.

Note that in the definition, all the base patterns in p need only behave the same at one perception bit. This is the minimum requirement. Of course this requirement can be higher. For example, to requiring to behave the same at all perception bits. These requirements all are subjective.

In a learning machine perception bits reflect how the learning machine perceives and processes pattern. Based on perception bits, we can understand subjective pattern. If the two base patterns behave the same at one perception bit, then they can be subjectively perceived identical. For any objective pattern, i.e. a set of base patterns p, if all of base pattern in p behave the same at one perception bit, p can be subjectively perceived as the same. This is where objective pattern connects to subjective pattern.

For a learning machine, the perception bits is changing with learning, their behavior changes, and their number changes.

Due to size, we skip details of operations of subjective patterns. But, they are super important. There are 3 subjective pattern operations ("¬", "+", "·"). Not like operations of objective pattern, which require no modifications inside the learning machine, operations of subjective pattern need to modify learning machine, often by adding perception bits and modify the behaviors.

3.3   X-form

There are 3 subjective operations ("¬", "+", "·") over subjective patterns. Their results are also subjective patterns. Naturally, we ask: if we apply the 3 subjective operations in succession, what will happen? Clearly, we will get an algebraic expression. But, what exactly is this algebraic expression? What should we do? What it can bring to us?

Definition 3.5 (X-form as Algebraic Expression). Suppose $g = \{p_1, p_2, \ldots, p_K\}$ is a set of variables, and we repeatedly apply 3 operators "¬", "+", "·" over $p_1, p_2, \ldots, p_K$, we then construct an algebraic expression E, we then call E as an X-form on g, which is written: $E = E(g) = E(p_1, p_2, \ldots, p_K)$.

But, we can view such algebraic expression as subjective pattern (under certain conditions):

Definition 3.6 (X-form as Subjective Pattern). Suppose $g = \{p_1, p_2, \ldots, p_K\}$ is a set of subjective pattern, and $E = E(g) = E(p_1, p_2, \ldots, p_K)$ is one X-form on g (as algebraic expression). If all operations in E indeed have support, this expression E is a new subjective pattern.

Further, such algebraic expression can be viewed as information processing:

Definition 3.7 (X-form as Information Processor). Assuming M is a learning machine, $g = \{p_1, p_2, \ldots, p_K\}$ is a set of subjective patterns subjectively perceived by M, and $E = E(g)$ is a X-form on g (as algebraic expression), then $E(g)$ is an information processor that processes information like this: when a basic schema p is put into M, since M perceives this pattern, the subjective patterns $p_1, p_2, \ldots, p_K$ forms a set of boolean variables, still written as: $p_1, p_2, \ldots, p_K$, and when this set of boolean variables is applied to E, the value of E is the output of the processor, and it is written as: $E(g)(p)$.

So, we now understand the meaning of X-form, in several steps. Why do we call as X-form? These expressions are mathematical forms and have rich connotations, but we don't know much about many aspects of such expressions, so following the tradition, we use X-form to name it. Following theorem connect objective pattern, subjective pattern and X-form. We skip the proof here.

Theorem 3.2 (Objective and Subjective Pattern, and X-form). *Suppose M is an N-1 learning machine. For any objective pattern p_o (i.e. a subset in PS_N^0), we can find a set of base pattern $g = \{b_1, b_2, \ldots, b_K\}$, and one X-form E on g, $E = E(g) = E(b_1, b_2, \ldots, b_K)$, so that M perceive any base pattern in p_o as E, and we write as $p_o = E(g)$. We say p_o is expressed by X-form $E(g)$. Moreover, we can also require that the number of base patterns less than N (i.e. $K < N$).*

According to the definition of the X-form, it is easy to see that if operations ("+" or "·") are used to join several X-forms, a new X-form is formed. Similarly, it is easy to see, if we do a suitable segmentation on an X-form, the segmented portion is also an X-form.

Definition 3.8 (Sub-form of X-form). Suppose E is an X-form, so E is an algebraic expression on $g = \{p_1, p_2, \ldots, p_K\}$, where $g = \{p_1, p_2, \ldots, p_K\}$ is a set of variables, E is an algebraic expression (with 3 subjective operations): $E = E(g) = E(p_1, p_2, \ldots, p_K)$. Then a sub-form of E, E_s, is an algebraic expression on g_s, where $g_s = \{b_{s_1}, \ldots, b_{s_J}\}$, $J \leq K$ is a subset of g, and $E_s = E_s(g_s) = E_s(b_{s_1}, \ldots, b_{s_J})$, and E_s is an appropriate sub-expression of E expression.

By definition, the sub-form is also an X-form.

We skip discussions on properties of X-form. But, we would like to point out this: an X-form can be thought as one information processor, and vice versa, a information processor can be expressed by at least one X-form. Importantly, we should note, a processor could be expressed by different X-forms. This simple fact is crucial.

3.4   Discussions About X-form

X-form is introduced in the process to understand subjective pattern. It turns out, X-form is a very good tool for this purpose. X-form is a clear mathematical object, playing multiple roles. First, it is one algebraic expression. Second, it is a subjective pattern. Third, it is an information processor. We use the same notation to denote the 3 meanings above. This is convenient.

We can see several aspects of X-form. On the one hand, it is an algebraic expression, so algebraic operations can be done, which are relatively easy. On the other hand, the X-form is associated with the perception bits of learning machine. Thus, we could not to change X-form arbitrarily, and changes of X-form requires modification of perception of learning machine. However, this two-sided nature makes X-form particularly good for learning machines. On the one hand, we can do algebraic operations on the X-form, while algebraic operations are relatively easy, and such easiness gives the learning machine good ways to do imagination and various reasoning. On the other hand, these operations, must be specifically placed inside the learning machine, and be realized by perception bits. And, one given X-form itself already illustrates how it can be specifically implementation inside a learning machine. This is very perfect.

As a comparison, we can see usual artificial neural network (ANN). ANN indeed is one information processor. But, it is very hard to do explicit algebraic operations on ANN, and it is hard to understand how perception of pattern is done inside ANN. It is very hard to see inside structure, and very hard to write down what is going on. Another comparison, we can see the motion of concepts with category theory or similar, where several operations can be performed on concepts. While such operations are good for reasoning, these operations are difficult to be directly related to internal structure of the learning machine. X-form can do all these easily. It provides us a powerful tool.

X-form can be thought of as a Boolean function. However, a Boolean function can be implemented in many different X-forms. In this sense, the X-form is a Boolean circuit. However, the X-form is different from the commonly defined

Boolean circuit. An X-form is act on patterns, not directly act on N-dim Boolean vector.

X-form itself is a very clearly defined logical relationship (this logical relationship can be very complicated if the X-form is long), so the X-form naturally is associated with logical reasoning within learning machine. X-form itself is conducting Boolean logic, so classical logic. However, more types of logical reasoning can be done, not have to be classic, e.g., various non-classical, probabilistic, fuzzy, etc. The work in these areas has yet to be unfolded.

In short, the X-form is an algebraic expression, subjective pattern and information processor, which can be easily understood, manipulated, and realized. It is network like, but much easier to handle than popular neural network. The learning is accommodated by X-forms, yes, many X-forms. Surely, X-form is a mathematical object that deserves a lot of researches.

The X-form is the core concept of our discussion, and play very crucial role in a learning machine. Thus we would like to explicitly propose a conjecture for learning machine.

Conjecture 3.9 (X-form conjecture). Any computing system, as long as it has flexible learning capabilities, has at least one X-form inside it, and learning is accommodated by the X-form.

4   Conceiving, Governing and Primary Consciousness

It is clear that X-form is the most important part of an universal learning machine, which is doing information processing, perceiving patterns, and accommodating learning. Naturally, we think that there will be many X-forms inside an universal learning machine, which forms conceiving space.

Definition 4.1 (Conceiving Space). The space formed by all X-forms in a learning machine M is called the conceiving space of M.

Learning is done on X-forms in conceiving space. But, what is promoting, regulating and implementing, governing those changes of X-forms? We introduce the concept of governing space.

Definition 4.2 (Governing Space). Inside a learning machine M, all mechanisms that promote, regulate, control the changes of X-forms inside conceiving space is called as governing space of the learning machine.

So what are inside the governing space? First, there are learning strategies and learning methods, which we will discuss in later sections. However, here we clearly point out that the first thing that governing space must have is its subjectivity. A learning machine must sens all kinds of inputs, and must handle these inputs accordingly. Using our term, it is acting according to subjective pattern. If there is no subjectivity, a learning machine is impossible to detect the input pattern, and it is impossible to actively control and modify the X-form in conceiving space. So how should we describe the subjectivity?

We can look deeper. In [9], the author phrased a term: primary consciousness, and well described it. We think that the term and its description are very good. So we will use this term. That is to say, in governing space, there is primary consciousness that is working to guide activities of learning machine. Refer to [9] for well phrased descritions. If a learning machine that does not have primary consciousness, it cannot be a universal learning machine.

However, even for primary, there is difference in capability. We believe that for the universal learning machine, it is important to understand minimal primary consciousness, which means that if primary consciousness is below a certain threshold, the learning machine cannot be universal. Understanding such threshold can help us to understand some important properties of a universal learning machine. We can note that the external environment of the learning machine is actually the base pattern space at the input space and feedback in the output space. Therefore, the primary consciousness should have sufficient awareness and correct response to this environment. In this way, primary consciousness should be expressed through some of the capabilities of the universal learning machine. That is to say, we can stipulate some abilities to reflect primary consciousness. These capabilities are crucial: (1) the ability to perceive the base pattern of the input. (2) The ability to detect X-forms within itself, and so on.

In Sect. 2, we discussed level of learning. By using conceiving space and governing space, it is very easy and clear to distinguish different level of learning.

5   Learning Dynamics and Data Sufficiency

Through the previous discussion, we have made clear that inside learning machine the information processing and pattern perception is done by X-form, and learning is developing and selecting a better X-form. We now turn to how to develop and select better X-form.

Learning is a motion of X-form, moving from one X form to another. Or, learning is the dynamics in the conceiving space. So let's discuss this dynamics. We only consider the N-1 learning machine M below.

Suppose M is an N-1 learning machine. The typical learning process is:

1. Set an objective pattern: $p_o \subset PS_N^0$;
2. Select a sample set $S_{in} \subset p_o$, usually S_{in} is a much smaller set than p_o. But in extreme cases, $S_{in} = p_o$;
3. Select another sample set $S_{out} \subset p_o^c$, i.e. all elements in S_{out} are not in p_o. The set S_{out} is usually much smaller than p_o^c, but in extreme cases, $S_{out} = p_o^c$ may appear.
4. Use the data in S_{in} and S_{out} to form the sample data sequence $b_i, i = 1, 2, \ldots$, $b_i \in S_{in}$ or $b_i \in S_{out}$.
5. However, feedback data (also known as labeled data) is also required. Data $b_i \in S_{in}$ or $b_i \in S_{out}$ could be labeled, or not labeled. Both cases are allowed, and also allowed sometime labeled, something not labeled. Thus, the data sequence is: $(b_i, o_i), i = 1, 2, \ldots$, if $b_i \in S_{in}$, o_i is 1 or $\varnothing$, if $b_i \in S_{out}$, o_i is 0 or $\varnothing$. Here $\varnothing$ stands for empty to indicate data is not labeled at this time.

6. Apply the data sequence continuously into M. There is no limit on how to enter, how long to enter, input frequency, repeat input, which parts to enter, etc.

Suppose the data sequence is $\{(b_i, o_i) \mid i = 1, 2, \ldots\}$, and the original X-form is E_0. We can consider the learning dynamics. We then can write down the learning in equations.

Starting from E_0, the first step would be:

$$E_1 = LM(E_0, b_1, o_1)$$

Here LM stands for learning method the specific how X-form changes. Note that we can write above way of function because we assume that the learning machine follows "a set of simple and fixed rules", without this, it is problematic to write the above equation.

This process continues, the data is entered in order, at the k step, we have:

$$E_k = LM(E_{k-1}, b_k, o_k) = LM(E_0, b_1, o_1, b_2, o_2, \ldots, b_k, o_k), k = 1, 2, \ldots \quad \text{(lm)}$$

We should note, the equation is very similar to equation in mathematical dynamics. Actually, they are dynamics of learning. We can say, the phase space is the conceiving space, and dynamics is written in learning methods in governing space. Of course, to make these statements mathematically precise, there are much more works to be done.

In the above equation, under the drive of data sequence, we hope eventually the X-form E_k would become what we want. There are several factors. One is data sequence, another is "smartness" of learning method. Here, we discuss data sequence. With learning driven by data, more data bring more driving. But it's better to use less data to do more, if possible. So, we need to know what is enough data. This leads to data sufficiency. Fortunately, the X-form itself gives a good description of enough data. Using X-form, we can define data sufficiency.

Definition 5.1 (Data Sufficient to Support). Suppose E is an X-form, and assume that there is a set of base patterns D, $D = \{b_j \mid j = 1, \ldots J\}$, if for any E subform of F, there is at least one data $b_j, 1 \leq j \leq J$ so that $F(b_j) = 0$, but $E(b_j) = 1$, we say that the data set D is sufficient to support X-form E.

Definition 5.2 (Data Sufficient to Bound). Suppose E is an X-form, and assume that there is a set of base patterns D $D = \{b_j \mid j = 1, \ldots J\}$, if for any X-form F that covers E (i.e. E is a sub-form of F), then there at least is a data $b_j, 1 \leq j \leq J$, so that $E(b_j) = 0$ and $F(b_j) = 1$, we say that the data set D is sufficient to bound X-form E.

We can see the meaning of data sufficiency: If for each sub-form of E, we have a data b_j that makes it in the supporting set of E, but not in the supporting set of this sub-form, the data set is sufficient to support. Also, if for each X-form with E as a sub-form, we have a data b_j outside the supporting set of E, but in the supporting set of the X-form, the data set is sufficient to bound.

6 Learning Strategies

In Eq. (lm), LM is actually formed by learning strategy and method. Strategy defines the framework while method defines more detailed part. There are multiple strategies and methods for different situations and requirements. We discuss some here. Due to restriction on size, we have to skip many details. Please refer to [5].

Strategy 1 - Embed X-forms into Parameter Space
The basic idea of this strategy is this: set a real Euclidean space $\mathbb{R}^U$, U is a huge integer, and then divide $\mathbb{R}^U$ into L mutually disjointed areas, then attach an X-form to each area, thus embedding L X-forms. Then introduce a dynamics in $\mathbb{R}^U$, which is equivalent to introducing dynamics on L X-forms. Dynamics in real European space is a very mature branch of mathematics, and we have a lot of tools to deal with this dynamics. In this way, we transform the learning dynamics into the dynamics in a real-European space.

More precisely, we can write down as below. Suppose $\mathbb{R}^U$ is a real Euclidean space, U is a huge integer, and then assume that we have somehow cut $\mathbb{R}^U$ to L pieces:

$$\mathbb{R}^U = \bigcup_{i=1}^{L} V_i, \quad V_i \cap V_{i'} = \varnothing \ \ \forall i \neq i'$$

Also, we preset L X-forms $E_i, i = 1, 2, \ldots, L$, and attach each E_i to V_i. Thus, for any point $x \in \mathbb{R}^U$, if $x \in V_i$, then at this point x, the corresponding X-form is E_i. We can write this X-form as $E_x = E_i$.

Then, using following loss function:

$$Lo(x) = \sum_{j=1}^{J} (E_x(b_j) - o_j)^2, \quad \forall x \in \mathbb{R}^U \tag{loss}$$

where E_x is the X-form corresponding to x. With these setting, we have theorem.

Theorem 6.1 (Embed into Parameter Space). *Suppose the X-form that we want is E^*, and assume that E^* is among $\{E_1, E_2, \ldots E_L\}$. More, suppose the data sequence D is sufficient to support and sufficient to bound E^*. Further assume that the learning dynamics is: seeking loss function Lo take the minimum value. Then, if $Lo(x_{k^*})$ reaches the minimum at some k^*, the X-form, E_{k^*} is the X-form we want to learn, i.e. $E^* = E_{k^*}$.*

We have some comments on this strategy: (1) Such a setting indeed is an effective learning machine. Deep learning is in fact using this strategy [3]. (2) Of course, the previous settings are ideal and are only for ease of discussion. However, deep learning fully demonstrates that this learning strategy and method is very effective to many situations. (3) This strategy is much more general then usual deep learning. But, it could not achieve universal learning machine.

Strategy 2 - Squeeze X-form from Inside and Outside to Higher Abstraction

This strategy can be summarized as: (1) Check the input space to see if necessary to include the input and whether necessary to exclude the input. (2) Squeeze current X-form to a higher level of abstraction and more generalization. (3) Select the best X-form from conceiving space.

The data requirements for this strategy are: sufficient to support and sufficient to bound the desired X-form, and all data is required to be labeled. So in data sequence $D = \{(b_j, o_j), j = 1, 2, \ldots\}$, all o_j are not empty. This strategy also has requirements for the ability of the learning machine.

Definition 6.1 (Capability 1 - to squeeze X-forms to higher abstractions). Could squeeze the current X-form E into another X-form E' with higher abstraction and more generalization. More precisely, squeeze will do the following: Suppose E is the X-form on g, $E = E(g)$, where g is a set of base patterns $g = \{b_1, b_2, \ldots, b_K\}$, then the squeezed X-form $E' = E'(g')$ should satisfy: 1. $g' \in g$, 2. $B \in B'$, where B is the supporting set for E, and B' is the supporting set for E'. If could find such an X-form, put it in conceiving space, otherwise take no action.

Theorem 6.2 (Squeeze from both inside and outside). *Suppose a learning machine M uses Strategy 2 to learn, and M has the ability 1. Assume further that the data sequence for driving learning is $D = \{(b_j, o_j), j = 1, 2, \ldots\}$, and D is sufficient to support and bound the X-form E^*. Then, starting from the empty X-form, if the data in D is fully input into M and repeat long enough (without missing any data, and can repeat), then M will eventually learn E^*.*

This theorem show that with Strategy 2, a learning machine with Capability 1, can become a universal learning machine.

Strategy 3 - Squeeze X-form from Inside to Higher Abstraction

Strategy 3 is: compressing the X-form from internal to high abstraction. We summarize this strategy as: (1) Check the input space to see if you need to include this input (a basic schema). (2) Compress the current X-form to a higher level of abstraction and more generalization, but not excessive. (3) Select the best X-form from the thought space. Just like strategy 2, strategy 3 can achieve universal learning machine.

Strategy 4 - Squeeze X-form from Inside to Higher Abstraction, Unsupervised

Both Strategy 4 and Strategy 3 squeeze the X-form from inside to high abstraction. However, Strategy 4 uses unsupervised data (at least partially), this is very desirable. But, using unsupervised data, we need to have a stronger capability. Just like strategy 2, strategy 4 can achieve universal learning machine as well.

Comments About strategies

Now, we know, in theory, universal learning machine is achievable. Although, the above strategies are more on theoretical side, and there is some distance to become practical. Nevertheless, this theoretical result is important, it show universal learning machine can be realized.

7   Engineering Model

To build an efficient universal learning machine, we need to have engineering model. We believe that there are many possible engineering models, by different technological means (such as hardware based, software based, biotech based, etc.). But these models must follow certain common principles.

One major part of a universal learning machine is its conceiving space that contains many X-forms. There are more than one ways to realize X-forms. For example, combination of inner product and nonlinear function could form X-forms [3]. A well-designed conceiving space should have properties below. (1) Should have enough expressive power for all X-forms. This is a very crucial. (2) Should have very efficient way to manipulate X-forms, for various activities, including connection, decomposition, combination, add on and decrease, logical reasoning, probability rationale, and so on. (3) Should have a clean structure, so that it can be easily explained and cooperated with other parts of the learning machine, especially with the governing space.

Learning machines should fully integrate logical reasoning and probability rationale, both are indispensable. Such integration needs to be done in conceiving space, i.e. on X-forms, as well as to be done in governing space, i.e. on learning mechanism.

According to [6], learning methods could be put into 5 major categories: (1) logical reasoning and deduction, (2) connectionism, (3) probability methods (specially Bayesian method), (4) analogy, (5) evolution. Governing space should have methods from all these 5 categories and integrate them naturally. Further, governing space can have a AGI module working inside it, such as NARS system [8]. Our study indicates that X-form is good for all of these. Inside governing space, primary consciousness should be established first so that learning machine can react to use these methods.

We have done some specific implementation of the engineering model, e.g. patent application [4].

References

1. Xiong, C.: Discussion on mechanical learning and learning machine. arxiv.org (2016). http://arxiv.org/pdf/1602.00198.pdf
2. Xiong, C.: Descriptions of objectives and processes of mechanical learning. arxiv.org (2017). http://arxiv.org/abs/1706.00066.pdf
3. Xiong, C.: What really is deep learning doing. arxiv.org (2017). http://arxiv.org/abs/1711.03577.pdf
4. Xiong, C.: Chinese patent application. Application # 201710298481.2. (in Chinese)
5. Xiong, C.: Principle, method and engineering model of universal learning machine. Researchage.net (2018). (in Chinese). https://www.researchgate.net/publication/323915356_jisuanjijinxingtongyongxuexideyuanlifangfahegongchengmoxing

6. Domingos, P.: The Master Algorithm, Talks at Google. https://plus.google.com/117039636053462680924/posts/RxnFUqbbFRc
7. Valiant, L.: A theory of the learnable. Commun. ACM **27** (1984). http://web.mit.edu/6.435/www/Valiant84.pdf
8. Wang, P.: A Logical Model of Intelligence: An Introduction to NARS. https://sites.google.com/site/narswang/home/nars-introduction?pli=1
9. Zhong, Y.: Mechanism-based artificial intelligence theory. CAAI Trans. Intell. Syst. **13**(1), 2–18 (2018). (in Chinese)
10. He, H.: Universal logic theory: logical foundation of mechanism-based artificial intelligence theory. CAAI Trans. Intell. Syst. **13**(1), 19–36 (2018). (in Chinese)
11. Wang, P.: Factor space-mathematical basis of mechanism based artificial intelligence theory. CAAI Trans. Intell. Syst. **13**(1), 37–54 (2018). (in Chinese)

An Improved CURE Algorithm

Mingjuan Cai[1] and Yongquan Liang[2(✉)]

[1] College of Computer Science and Engineering, Shandong Province Key
Laboratory of Wisdom Mine Information Technology,
Shandong University of Science and Technology, Qingdao 266590, China
[2] College of Computer Science and Engineering,
Shandong University of Science and Technology, Qingdao 266590, China
lyq@sdust.edu.cn

Abstract. CURE algorithm is an efficient hierarchical clustering algorithm for large data sets. This paper presents an improved CURE algorithm, named ISE-RS-CURE. The algorithm adopts a sample extraction algorithm combined with statistical ideas, which can reasonably select sample points according to different data densities and can improve the representation of sample sets. When the sample set is extracted, the data set is divided at the same time, which can help to reduce the time consumption in the non-sample set allocation process. A selection strategy based on partition influence factor is proposed for the selection of representative points, which comprehensively considers the overall correlation between the data in the region where a representative point is located, so as to improve the rationality of the representative points. Experiments show that the improved CURE algorithm proposed in this paper can ensure the accuracy of the clustering results and can also improve the operating efficiency.

Keywords: Clustering algorithm · CURE algorithm · Sampling
Representative point

1 Introduction

Clustering is a basic task of data mining and is also an unsupervised learning process. The goal of clustering is to gather the n objects in a given d-dimensional space into k clusters according to a specific data metrics, and to make the objects in a cluster have a maximum similarity degree, while the objects between the clusters have a minimum similarity degree [1]. The existing clustering methods are mainly divided into four categories (Berkhin 2006): partitioning method, hierarchical method, density method and model-based method [2].

Each algorithms has its own advantages and disadvantages when solving problems. The CURE (Clustering Using REpresentatives) algorithm is an agglomerative hierarchical clustering method [3, 4]. In recent years, researchers have proposed many novel algorithms from different angles to improve the CURE algorithm. For example: Kang et al. introduced the CURE algorithm in detail in the literature [5] and proposed an improved algorithm. Shen proposed an improved algorithm K-CURE [6]. Wu et al. proposed an improved algorithm RTCURE [7] based on information entropy for measuring inter-class distances. Inspired by these, a new improved algorithm

Z. Shi et al. (Eds.): ICIS 2018, IFIP AICT 539, pp. 102–111, 2018.
https://doi.org/10.1007/978-3-030-01313-4_11

ISE-RS-CURE (CURE algorithm based on improved sample extraction and repre-sentative point selection) is proposed to optimize the sample extraction and represen-tative point selection process of CURE algorithm in this paper. Experiments show the algorithm has better performances as well as low time complexity. ·

2   Related Work

The CURE algorithm can find arbitrarily shaped clusters and is insensitive to noise points. It is different from the traditional clustering algorithm using a single particle or object to represent a class, while choosing a fixed number of representative points to represent a class [8]. This intermediate strategy [9] based on particle and representative object method can detect clusters of different shapes and sizes, and is more robust to noise, overcoming the problem that most clustering algorithms are either only good at dealing with clusters of spherical and similar sizes, or they are relatively fragile when dealing with isolated points.

The CURE algorithm includes both the hierarchical part and the divided part, which overcomes the disadvantage of using a single clustering center tend to discover spherical clusters. A large number of experiments and experiments have proved that the CURE algorithm is effective. Under normal circumstances, the value of the contraction factor is between 0.2 and 0.7, and the number of points larger than 10 can get the correct clustering result [10].

Figure 1 shows the basic steps of the CURE algorithm. The data set is listed in Fig. 1(a). For large-scale data, it is usually obtain a sample set by random sampling. Figure 1(b) shows three clusters of current clustering, expressed in three colors respectively. Figure 1(c) shows the representative points selected from each cluster are shown by the black point. Figure 1(d) shows after "shrinkage" of the representative points, the two clusters with highest similarities are merged. Then, reselect the repre-sentative points of the new cluster and repeat the two processes of "shrinking" the representative points and merging the clusters until the preset number of clusters is

Fig. 1. The basic steps of CURE algorithm

reached. If a class grows slowly during clustering or contains very little data at the end of the cluster, it is treated as noise.

3 Improved CURE Algorithm

3.1 Sample Extraction

The traditional CURE algorithm uses random sampling to obtain sample sets, which is ideal for uniformly distributed data sets; however it may result in incorrect clustering results for non-uniformly distributed data [11]. This paper uses a sample extraction algorithm, which combines statistical ideas to ensure that the data set can be processed more efficiently and the sample set can describe the dataset more accurately. Before introducing the algorithm, we need to establish the mathematical statistics model of the data set first, and then use the merge decision criterion based on independent finite difference inequalities to complete the data extraction.

Definition 1. $A = \{A_1, A_2, \ldots, A_n\}$ is the feature set of data set. Any feature A_i can be represented by m independent random variables, where $1 \leq i \leq n$, $m > 0$. Suppose the range of any feature is $[L_i, R_i]$, and after normalization, the range of the feature is changed to $[0, h_i]$, where $h_i \leq 1$. Therefore, the range of values for m independent random variables should also be $[0, h_i]$, and the range of each independent random variable is $[0, h_i/m]$.

When all data are represented by multiple sets of random variables, the statistical model of the data set is established. In the statistical model, sampling of data points is independent of each other, and no distribution of m random variables is required. The criterion for clustering merged data points is mainly derived from the independent finite difference inequality.

Theorem 1. Let $X = (X_1, X_2, \ldots, X_m)$ be a set of independent random variables. The range of X_i is $l_i (1 \leq i \leq m)$. Suppose there is a function F defined in $\prod_i X_i$. When the variables X and X' only differ in the k-th condition, $|F(X) - F(X')| \leq r_k$ is satisfied. There has

$$P(F(X) - E(F(X)) \geq \tau) \leq exp\left(-2\tau^2 \Big/ \sum_k (r_k)^2\right) \tag{1}$$

Definition 2. Suppose q is the number of features of data items in the data set, and any feature can be represented by m independent random variables. If there exist a class C, the number of data points in class C is represented by $|C|$, the expectation of expected sum of m random variables of related data points in class C is expressed by $E(C)$, (C_1, C_2) is a combination of classes, C_{1c} and C_{2c} represent the core points of classes C_1 and C_2. The merging criterion is formula [2], where $0 < \delta \leq 1$.

$$|(C_{1c} - C_{2c}) - E(C_{1c} - C_{2c})| < \sqrt{\frac{1}{2mq}\left(\frac{|C_1| + |C_2|}{|C_1||C_2|}\right) \ln \delta^{-1}} \tag{2}$$

Proof: Firstly, prove

$$|(C_{1c} - C_{2c}) - E(C_{1c} - C_{2c})| \geq \sqrt{\frac{1}{2mq}\left(\frac{|C_1| + |C_2|}{|C_1||C_2|}\right)} \ln \delta^{-1} \tag{3}$$

conforms to Theorem 1.

Let $\tau = \sqrt{\frac{1}{2mq}\left(\frac{|C_1| + |C_2|}{|C_1||C_2|}\right)} \ln \delta^{-1}$, and there have

$$max\left(\sum_k (r_k)^2\right) = mq|C_1| \cdot max(r_{C_1})^2 + mq|C_2| \cdot max(r_{C_2})^2 = \frac{1}{mq}\left(\frac{|C_1| + |C_2|}{|C_1||C_2|}\right) \tag{4}$$

then

$$P(|(C_{1c} - C_{2c}) - E(C_{1c} - C_{2c})| \geq \tau) \leq exp\left(-2\tau^2 \Big/ \sum_k (r_k)^2\right)$$

$$\leq exp\left(-2\tau^2 \Big/ max\left(\sum_k (r_k)^2\right)\right) = \delta \tag{5}$$

From formula [4] we can see that the formula [3] conforms to Theorem 1. So when δ approaches 0 (δ = 0.0001 in this paper), the probability of $|(C_{1c} - C_{2c}) - E(C_{1c} - C_{2c})| < \tau$ is close to 1.

Let $b(C_1, C_2) = \sqrt{\frac{1}{2mq}\left(\frac{|C_1| + |C_2|}{|C_1||C_2|}\right)} \ln \delta^{-1}$, and if C_1 and C_2 belong to the same class, then $E(C_{1c} - C_{2c}) = 0$. It can be concluded that if the class combination (C_1, C_2) satisfies inequality $|(C_{1c} - C_{2c})| < b(C_1, C_2)$, it can be merged. Therefore, the merging criterion is defined as: For the class combination (C_1, C_2), when each feature satisfies $|C_{1c} - C_{2c}| < b(C_1, C_2)$, the C_1 and C_2 are merged; otherwise the merging is not performed.

3.2 Representative Point Selection

The reasonable selection of representative points can directly affect the quality of clustering result [12]. A representative point selection method based on the distance influence factor is proposed in the literature [13], the appropriate representative point can be selected in different regions of the cluster according to the density distribution of the cluster, but the distance influence factor ignoring the influence of the entire district where the representative point is located. In this paper, a representative point selection method based on Partition Influence Weight (PIW) is proposed, which takes into account the influence of the partition where the representative point is located in the cluster, can effectively eliminates noise points and the appropriate representative points can be selected according to the density distribution of the cluster.

Definition 3. $C = \{d_1, d_2, \ldots, d_n\}$ is a cluster in data set, the representative point in the cluster is d_{ri} ($0 < i \le \sqrt{|C|}$), and C_c is the core point. Under the minimum distance between the data point and the representative point, a cluster partition $\{C_1, C_2, \ldots\}$ is obtained. Each of the segments C_i has a one-to-one correspondence with the d_{ri}. The representative point d_{ri} has a weight of

$$PIW(d_{ri}) = \frac{n}{m} \cdot \frac{\sum_{j=1}^{m} d(d_j, C_c)}{\sum_{i=1}^{n} d(d_i, C_c)} \cdot dist(d_{ri}, C_c) \tag{6}$$

Among them, n is the number of data items in the cluster C, m is the number of data items in the block C_i, and $d(d_i, C_c)$ is the distance from the core point C_c to each data item in the cluster C, and $d(d_j, C_c)$ is the distance from the core points C_c to the data items in the block C_i.

In the process of selecting the representative point, the $PIW(d_{ri})$ value is compared with the threshold value η, and the adjustment of the representative point selection is performed until the proper representative point is selected. The initial threshold is set as $\eta = \frac{n}{5m} \cdot \frac{\sum_{j=1}^{m} d(d_j, C_c)}{\sum_{i=1}^{n} d(d_i, C_c)} \cdot dist(d_{ri}, C_c)$.

3.3   The ISE-RS-CURE Algorithm

The ISE-RS-CURE algorithm firstly uses the sample extraction algorithm based on statistical ideas is used to extract samples, in which the merge criterion is used instead of the distance threshold. Removing the representative points with count less than 2 can effectively eliminate the noise points and reduce the interference of noise points and abnormal points. This sample extraction algorithm carries out a rough partition of the sample set while obtaining the sample set. This information can be used to complete the final clustering result during the stage of labeling the non-sample data, which can effectively improve the efficiency of the algorithm. The partition strategy in the representative point selection algorithm guarantees the distribution of representative points. The partition influence factor can be used to continuously adjust the selection of representative points. The noise points can be eliminated so that the representative point set can better reflect the shape size and density information of the data set, and avoid the phenomenon that the representative point gathers and the noise point becomes the representative point. The pseudo code for the ISE-RS-CURE algorithm is:

Algorithm 1. ISE-RS-CURE algorithm

Input: data set D; the number of independent random variables m

Output: Clustering results

//**Phase 1:** Get a sample set

1. Initialize the sample set S

2. for each data item x in data set D do

3. for each sample point s in sample set S do

4.   if x satisfy: $s_i \in S$, s_i located C_i, for all features there is $|c_{ik} - x_k| < b(C_i, x)$, then $C_i = C_i \cup \{x\}, D = D - \{x\}$

5.   else $S = S \cup \{x\}, D = D - \{x\}$

6. end for

7. end for

8. Delete the s_i from S, where $|C_i| < 2$, and mark s_i as noise point

//**Phase 2:** Clustering stage

9. Initialize the representative set R

10. Each item in sample set S is treated as a class and aggregated hierarchical clustering

11. for each cluster C_i do

12. random select d_i, where $d_i \in C_i$, if $d(d_i, C_c) = min\left(max\left(d(d_i, C_c)\right)\right)$, let $C_c \leftarrow d_i$, $R_i = R_i \cup \{d_i\}$. Cycle $\sqrt{|C_i|}$ times.

13. divide the cluster C_i according to R_i, and get $\left\{C_{i1}, C_{i2}, \cdots, C_{i\sqrt{|C_i|}}\right\}$

14. calculate the $PIW(d_{rij})$ for each partition C_{ij}

15. if $PIW(d_{rij}) \leq \eta$ then

16.   Delete d_{rij}, select a new representative point and update R_i

17. end for

18. "shrink" the representative point to the cluster center point

19. Merge the two clusters with the highest similarity and return to **step 11** until the number of target clusters is reached

20. Assigning non-sample data items to different clusters in data sets

---

In the Step (16) the selection strategy of the new representative point is: in the cluster where the representative point d_{rij} is located, a data item that has not been selected as a representative point is randomly selected as a new representative point, and the value of $PIW(d_{rij})$ is recalculated until all the weights of the representative points are all greater than η. The sample extraction algorithm requires only one scan of the data set. Therefore, for a given data set of size n, the time complexity is only $O(n)$, and the advantage is obvious when dealing with large-scale data sets. The representative point selection algorithm needs to calculate the distance metric matrix of the sample set. The time complexity is $O(m^2)$, and m is the number of sample set data points. Thus, the time complexity of the ISE-RS-CURE algorithm is $O(n + m^2)$ and the space complexity is $O(n)$.

4   Experiment Analysis

In order to test the actual clustering performance of the improved CURE algorithm presented in this paper, the experimental part compares the traditional CURE algorithm and the ISE-RS-CURE algorithm by using the synthetic data set and the real data set respectively.

4.1   Experiments on Synthetic Data Sets

Experiment 1: In order to compare the ISE-RS-CURE algorithm and the traditional CURE algorithm in the accuracy of the clustering results, the experiment uses an synthetic data set D1 (as shown in Fig. 2). The data distribution of data set D1 is complex, which contains 6 clusters with different shape size and large location difference, and dense noise points with cosine distribution and a large number of scattered noise points. The total number of data points is 8000 and the number of attributes is 2. For ease of discussion, the six major clusters clustered are numbered as ① to ⑥ respectively.

  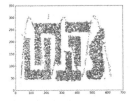

Fig. 2. Data set D1   **Fig. 3.** CURE clustering results   **Fig. 4.** ISE-RS-CURE sample set

 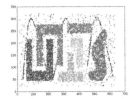

Fig. 5. Sample set clustering results   **Fig. 6.** ISE-RS-CURE clustering results

Figure 3 shows the clustering results of the traditional CURE algorithm, it can be seen that due to the influence of cosine noise points, cluster ① and cluster ② are different from ideal clusters, and ④⑤⑥ is also disturbed to varying degrees. Figure 4 is a sample set obtained by the sampling algorithm in ISE-RS-CURE algorithm. It can be seen that the sample set conforms to the distribution of the original data set, and the noise data with lower density is filtered out. The sample set has a higher representative. Figure 5 shows the clustering result of the sample set. Since the noise points have less interference, the correct clustering results can be displayed for the six major clusters.

Figure 6 shows the final clustering result of the ISE-RS-CURE algorithm. Compared with the traditional CURE algorithm, the ISE-RS-CURE can get clustering results that are closer to the real cluster, so it has significantly improved the clustering accuracy.

Experiment 2: In order to compare the run-time performance of the traditional CURE algorithm and the ISE-RS-CURE algorithm, this experiment uses an synthetic data set D2 (as shown in Fig. 7). Data set D2 is a data set of 3 round clusters composed of random numbers. Because the data set is distributed evenly and there is no noise point, the traditional CURE algorithm and the improved CURE algorithm can both get the correct clustering results. In the experiment, the traditional CURE algorithm uses random sampling to obtain the same size sample set. The number of data points is increased from 5000 to 25000, and comparing the running time of the two algorithms.

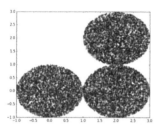   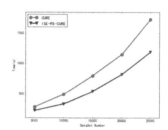

Fig. 7. Data set D2   **Fig. 8.** Time contrast of two algorithms

It can be seen from Fig. 8 that with the increase of data number, the execution time of the traditional CURE algorithm increases exponentially. Although the execution time of the improved algorithm also increases exponentially, it obviously slows down; the execution time of the ISE-RS-CURE algorithm is less than the execution time of the traditional algorithm, and as the data volume increases, the gap gradually increases. Therefore, the ISE-RS-CURE algorithm has an improvement in execution time compared with the traditional CURE algorithm.

4.2   Experiments on the Real Data Sets

This experiment selects 5 data sets in the UCI machine learning database [14], tests the traditional CURE algorithm and the ISE-RS-CURE algorithm, and evaluates it with 3 evaluation criteria. The basic attributes of the dataset are display in Table 1.

Table 1. Data set table

Data set	Size	Attribute	Cluster
Wine	178	13	3
Seeds	210	7	3
Ionosphere	351	34	2
Waveform	5000	31	3
Segmentation	2310	19	7

The clustering results are compared and analyzed with 3 evaluation indexes: CA, ARI and NMI [15]. The results of each data set and corresponding evaluation indicators are recorded in Table 2.

Table 2. Comparison of results of evaluation indicators on UCI datasets

Data set	Algorithm	CA	ARI	NMI
Wine	CURE	0.7394	0.5271	0.7058
	ISE-RS-CURE	**0.7401**	**0.5389**	**0.7125**
Seeds	CURE	0.6910	0.5936	0.6773
	ISE-RS-CURE	**0.7028**	**0.6192**	**0.6854**
Ionosphere	CURE	0.6283	0.7492	0.6787
	ISE-RS-CURE	**0.6398**	**0.7635**	**0.7021**
Waveform	CURE	0.5862	0.3368	0.4920
	ISE-RS-CURE	**0.6750**	**0.4068**	**0.5473**
Segmentation	CURE	0.6449	0.4839	0.6281
	ISE-RS-CURE	**0.7061**	**0.5325**	**0.6657**

From Table 2, we can see that for the five different data sets, the ISE-RS-CURE algorithm performs better than the traditional CURE algorithm on the 3 clustering evaluation indexes of CA, ARI and NMI. Especially in the large amount of data and attribute multiple data sets, experiments show that the ISE-RS-CURE algorithm has more obvious advantages, which further proves the effectiveness of the ISE-RS-CURE algorithm in clustering.

5   Conclusion

This paper proposes an ISE-RS-CURE algorithm for the disadvantages of the traditional CURE algorithm, which sample set can't reflect the data distribution and Representative points are less representativeness. The ISE-RS-CURE algorithm uses a combined decision criterion based on statistical ideas for sample extraction and the representation point selection strategy based on the partition influence weight. It can not only obtain the set of samples which conform to the data set distribution, but also can gather arbitrary shape clusters and have better robustness. Experimental results show that the ISE-RS-CURE algorithm can effectively improve the accuracy and efficiency of the algorithm.

References

1. Han, J., Kamber, M.: Data Mining: Concepts and Techniques, 3rd edn. China Machine Press, Beijing (2012)
2. Niu, Z.-H., Fan, J.-C., Liu, W.-H., Tang, L., Tang, S.: CDNASA: clustering data with noise and arbitrary shape. Int. J. Wirel. Mob. Comput. **11**(2), 100–111 (2016)

3. Guha, S., Rastogi, R., Shim, K.: CURE: an efficient clustering algorithm for large databases. In: ACM SIGMOD International Conference on Management of Data, pp. 73–84. ACM (1998)
4. Guha, S., Rastogi, R., Shim, K., et al.: CURE: an efficient clustering algorithm for large databases. Inf. Syst. **26**(1), 35–58 (2001)
5. Kang, W., Ye, D.: Study of CURE based clustering algorithm. In: 18th China Conference on Computer Technology and Applications (CACIS), vol. 1. Computer Technology and Application Progress, pp. 132–135. China University of Science and Technology Press, Hefei (2007)
6. Jie, S., Zhao, L., Yang, J., et al.: Hierarchical clustering algorithm based on partition. Comput. Eng. Appl. **43**(31), 175–177 (2007)
7. Wu, H., Li, W., Jiang, M.: Modified CURE clustering algorithm based on entropy. Comput. Appl. Res. **34**(08), 2303–2305 (2017)
8. Wang, Y., Wang, J., Chen, H., Xu, T., Sun, B.: An algorithm for approximate binary hierarchical clustering using representatives. Mini Micro Comput. Syst. **36**(02), 215–219 (2015)
9. Fray, B.J., Dueck, D.: Clustering by passing messages between data points. Science **315** (5814), 972–976 (2007)
10. Jia, R., Geng, J., Ning, Z., et al.: Fast clustering algorithm based on representative points. Comput. Eng. Appl. **46**(33), 121–123+126 (2010)
11. Zhao, Y.: Research on user clustering algorithm based on CURE. Comput. Eng. Appl. **11**(1), 457–465 (2012)
12. Shao, X., Wei, C.: Improved CURE algorithm and application of clustering for large-scale data. In: International Symposium on it in Medicine and Education, pp 305–308. IEEE (2012)
13. Shi, N., Zhang, J., Chu, X.: CURE algorithm-based inspection of duplicated records. Comput. Eng. **35**(05), 56–58 (2009)
14. Lichman, M.: UCI machine learning repository [EB/OL] (2013). http://archive.ics.uci.edu/ml.2018/02/24
15. Pengli, L.U., Wang, Z.: Density-sensitive hierarchical clustering algorithm. Comput. Eng. Appl. **50**(04), 190–195 (2014)

Data Intelligence

D-JB: An Online Join Method for Skewed and Varied Data Streams

Chunkai Wang[1,2(✉)], Jian Feng[1], and Zhongzhi Shi[2]

[1] Post-Doctoral Research Center, China Reinsurance (Group) Corporation,
Beijing, China
{wangck,fengj}@chinare.com.cn

[2] Institute of Computing Technology, Chinese Academy of Sciences, Beijing, China
shizz@ict.ac.cn

Abstract. Scalable distributed join processing in a parallel environment requires a partitioning policy to transfer data. Online theta-joins over data streams are more computationally expensive and impose higher memory requirement in distributed data stream management systems (DDSMS) than database management systems (DBMS). The complete bipartite graph-based model can support distributed stream joins, and has the characteristics of memory-efficiency, elasticity and scalability. However, due to the instability of data stream rate and the imbalance of attribute value distribution, the online theta-joins over skewed and varied streams lead to the load imbalance of cluster. In this paper, we present a framework D-JB (Dynamic Join Biclique) for handling skewed and varied streams, enhancing the adaptability of the join model and minimizing the system cost based on the varying workloads. Our proposal includes a mixed key-based and tuple-based partitioning scheme to handle skewed data in each side of the bipartite graph-based model, a strategy for redistribution of query nodes in two sides of this model, and a migration algorithm about state consistency to support full-history joins. Experiments show that our method can effectively handle skewed and varied data streams and improve the throughput of DDSMS.

Keywords: Distributed data stream management system
Online join · State migration · Bipartite graph-based model

1 Introduction

Nowadays, with the increasing number of data types and the emergence of data intensive applications, the number and speed of data increase rapidly. It makes that data stream real-time analysis and processing has become one of the hottest research areas. So, distributed data stream management systems (DDSMS) are widely used in real-time processing and query analysis of large-scale data streams.

Z. Shi et al. (Eds.): ICIS 2018, IFIP AICT 539, pp. 115–125, 2018.
https://doi.org/10.1007/978-3-030-01313-4_12

In applications such as analytic over microblogs, monitoring of high frequency trading and real-time recommendation, it often involves joins on multiple data streams to get the query result and to maintain large state for full-history query requests based on the full-history data [1,2]. In these applications, data rate tends to fluctuate and the distribution of attribute values is also imbalance. It makes the join operation over skewed and varied streams prone to cluster load imbalance. The phenomenon leads to the decrease of query efficiency and the increase of computation cost in the cloud environment. Due to the imbalance of data rate and distribution, it causes attribute value skew (AVS) [3]. Due to data partition, it causes tuple placement skew (TPS) [3]. So, how to deal with the efficient joins over skewed and varied streams and the load balance of clusters is the focus of our attention.

In order to design an efficient distributed stream theta-join processing system, there are several models designed for join operator over data streams. The join-matrix model [4] and the join-biclique model [5] are two representative approaches to deal with the scalable join processing. For supporting arbitrary join-predicates and coping with data skew, the join-matrix model uses a partitioning scheme to randomly split the incoming data stream into a non-overlap substreams. As a representative of alternative model, the join-biclique model is to organize the processing units as a complete bipartite graph (a.k.a biclique), where each side corresponds to a relation. Two streams are divided into the different side. And, according to the key-based partitioning method (such as, hash function), tuples are distributed to the different nodes for storing in the same side. At the same time, tuples also are sent to the opposite side to do the join operation using the same hashing strategy. After obtaining the results, these tuples are discarded.

Join-matrix and join-biclique models can effectively deal with the online join operation of distributed data streams, but are faced with the following problems and challenges.

1. The join-matrix model needs high memory usage to replicate and store in an entire row or column. Although join-biclique model has the strength of memory-efficiency, it cannot dynamically adjust the distribution of processing units based on skewed streams.
2. Due to the inconsistency of streams distribution, the balance of DDSMS is lost. It leads to performance degradation of DDSMS.
3. It is necessary to have a good salability of DDSMS for join operation. When the pressure of a node is too large/small, the cluster size can dynamically scale out/down according to its application workloads.

So, in this paper, we propose the mixed workload partitioning strategy D-JB for handling the skewed online join based on the join-biclique model, so as to achieve load balancing and high throughput of DDSMS. The contributions of our work are summarized as follows:

1. We propose a mixed key-based and tuple-based partitioning scheme to handle skewness in each side of the join-biclique model, and a normalized

optimization objective by combined with the different cost types involved in the dynamic migration strategy.

2. We present a strategy for redistribution of processing units in two sides of this model. The load balance of D-JB is implemented by repartitioning query nodes logically. And, we prove the efficiency and feasibility by using the different query tasks.

3. We use the operator states manager to migrate processing units between different nodes to ensure the consistency and scalability of the full-history join operation.

The rest of the paper is organized as follows. Section 2 surveys the related work. Then, there are the preliminaries in Sect. 3. And, in Sect. 4, we give the architecture of D-JB and describe details of data migration. Section 5 gives the results of our experiment evaluation. Finally, Sect. 6 concludes this paper.

2 Related Work

In recent years, there has been a lot of research work on the online join operation for skew resilience.

Online joins often require un-blocking tuple processing for getting query results in real-time. The join-matrix and join-biclique models are the most popular research models in parallel and distributed environment. Intuitively, the join-matrix model design a join between two data stream R and S as a matrix, where each side corresponds to one relation. In data stream processing, DYNAMIC [4] supports adaptive repartitioning according to the change of data streams. To ensure the load balancing and skew resilience, Aleksandar et al. [6] proposed a multi-stage load-balancing algorithm by using a novel category of *equi-weight* histograms. However, [4,6] assumes that the number of partitions must be 2^n. So, the matrix structure suffers from bad flexibility. For reducing the operational cost, Junhua et al. [7] proposed the cost-effective stream join algorithm by building irregular matrix scheme. However, The basic model is also matrix-based, data redundancy is still more. JB [5] can save resource utilization significantly. And in order to solve the problem of load imbalance by key-based partitioning, it adopt a hybrid routing strategy, called *ContRand* [5], to make use of both the key-based and random routing strategies. *ContRand* logically divide processing units into disjoint subgroups in each side, and each subgroup contains one or more units. The key-based (content-sensitive) routing strategy is used between the subgroups, and the tuple-based (content-insensitive) routing strategy is used in each subgroup. However, this strategy requires the user to define parameters of subgroupings, and cannot be adjusted dynamically according to the dataflow. Moreover, if the data stream is too skew, it will cause a key to be overloaded in a processing unit and exceed the upper bound of imbalance tolerance. So, we need to partition the tuples with the same key into several processing units. In this case, the problem cannot be reduced to Bin-packing problem [8] in paper [9], which is not considered that a processing unit exceeds the threshold storing tuples with the same key. So, In this paper, we propose the mixed workload partitioning strategy for handling the skewed and varied online join based on the join-biclique model.

3   Preliminaries

We gives the relevant preliminaries for the full-history online join operation in this section.

3.1   Concept Description

In order to make clear the optimization target, the notations involved in our proposed model are listed in Table 1.

Table 1. Table of Notations

Notations	Description
K_{tup}	the key of a tuple
pu	the processing unit (called *task instance* in Storm [10])
m, n	the number of pu in two sides of B
$L(pu)$	the total workload in pu
$\theta(pu)$	load balance factor of pu
θ_{max}	the maximum bound of imbalance factor

Definition 1. At time t, the **load balance factor** of a pu is defined as:

$$\theta_t(pu) = \left| L_t(pu) - \bar{L}_t \right| / \bar{L}_t \tag{1}$$

where $\bar{L}_t$ represents the mean of total workloads in PU. $\bar{L}_t$ is defined as:

$$\bar{L}_t = \sum_{pu=1}^{N_{PU}} (L_t(pu))/N_{PU} \tag{2}$$

So, the pu is relatively balanced at time t, if $\theta_t(pu) \leqslant \theta_{max}$.

Definition 2. There are three types of migration at time t, when $\exists pu$ ($pu \in PU$) $\theta_t(pu) > \theta_{max}$. (1) *data immigration*. Tuples with the same K_{tup} at different pus merge to the starting pu. (2) *data emigration*. All tuples with the same k are migrated to other pu. (3) *data splitting*. Some tuples with the same k are migrated to other pu, the other tuples are not moved.

Definition 3. According to the distribution and skewness of data streams, we need to design the migration strategy dynamically. It involves three types of costs: (1) routing cost $C_{routing}$ is the cost of maintaining the routing table for mapping relationships between K_{tup} and pus. (2) duplication cost $C_{duplication}$ is the cost of replicating tuples with the same K_{tup} after data splitting. (3) migration cost $C_{migration}$ is the cost of migrating tuples from a pu to the other.

From Definitions 2 and 3, it is known that *data immigration* involves $C_{migration}$; *data emigration* involves $C_{routing}$ and $C_{migration}$; *data splitting* involves $C_{routing}$, $C_{duplication}$ and $C_{migration}$.

3.2 Optimization Objective

Let $F = (f_1, f_2, f_3, ...)$ be the set of all migration functions at time t. The migration cost by using each function f_i can be defined as follows.

$$C_t(f_i) = \alpha * C_{routing}(f_i) + \beta * C_{duplication}(f_i) + \gamma * C_{migration}(f_i) \qquad (\alpha + \beta + \gamma = 1) \ (3)$$

where, α, β and γ are the weight of three costs respectively. In order to determine the specific weights, we use the consumed time to calculate. For detail, the hash routing time T_{hash} affects $C_{routing}$; one tuple transferring time T_{trans} affects $C_{duplication}$ and $C_{migration}$. So, the total time $T_{total} = T_{hash} + m * T_{trans} + n * T_{trans}$, where, m is the number of duplication tuples, n is the number of migration tuples. Finally, $\alpha = T_{hash}/T_{total}$; $\beta = m * T_{trans}/T_{total}$; $\gamma = n * T_{trans}/T_{total}$.

The optimization objective is as below:

$$\min_{f_i \in F} C_t(f_i)$$
$$s.t. \quad \theta_t(pu) \leq \theta_{max}, \forall pu \in PU. \tag{4}$$

The target is to minimize the total costs, while meeting the constraint on load balance factor. It involves the range of K_{tup}, the number of PU and the maximum bound of imbalance factor θ_{max}, which is a combinatorial NP-hard problem. And, this problem is more complex than Bin-packing problem, due to the data inside a pu may be split. Therefore, in the next section, we set up a number of heuristics to optimize it.

4 D-JB Model Design

This section gives the architecture of D-JB and the algorithms of data migration.

4.1 D-JB Architecture

The architecture of D-JB is shown in Fig. 1, we design the *controller* on Storm by using the join-biclique model. The basic process of the workflow is as follows.

(1) Firstly, data stream R (resp. S) are partitioned by the key-based hash function, stored to n (resp. m) *pus*, and sent to the opposite side for online join operation.

(2) At each time interval (the setting is 5 s in our experiment), we periodically monitor the statistical information of each *pu* load on both sides of B, and collect the information to *controller*. And, we develop migration strategies based on the heuristics (see in Sect. 5.1).

(3) Then, new data streams are temporarily stored in Kafka [11] and postponed online joins before data migration. Meanwhile, we migrate data streams and the state based on migration strategies, and update the routing table synchronously.

(4) Finally, we continue to send data streams and do the online join operation.

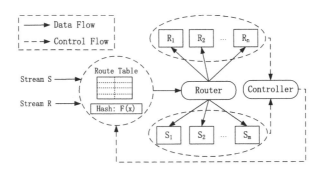

Fig. 1. architecture of D-JB

4.2   Algorithms of Data Migration

The problem of minimizing migration cost is the NP-hard problem. So, we need to set some heuristics to optimize the objective function. There are two migration strategies: the data migration strategy in one side of B, called *internal-side migration* (ISM); and the logical migration strategy in two sides of B, called *side-to-side migration* (S2SM).

Heuristics. At time t, it is assumed that the load of pu exceeds the upper limit of the non-equilibrium factor, which is $L_t(pu) > (1+\theta_{max})^* \bar{L}_t$, or the load of pu is lower than the lower limit of the non-equilibrium factor, which is $L_t(pu) < (1-\theta_{max})^* \bar{L}_t$. In order to satisfy the balance of each pu and minimize data migration, heuristics are as follows.

H1. For data emigration, we can directly migrate keys, if the threshold requirements can be met after migrating these keys, and record them in the routing table.

H2. For data emigration, if the threshold requirements cannot be met after migrating some keys, we need to partition the keys with a larger number of tuples. Then, we migrate some tuples, and record them in the routing table.

H3. For data immigration, if there are records in the routing table, we need to merge data into the key-based hashing pu, and then clear the records in routing table.

According to the above heuristics, we propose the process of moving out tuples, which called $MoveOut(PU_{out}, PU_{in}, RT)$. Firstly, we judge the range of key values of moving out tuples and determine the processing units waiting to move in tuples. Then, for each emigration key, we move out tuples by H1 and H2, and update the routing table.

Next, we propose the process of moving in tuples, which called $MoveIn(PU_{in}, PU_{out}, RT)$. Firstly, we judge the range of key values of moving in tuples and determine pus waiting to move out tuples. Then, for each immigration key, we move in tuples by H3, and update the routing table.

ISM. At time t, in order to satisfy the balance of one side of B and minimize the data migration, the ISM algorithm is described in Algorithm 1. Firstly, we compute the each $L_t(pu)$ and $\bar{L}_t$ at time t (line 1–4). Then, for the pu that needs to emigrate data, we call the $MoveOut$ algorithm (6–8 lines). Finally, for the pu that needs to immigrate data, we call the $MoveIn$ algorithm (9–11 lines).

Algorithm 1. ISM Algorithm.

Require:
    Processing units PU in one side of B;
    Routing table RT;
    The threshold of imbalance factor θ_{max};
Ensure:
    The Migration Plan MP;
 1: **for** $(i = 1; i < Number\_of\_PU; i++)$ **do**
 2:    Computing $L_t(pu_i)$;
 3: **end for**
 4: Computing $\bar{L}_t$;
 5: **for** $(i = 1; i < Number\_of\_PU; i++)$ **do**
 6:    **if** $(L_t(pu_i) > (1+\theta_{max})*\bar{L}_t)$ **then**
 7:        $MoveOut(PU, PU, RT)$;
 8:    **end if**
 9:    **if** $(L_{t_{pu_i}} < (1-\theta_{max})*\bar{L}_t)$ **then**
10:        $MoveIn(PU, PU, RT)$;
11:    **end if**
12: **end for**

S2SM. Because the stream rate is often dynamic, which results in a large gap between two sides and affects the throughput of DDSMS. In this section, we design the S2SM algorithm for dynamically changing data distribution on both sides. The overall migration is shown in Algorithm 2.

Algorithm 2. S2SM Algorithm.

Require:
> Processing units PU_m and PU_n in two sides of B;
> Routing table RT;
> The threshold of imbalance factor θ_{max};

Ensure:
> The Migration Plan MP;
> 1: **for** $(i = 1; i < (Number\_of\_PU_m + Number\_of\_PU_n); i++)$ **do**
> 2:   Computing $L_t(pu_i)$;
> 3: **end for**
> 4: Computing $\bar{L}_{t_m}$, $\bar{L}_{t_n}$ and $\bar{L}_t$;
> 5: Choosing the PU_{out} and PU_{in} based on θ_{max}
> 6: **for** $(j = 1; j < Number\_of\_PU_{out}; j++)$ **do**
> 7:   **if** $(L_t(pu_j) > (1+\theta_{max})*\bar{L}_t$) **then**
> 8: $MoveOut(PU_{out}, PU_{in}, RT)$;
> 9:   **end if**
> 10: **end for**
> 11: **for** $(k = 1; k < Number\_of\_PU_{in}; k++)$ **do**
> 12:   **if** $(L_t(pu_j) < (1-\theta_{max})*\bar{L}_t$) **then**
> 13: $MoveIn(PU_{in}, PU_{out}, RT)$;
> 14:   **end if**
> 15: **end for**

Firstly, at time t, we compute the workload of each pu, and the average workload in each side and the whole cluster (line 1–4). Then, we choose the side of data emigration and the side of data immigration (line 5). Finally, for the side of data emigration, we judge pus that need to move out tuples and call the *MoveOut* algorithm (line 6–10). For the side of data immigration, we judge pus that need to move in tuples and call the *MoveIn* algorithm (line 11–15).

State Migration. BiStream [5] adopts a *requesting* phase and a *scaling* phase to adaptively adjust the resource management. However, it can only be adjusted in the window-based join model. In this section, we introduce the Tachyon [12] as a in-memory file server to store the state information, and use the operator states manager (OSM) [13] to achieve live migration from different nodes. This help us to complete the adaptive resource management in the full-history join model.

5   Evaluation

5.1   Experimental Setup

Environment. The testbed is established on a cluster of fourteen nodes connected by a 1 Gbit Ethernet switch. Five nodes are used to transmit data source through Kafka. One node serves as the nimbus of Storm, and the remaining eight nodes act as supervisor nodes. Each data source node and the nimbus node have a Intel E5-2620 2.00 GHz four-core CPU and 4 GB of DDR3 RAM. Each supervisor node has two Intel E5-2620 2.00 GHz four-core CPU and 64 G of DDR3 RAM. We implement comprehensive evaluations of our prototype on Storm-0.9.5 and Ubuntu-14.04.3.

Data Sets. We use TPC-H benchmark [14] to test the proposed algorithms, and generate the TPC-H data sets using the *dbgen* tool shipped with TPC-H benchmark. All the input data sets are pre-generated into Kafka before feeding to the stream system. We adjust the data sets with different degrees of skew on the join attributes under *Zipf* distribution by choosing a value for skew parameter z. By default, we set $z = 1$, and generate 10 GB data.

Queries. We employ three join queries, namely two equi-joins from the TPC-H benchmark (Q3 and Q5) and one synthetic band-join (*Band*) is used in [4,5]. The *Band* are expressed as follows:
    SELECT *, FORM LINEITEM L1, LINEITEM L2
        WHERE ABS(L1.orderkey-L2.orderkey) <= 1
        AND (L1.shipmode='TRUCK' AND L2.shipinstruct='NONE')
        AND L1.Quantity > 48

Models. We use three algorithms for evaluating the query performance, namely D-JB, JB [5] and JB6 [5]. D-JB is proposed in this paper. JB divides the *pus* on average, and each half of the *pus* corresponds to a data stream. JB6 means there are 6 subgroups in each side for random routing.

5.2 Throughput and Latency

We compare the throughput and latency among the different models by using queries of Q3, Q5 and *Band*. As shown in Fig. 2(a), the throughput of D-JB is largest than JB and JB6. However, the throughput of JB is lowest due to it needs to do the whole network broadcast operation, the communication is too large. Figure 2(b) shows that the latency of D-JB is lowest, and the latency of JB is highest.

5.3 Scalability

When the cluster scales out, we further study the scalability of D-JB, JB and JB6. Since JB and JB6 do not support scaling out dynamically for full-history online join, we use Taychon to implement these models, which can scale out processing nodes without restarting topologies in Storm. As shown in Fig. 3(a), (b) and (c), the run time of D-JB is the shortest in these models, and the scalability of D-JB is the strongest. Moreover, since Q5 involves the largest number of data streams and the most complex join operation, the run time of Q5 is more than Q3 and *Band*. And, because of *Band* involves only one data stream, it can get the minimum run time.

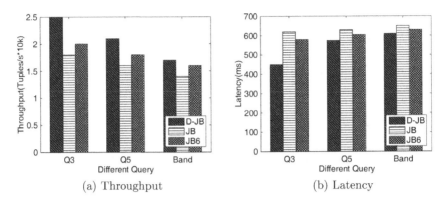

(a) Throughput   (b) Latency

Fig. 2. Throughput and latency

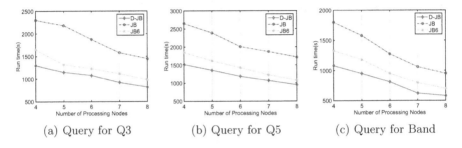

(a) Query for Q3   (b) Query for Q5   (c) Query for Band

Fig. 3. Runtime

6   Conclusion

In this paper, we propose online join method for skewed and varied data streams. Based on the join-biclique model, we give a mixed key-based and tuple-based partitioning scheme to handle data skew in one side of the join-biclique model, and present a strategy for redistribution of processing units in two sides of this model. Finally, we design a migration algorithm about state consistency to support full-history joins and adaptive resource management. Our experimental results demonstrate that our proposed framework can be better to deal with data skew, significantly improve the throughput and reduce the latency of DDSMS.

References

1. Hwang, J., Balazinska, M., Rasin, A., et al.: High-Availability algorithms for distributed stream processing. In: Proceedings of the 21st International Conference on Data Engineering, pp. 779–790. IEEE Press (2005)
2. Fernandez, R., Migliavacca, M., Kalyvianaki, E., et al.: Integrating scale out and fault tolerance in stream processing using operator state management. In: Proceedings of the 2013 ACM SIGMOD International Conference on Management of Data, pp. 725–736. ACM Press (2013)

3. Walton, C., Dale, A., Jenevein, R.: A taxonomy and performance model of data skew effects in parallel joins. In: Proceedings of the 17th International Conference on Very Large Data Bases, pp. 537–548. ACM Press (1991)

4. Elseidy, M., Elguindy, A., Vitorovic, A., et al.: Scalable and adaptive online joins. In: the VLDB Endowment, vol. 7(6), pp. 441–452 (2014)

5. Lin, Q., Ooi, B.C., Wang, Z., et al.: Scalable distributed stream join processing. In: ACM SIGMOD International Conference on Management of Data, pp. 811–825. ACM Press (2015)

6. Vitorovic, A., Elseidy, M., Koch, C.: Load balancing and skew resilience for parallel joins. In: IEEE International Conference on Data Engineering, pp. 313–324. IEEE Press (2016)

7. Fang, J., Zhang, R., Wang, X., et al.: Cost-effective stream join algorithm on cloud system. In: the 25th ACM International on Conference on Information and Knowledge Management, pp. 1773–1782. ACM Press (2016)

8. Narendra, K., Richard, K.: An efficient approximation scheme for the one-dimensional bin-packing problem. In: 23rd Annual Symposium on Foundations of Computer Science, pp. 312–320. IEEE Press (1982)

9. Fang, J., Zhang, R., Wang, X., et al.: Parallel stream processing against workload skewness and variance. In: CoRR abs/1610.05121 (2016)

10. Toshniwal, A., Taneja, S., Shukla, A., et al.: Storm@twitter. In: ACM SIGMOD International Conference on Management of Data, pp. 147–156. ACM Press (2014)

11. http://kafka.apache.org/

12. Li, H., Ghodsi, A., Zaharia, M., et al.: Tachyon: memory throughput I/O for cluster computing frameworks. In: LADIS (2013)

13. Ding, J., Fu, T., Ma, R., et al.: Optimal operator state migration for elastic data stream processing. In: HAL - INRIA, vol. 22(3), pp. 1–8 (2013)

14. http://www.tpc.org/tpch

The Application of Association Analysis in Mobile Phone Forensics System

Huan Li[✉], Bin Xi, Shunxiang Wu, Jingchun Jiang, and Yu Rao

Department of Automation, Xiamen University,
No. 422, Siming South Road, Xiamen 361005, Fujian, China
1748468754@qq.com, 945366381@qq.com, 360584748@qq.com,
bxi@xmu.edu.cn, wsxl009@163.com

Abstract. As the connection between human life and smart phones becomes close, mobile devices store a large amount of private information. Forensic personnel can obtain information related to criminal through mobile phones, but the traditional mobile phone forensics system is limited to a simple analysis of the original information and can't find the hidden relationship between data. This article will introduce a method based on K-means clustering and association rule mining to improve the traditional forensics system. Through the cluster analysis of basic information, the relationship between suspects and their contactors can be explored. At the same time, association rules mining can be used to analyze the behavior of suspects so as to predict the time and contact of each event. Help law enforcement agencies find evidence hidden behind the data and improve the efficiency of handling cases.

Keywords: Mobile phone forensics system · K-means clustering
Association rules · Apriori algorithm

1 Introduction

With the rapid development of communication technology, smart mobile devices have become the main communication tools of the society. Many criminal activities using smart devices have also been brought out. It is common to forge information and transmit viruses [1]. At the same time the digital forensics technology changed from computer to mobile device, especially mobile digital forensics technology based on Android and IOS operating systems. As mobile phone forensics plays an increasingly important role in handling of cases, extracting effective information from devices is particularly vital for improving the efficiency of police's work [2]. However, the traditional method of forensics has not been applied to the needs of mobile devices, mass data, and intelligent analysis [3]. Currently the digital forensics research should focus on the analysis of smart mobile devices and mass data, and provide the judicial department with a complete package for evidence collection, evidence analysis, and evidence reporting [4]. In modern forensics systems, more hidden evidence can be found on the existing basis through clustering and association rules mining. Clustering is an unsupervised learning. By classifying similar objects into similar classes, all objects are divided into several categories. Entities within a cluster are similar, and the

Z. Shi et al. (Eds.): ICIS 2018, IFIP AICT 539, pp. 126–133, 2018.
https://doi.org/10.1007/978-3-030-01313-4_13

entities of different clusters are not similar. A class cluster is a collection of test points in the space, the distance between any two points in the same cluster smaller than the distance of any two points between different clusters. Using this method, the person associated with the suspect can be distinguished and the scope of the investigation can be narrowed. Association rule mining is a means of discovering hidden relationships in data sets. It can help us to extract valuable related information between transactions. Through the analysis of these information, the suspect's characteristics and daily life habits can be portrayed, thus providing more clues for investigators.

2   The Analysis of Intimacy Based on Clustering

In the traditional mobile phone forensics system, the user's basic information is simply listed, and the relationship between the contactors is hidden. In order to improve this, the clustering algorithm can be applied to the forensic work. So the forensic system becomes more effective.

2.1   Traditional Clustering Method

K-means clustering algorithm is one of the most classic and widely used clustering algorithm [5]. It divided data objects into clusters based on similarity. The concert steps of the algorithm are as follows:

- Select initial cluster centroids from data sets randomly;
- Calculate the distance of point to the centre of each cluster, and assign it to the nearest one;
- Re-calculate the centre of every cluster, if no one has changed then go to step four, otherwise go to step two;
- Get the results of clustering.

The result of this algorithm is mainly dependent on the initial cluster centroid, so the choice of the initial cluster centroid may have a great impact on the final result. Once the initial centre is chosen improperly, it maybe lead to an incorrect result. At the same time, due to the existence of isolated points, it will affect the calculation of the cluster centre after completion of the iteration. If choosing noise data far from the data-intensive area, it will cause the formed cluster centre to deviate from the real data-intensive area and reduce the accuracy of clustering.

2.2   An Improved K-Means Algorithm
2.2.1   The Choice of Initial Cluster Centre

The traditional algorithm does not consider the distribution characteristics of the sample when selecting the initial point, so there will be a large deviation in the random selection, which leads to the clustering results do not meet expectations. In this paper, we will determine the classification of the categories and the choice of the initial centre point according to the sample distribution characteristics. As shown in the Fig. 1, It's obvious that K = 3, and then calculate the average of these three cluster.

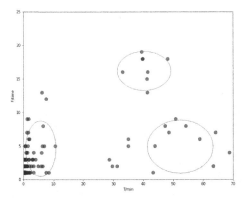

Fig. 1. The initial number of clustering

2.2.2 The Pre-processing of the Isolated Points

Due to the sensitivity of K-means to isolated points, the number of data in different cluster may vary greatly, and it's one of factors that affect the final results. To solve the problem, this paper proposes a method based on KNN algorithm to reduce the impact of isolated points. KNN algorithm is used to detect isolated points as shown in Fig. 2. For numerical data, it is a relatively common algorithm. However, in large-scale data sets, it has a large amount of calculations and high algorithm complexity.

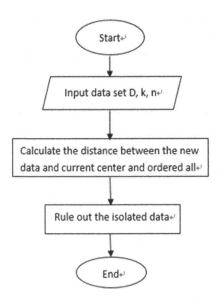

Fig. 2. The detection of the isolated points

3   An Improved Association Analysis Algorithm

Association rule mining is a kind of data mining technology [6]. It can discover the relationship between items or attributes in a data set. These relationships are unknown in advance and cannot be directly derived from the operation of the database. In general, correlation analysis is used to find meaningful relationships hidden in large data sets. The links found can be expressed in the form of association rules or frequent item-sets. These rules can be used to assist people in market operations, decision support, and business management.

3.1   The Basic Concepts of Association Rule Mining

Suppose that $D = \{d1, d2, \ldots, dn\}$ is a set of all transactions, N represents the total number of transactions, and $I = \{i1, i2, \ldots, im\}$ is a set of all items in the data set, A and B are a subset of I, and λ (A) = $|\{di| A \subseteq di, di \in D\}|$ indicates the number of transactions that include A, A $\rightarrow$ B support:

$$s(A \rightarrow B) = \frac{\lambda(A \cup B)}{N} \tag{1}$$

$s(A \rightarrow B)$ determines how often a rule can be used for a given dataset. If $s(A \rightarrow B)$ is greater than the initially set threshold, $(A \rightarrow B)$ is a frequent set.

Based on frequent sets can calculate the corresponding confidence:

$$c(A \rightarrow B) = \frac{\lambda(A \cup B)}{\lambda(A)} \tag{2}$$

$c(A \rightarrow B)$ determines the reliability of the rule. If $c(A \rightarrow B)$ meets the minimum confidence requirement, the rule is considered to be reliable.

Based on the above analysis, association rule mining can be divided into two steps:

- Generate frequent item sets that meet the minimum support requirements;
- Extract the rules that satisfy the minimum confidence requirement from frequent item-sets.

3.2   Apriori Algorithm and Its Improvement

Apriori algorithm is an algorithm used for association mining [7]. The principle is: If an item-set is frequent, all its subsets must be frequent. The Apriori algorithm uses a layer-by-layer iterative method to generate frequent sets. Starting from the 1-item set, pruning is based on the support degree to find a frequent 1-item set L1. A candidate 2-item set C2 is generated from L1, and then C2 is pruned by support to obtain a frequent 2-item set L2. In sequence, C3 is obtained from L2, and L3 is obtained from C3 pruning until no new frequent item-sets are generated. After that, rule extraction is performed for all frequent item-sets one by one: Initially, the extraction rule contains only one item that satisfies the requirement for confidence, and then uses a back piece rule to generate two back items... Until Lk items are created [8]. However when a candidate K-item set is first generated from a frequent $(K - 1)$-entry set, a $(K - 1)$-item set pairwise join method is used, so the number of candidate set entries grows exponentially, and this part of the data will taking up a lot of memory affects the performance of the machine.

Secondly, for the newly generated candidate K-item set, it is necessary to re-scan the database to count its support count, so as to determine the frequent K-item sets. Assume that the number of transactions in the data set is N and the number of candidate K-item sets is M. For frequent K-item sets, it is necessary to scan the MN database, which will consume a lot of time during the entire rule generation process, which will seriously affect the efficiency of the algorithm [9]. In view of the above deficiencies, this article adopts an improved Apriori algorithm to improve the mining efficiency, uses a vertical structure to represent database data, marks each item contained in the transaction with 1 and uncontained items with 0, thus the original transaction database. Table 1 is converted to Boolean matrix Table 2.

Table 1. Transaction data

TID	Item_sets
1	I1, I2, I5
2	I2, I4
3	I1, I3
4	I3, I5
5	I1, I2, I3, I4

4 Experiment and Results

4.1 The Results of the Intimacy Analysis

The improved K-means algorithm is used to cluster the call records in the forensics system. After clustering, the results of classifying some clusters based on experience are as follows (Fig. 3):

The abscissa indicates the call duration, the ordinate indicates the number of calls, the red dots indicate strangers, the black dots indicate classmates, the green dots indicate family members, and the blue dots indicate friend relationships.

After comparing the experimental results with the real situation, it was found that the accuracy of the clustering of family and friends is high, and the accuracy of the relationship between strangers and classmates is relatively low.

4.2 The Results of Association Rule Mining

In the mining of association rules, the chat records of social software are used as the original data set, and the improved Apriori algorithm is used to extract the rules that satisfy the confidence requirement [10, 11].

Table 2. Boolean data

ID	I1	I2	I3	I4	I5	Length
1	1	1	0	0	1	3
2	0	1	0	1	0	2
3	1	0	1	0	0	2
4	0	0	1	0	1	2
5	1	1	1	1	0	4
	3	3	3	2	2	

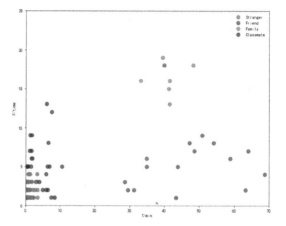

Fig. 3. The result of clustering (Color figure online)

The total number of chat records in this experiment is 3541. Each record contains 5 attributes: app_name, account, friend_account, friend_nickname, and send_time. The total number of mining rules is 93. Table 3 shows the filtered rules.

Table 3. Association rules

No	Rule	Confidence
1	21 → WeChat	0.957
2	veter** → QQ	0.756
3	Adam**, silv** → 10	0.960
4	QQ, 363****56, 16 → 283****06	0.718
5	WeChat, Adam**, Xia** → 18	0.704
...	...	...

- The first rule indicates that users usually use WeChat to contact people around 9:00 pm;
- The second rule indicates that users and nicknames veter** are usually contacted by QQ;

- The third rule shows that Adam** and Silv** generally communicate during the day;
- The fourth rule indicates that users with an account number of 363****56 generally use QQ in the afternoon and users with an account number of 283****06 to contact;
- The fifth rule states that the user whose account is Adam** normally uses WeChat in the afternoon to contact the user whose account is Xia**.

Through the excavation of rules, criminal investigators can better understand the suspect's daily life habits and characteristics, so as to discover important clues hidden in the mobile phone, speed up the process of handling cases.

5 Conclusions

In this paper, the improved K-means algorithm and Apriori algorithm are applied to the mobile forensics system, thus making up for the insufficiency of traditional forensics equipment for data association analysis. According to the intimacy, different contactors are clustered, and then the clustering results are properly classified to obtain the relationship between different people. In the mining of association rules, high-confidence rules reflect the user's daily habits and characteristics [12], while low confidence indicates that the relationship between transactions is not tight, which means that there may be abnormal activities, so these two aspects All need attention. The new forensic system can provide judicial personnel with more valuable information, help them to further understand the suspects and improve the efficiency of handling cases.

References

1. Li, Z.: Design and Implementation of a Mobile Digital Forensic System. Dissertations, Xiamen University (2017)
2. Yang, Z., Liu, B., Xu, Y.: Research status and development trend of digital forensics. Sci. Res. Inf. Technol. Appl. **6**(1), 3–11 (2015)
3. Pollitt, Mark: A history of digital forensics. In: Chow, K.-P., Shenoi, Sujeet (eds.) DigitalForensics 2010. IAICT, vol. 337, pp. 3–15. Springer, Heidelberg (2010). https://doi.org/10.1007/978-3-642-15506-2_1
4. Gao, C.: Analysis of computer forensics technology. Inf. Syst. Eng. **2**, 19 (2017)
5. Li, J.T., Liu, Y.H., Hao, Y.: The improvement and application of a K-means clustering algorithm. In: IEEE International Conference on Cloud Computing and Big Data Analysis, pp. 93–96. IEEE (2016)
6. Zhang, Tao: Association rules. In: Terano, Takao, Liu, Huan, Chen, A.L.P. (eds.) PAKDD 2000. LNCS (LNAI), vol. 1805, pp. 245–256. Springer, Heidelberg (2000). https://doi.org/10.1007/3-540-45571-X_31
7. Yu, H., Wen, J., Wang, H., et al.: An improved Apriori algorithm based on the Boolean matrix and Hadoop. Proc. Eng. **15**(1), 1827–1831 (2011)
8. Zhang, T., Yin, C., Pan, L.: Improved clustering and association rules mining for university student course scores. In: International Conference on Intelligent Systems and Knowledge Engineering, pp. 1–6 (2017)

9. Chen, X., Dong, X., Ma, Y., Geng, R.: Application of non-redundant association rules in university library. In: Yu, W., Sanchez, E.N. (eds.) Advances in Computational Intelligence. AINSC, vol. 116. Springer, Heidelberg (2009). https://doi.org/10.1007/978-3-642-03156-4_43

10. Chen, Z., Song, W., Liu, L.: The application of association rules and interestingness in course selection system. In: IEEE International Conference on Big Data Analysis, pp. 612–616. IEEE (2017)

11. Raj, K.A.A.D., Padma, P.: Application of association rule mining: a case study on team India. In: International Conference on Computer Communication and Informatics IEEE, pp. 1–6 (2013)

12. Wang, H., Liu, P., Li, H.: Application of improved association rule algorithm in the courses management. In: IEEE International Conference on Software Engineering and Service Science, pp. 804–807. IEEE (2014)

How to Do Knowledge Module Finishing

Shunpeng Zou[1,2], Xiaohui Zou[1,2(✉)] (ID), and Xiaoqun Wang[2]

[1] China University of Geosciences (Beijing),
29 Xueyuan Road, Beijing 100083, China
407167479@qq.com, 949309225@qq.com,
zouxiaohui@pku.org.cn
[2] Peking University Teacher Teaching Development Center,
Room 415, Beijing 100871, China
xqwang@pku.edu.cn

Abstract. The aim is to introduce a simple and unique tool to realize the ergonomics of man-machine collaborative research on "governance of the country". This method: First, the original text is imported, and then, a Chinese character board is generated, and then the language points, knowledge points, and original points are matched. The feature is human-machine collaboration. The result is that the Chinese character is tessellated, the Chinese chess spectrum is made, and the original chess is played. The knowledge module can be used for finishing. Its significance is: the language board, the knowledge game, and the original board game are digital, structured, and formal indirect ways to achieve automatic calculation: language points, knowledge points, and original points. The five menu-based knowledge modules of this example can be layered and distributed, and at least sixteen knowledge modules can be finished. Man-machine can be repeated calls. This is a contribution to the interpretation or understanding of texts.

Keywords: Linguistic cognition · Brain-machine integration
Indirect formalization · Computer-aided teaching
Knowledge module finishing

1 Introduction

Through research, we have found that due to language understanding, knowledge expression, and habits of people's thinking, etc., what people desperately need is not only interpersonal academic communication but also the intelligent ability of human-computer interaction, especially in this area of ideology and politics education informatization. This paper takes computer-assisted study of the essence of Xi Jinping's thoughts on ruling and managing the country as an example. It attempts to explain how man-machine combination can be used as a new method of knowledge module finishing to help understand the basic principles and core strategies. For example, all of them are interpreting Xi Jinping's thoughts on ruling the state, but there are many versions of interpretations by experts. Especially in the social environment where computers and smart phones and other types of artificial intelligence terminals are popular today, they are more prone to the appearance of similar blind people's

Z. Shi et al. (Eds.): ICIS 2018, IFIP AICT 539, pp. 134–145, 2018.
https://doi.org/10.1007/978-3-030-01313-4_14

images. This is not conducive to the unity of one mind, one knowledge and action, and then concentrates on working together to make the nation a good game. Regardless of the bottom-up or the top-down, the key lies in: How to achieve the goal of swiftly realizing the nation's ups and downs and steadily realizing the goal of ruling the country? Computer-assisted research on Xi Jinping's idea of governing the country is helpful for us to give full play to the comprehensive advantages of human-computer-brain collaborative processing of information.

Therefore, the purpose of this paper is to use a simple and unique method and tool to collaboratively study the essence of Xi Jinping's thoughts on governing the country, which can be summed up briefly as a computer-aided study of the essence of ruling the country, which is characterized by a reasonable division of labor complementary advantages.

2 Method

2.1 The Scientific Principles on Which This Method Is Based

The method adopts the indirect formalization method and the indirect calculation model. The principle of logic by order and position, the principle of bilingual mathematics and the principle of generalized translation, which are the three basic laws of the digital and textual chessboard, are the scientific support. Among them, the order and position of the numbers and the order and position of the words combine to form the double-chess board principle. It explains the specific order and position of all the structural units of Chinese on the board. This determines the most basic sequencing positioning development environment and calling platform for human-computer interaction, and any piece of text, any sentence, any phrase or even any word must be just a series of choices. It is not only theoretically applicable to the intelligent analysis of any text, but also in practice, through a series of targeted large and small double chessboards, and the corresponding knowledge module for the specific text. Because it is limited to plain text analysis, its scalability reaches its limit; because it is limited to Chinese examples, it can reach its limits. In other words, any discourse fragment or text, which contains language points, knowledge points and original points (referred to as "three points") and even featured software modules, such as: knowledge maps and mind maps, can be recorded from the menu of the "three points" come from the combination and construction. Since the task of this article is limited to demonstrating the method with its characteristics through a typical example, the explanation is not extended.

2.2 Taking the Formal Study of the Ideology of "Ruling the Country" as an Example

Taking the formal study of the ideology of "Ruling the country" as an example, demonstrating new methods and tools.

One of the authors took the case of "big data processing and knowledge management based on double- chessboard: taking abstracts of papers and patents as an example" and "taking an abstract of the organizational profile and personal profile in the thinking and exploration of various high-end think tanks as an example" In practice, we promote the popularization of big data processing and knowledge management based on double- chessboards. Such new methods and tools for human-computer interaction have made great progress. Therefore, the author here introduces the Marxist Academic Forum and tries to make a set of typical examples in the study of Xi Jinping's themes, main lines, and scientific theoretical systems [1, 2]. Its popularization involves the application of a series of "diversity" models. Its empirical analysis has proved that macroscopic models and microscopic models can be used not only for the "involved research object" of the "involved system" discipline. The four theoretical questions, interlocking disciplines, conceptual systems, and methodological systems, give clear explanations, and then can use indirect formalization and indirect calculations of system engineering solutions to deal with terminological descriptions at all levels. The previous examples of the author reveal that the research object of the (community) teaching management formed by the "diploid" model of university education and management is "the discussion of advanced talent training and advanced knowledge", and the nature of the discipline is "the overlapping of teaching and management". "The concept system and method system relate to a series of activities "classification" and "cases" formed by the "teaching" and "management" of the campus classroom. Among them, the rough concept can get the basic concepts involved in macro models. The framework can then be further subdivided to obtain the basic terms involved in the microscopic model [3]. This will further systematize Xi Jinping's idea of ruling the state, so that further formation of a universal scientific theory system through computer-aided research is helpful. For the time being, it is not necessary to construct an expert knowledge system for human-computer interaction that requires a human-computer interaction native language environment that is compatible with personalization and standardization [4]. Only a set of specific and operable embodiments is cut in.

As shown in Fig. 1, one prominent line is to uphold and develop the main line of socialism with Chinese characteristics. Below the speech fragment there is a series of digits that record the selected word group (there are several) of the user, visible below, and formatted and highlighted the "task, article, main line, feature". These are formalized, digitized, and formatted textual texts, that is, the ideological and political education information methods described in this article. They are to achieve this step by step:

The first step is to enter the "Add Article" interface we provide (http://kb2.sloud.cn/article/add)

The second step, fill in the "Article Content" and "Title" (see the following example)

The third step, release (visible)

As shown in Figs. 1 and 2, the first step, the second step and the third step are all extremely silly and easy to understand (the pupils can also quickly understand and learn how to operate). From Fig. 3, after the text is imported, the system will automatically count the total number of characters and their non-repeated number of words (in this case, the non-repetitive Chinese characters, that is, the smallest language structure form unit of Chinese or Chinese) is the most basic language point. From

Fig. 1. "One main line" example of knowledge module diagram

Fig. 4, the language points and knowledge points implied in the text can be highlighted by means of selection and clicking on the Chinese character board. For example, the "mainline" is where the matching of language points and knowledge points or even original points overlaps. Another example is that "ism" is very interesting. At this point, the language points and knowledge points and even the original points are only similarly matched, similar, or similar. What's more interesting is that each person's choice or thinking can be different, but they must be limited to the range of the double word board; the machine can also have its own algorithm or choice. The most interesting thing is that not only man-machines can have their own choices, but also, the combination of man-machine formation and the dual-brain collaborative selection, that is, intelligence, is better than pure human brain intelligence and purely computer-based intelligence in both physics and practice. Smart has more advantages. The unique advantages of the discovery and invention of the double-word board are specifically described in another book.

Fig. 2. The window for adding articles is the man-machine interface

Fig. 3. Fill in or import the text of "Title" and "Article Content" in the window of the increase article

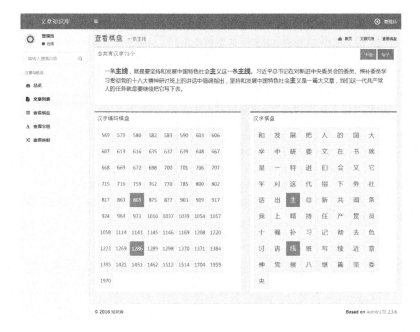

Fig. 4. Check the board

With such knowledge module finishing methods and tools, we can better discuss the differences and links between their respective thinking, understanding and understanding. For example, we can better understand "a main line."

We can do an in-depth analysis and discussion on that sentence itself, and we can continue to analyze and think in depth on the basis of the intensive processing of basic knowledge modules.

As shown in Fig. 5, the essence of two centuries is highlighted or expressed or understood even more through a series of word combinations that are related, similar or similar to each other. The most crucial is the human-computer interaction process. In terms of individuals and groups, human-computer interaction can be easily achieved, and even human-computer interaction and interpersonal networking can be achieved. The combination of such intellectual activities and artificial intelligence is far more than simply reading texts, even graffiti discussions on paper or chalkboard, and is more flexible and easier to retain, share, and have standard measurability. Readers can compare themselves.

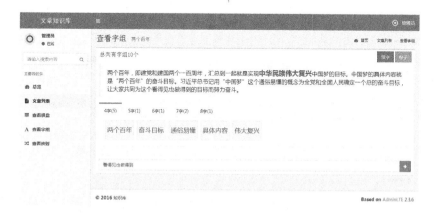

Fig. 5. "Two hundred years" example of knowledge module diagram

As shown in Fig. 6, in addition to the features illustrated in Figs. 1 and 5 above, the author also emphasizes: "Wealth, Democracy, Civilization, Harmony," and "Freedom, Equality, Justice, the Rule of Law," and "Patriotic and Professionalism." Each of the three sequences of honesty and friendliness can be grouped together into three categories for an overall explanation and a corresponding knowledge module. Each keyword can be further explained individually and individually. And make more specific knowledge modules. This section is left to the reader to do it and make a corresponding comparison.

As shown in Fig. 7, each of the four comprehensive aspects is not only an organic whole, but also has rich content in every aspect. Therefore, not only can it be made of four separate items, it is conducive to further study and understanding. The knowledge module, but also can take advantage of the situation, the content of each aspect is a system of combing, making a series of knowledge modules.

This is more meaningful than simply staying in these words. Ask the reader to try and compare.

As shown in Fig. 8, this is just a screenshot. Readers can further review the knowledge modules in many aspects and even in combination with the aforementioned suggestions.

The above five aspects are interlocking. The main line for the development of socialism with Chinese characteristics must further clarify its direction, that is, realize

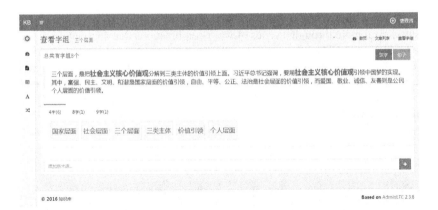

Fig. 6. Shows an example of a three-level knowledge module

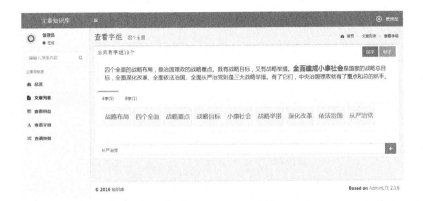

Fig. 7. An example of a "four comprehensive" knowledge module

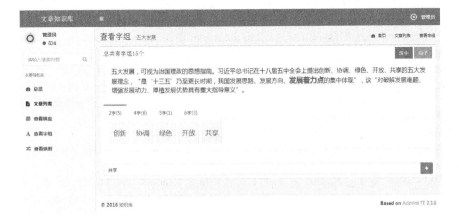

Fig. 8. Shows an example of a "five developments" knowledge module

the Chinese dream. With this goal, it is necessary to further clarify the value orientation of achieving the goal, that is, the core values of socialism; with value guidance, we must further clarify the current The strategic focus in the future and in a period of time will be four comprehensive strategic layouts; to achieve this strategic layout, the five major development concepts will be fulfilled [5].

2.3   Expert Knowledge of Human-Computer Collaboration System Platform

Expert knowledge of human-computer collaboration System engineering development platform: Creating a Chinese development environment compatible with personalization and standardization.

The author's further simplification and combing can be referred to as the ideological essentials of the "Governing the State and Managing the State" (enriched version).

Contains 116 non-repeating characters after clarifying the need to develop the main line of socialism with characteristics, we must further clarify its direction. The goal of achieving the great rejuvenation of the Chinese nation in two centuries is to realize the Chinese dream, and it is necessary to realize it clearly. Value orientation guides the values of the three levels of the country, society, and individual, that is, the socialist core values. Further, we must also clarify the strategic priorities for the current and future period, that is, four comprehensive strategic layouts that cover the building of a well-to-do society. Strategic goals, comprehensive deepening of reforms, comprehensively governing the country according to law, and comprehensively rigorously governing the party's three strategic initiatives requires the realization of this strategic layout. In the end, we must not only rely on but also honor the five development concepts of "innovation, coordination, greenness, openness, and sharing" [6].

It was introduced into the expert knowledge system engineering development tool platform for human-robot collaboration, and the big data processing and knowledge management based on the double-word chessboard [8] can be implemented by following the rules of "sequencing and positioning, synonymous juxtaposition and consent" [7]. Specific practices and explanations:

The import can instantly generate the summary-specific double-word board, which is characterized by a non-repetitive semi-structured Chinese chess board (digitized) and its mapped purely structured digital board (numbers) and the unstructured summary (data) All at a glance, so that party and government cadres, experts, scholars, and teachers and students can work together to further refine their knowledge, without having to use any other programming language development environment based on the US standard information exchange code. Based on the "two broad categories of formalization strategies," we have created a specific implementation tool for the dual formal approach: the generalized bilingual information processing platform:

As shown in Fig. 9, this example is actually a sample version of the super silly programming development environment. It can not only automatically generate a double-word board, the unstructured abstract is the product of the human brain, that is, we further comb and refine on the basis of expert knowledge, which can not avoid the common disadvantages of the natural person [9, 10]. By the computer by importing

226 characters and non-repeating Chinese characters 116 that is the summary, the summary automatically generates the abstract specific double-word board namely the semi-structured Chinese character board (matrix) and its mapped purely structured digital board (Matrix) is usually difficult for natural people to do. This lays the foundation for human-computer collaboration and even man-machine collaboration - super silly programming development environment platform. Furthermore, we can use the large-scale chess game of "individual governance" as an example to carry out a large-scale cooperation of all parties, the whole army, and all walks of life in the country, and lay down a simplified version of the reference frame that can be considered comprehensively. Among them, because the logical sequence order that the numerical chessboard and the Chinese character board follow is the same, there is a one-to-one correspondence between the two symmetrical matrices. Thus, the basis for the formal reasoning of the size premise is already in place. The rest is to achieve computer-aided research through the super-stupid programming development of the concrete double-word board. Through man-machine collaboration and interpersonal negotiation, the feature set implied by the abstract is extracted, and then the reference frame of reference is established.

The result is: the tessellation of the Chinese text and the chess genre of the Chinese text and the spirit of the ideological content, ensuring the essence of the abstract, i.e. the core idea and the style and characteristics of its expression, etc. Under the overall pattern, it is accurate and accurate.

The specific practices and steps are as follows:

As shown in Fig. 10, the "one main line, two hundred years, three levels, four comprehensive, and five major developments" in the abstract are identified by the user or the system, respectively, and then, through the "add new terms" approach, Structured text (summary) establishes a mapping relationship with structured text (the set of menus that match the language points and knowledge points in the box and the original points).

If the way that natural persons help computers do precise learning is to teach them how to choose a group of words, the reverse is that computer-assisted natural persons learn and study. Here, the user (natural person) finds a way to add a new term in the abstract, and lets the system (computer) react "a main line" (in the box) to establish a mapping relationship. Segmentation and tagging can be implemented on this hyper-silent programming development environment platform using only the native language (Chinese). In the same way, it can also be found in the abstract that "two hundred years, three levels, four comprehensive, and five major developments" are added one by one into the box and formalized. The main purpose is to tell the process of segmentation and labeling. The machine implements a computer-assisted set up of a series of mapping relationships.

Of course, we cannot demand that computer systems truly understand like natural persons: This "one line" is to continue to follow the path of socialism with Chinese characteristics. However, we can pass this keyword—whether it is a term or a common saying., Make a hyperlink to make it look like it really understands that "a main line" is to continue to follow the path of socialism with Chinese characteristics. Such a path or algorithm can all be told by the user and reused by the system.

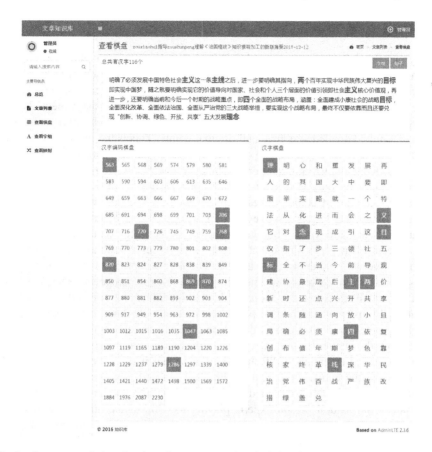

Fig. 9. Summary of the chessboard structure and principle of the analysis of the general knowledge module

Fig. 10. Summary of the general knowledge module in "Governing the State"

3   Result

Simply put, this is how we can complete this kind of knowledge module finishing by human-robot collaboration. Attentive readers will remember and discover that we have said "respectively by the user or system to find out" such words. Its meaning is that it can be used as a way to help people, but it can also use machine-assisted methods, and can also use machine learning methods to collaboratively complete the matching of language points, knowledge points, and original points. This is the process of human-machine collaboration. The matching of terms and colloquialisms matches the language points, knowledge points, original points, and even matches with grammar, semantics, and pragmatics. Not only is it related to the double-word board that is composed of digits and words, but it is also linked to the "Chinese chess spectrum" consisting of bilingual Chinese and English. We regard terminology and colloquialism as alternative bilingualism. The three types of bilingual matching process are extensions of this study. In the process of man-machine collaboration, once the user (natural person) selected "one main line, two hundred years, three levels, four comprehensive, and five major developments" from the above abstract (unstructured), the computer system immediately They are all structured and formalized to lay the foundation for further automated invocation. In this process, the two parties are quite simple, but they are very useful. Embedding other methods (such as deep learning) will be more effective.

Although the above illustration is only a series of computer-assisted studies based on the ideological essentials of "Ruling the State," but it can at least prove that the combination of the human brain and the brain can indeed use super-stupid programming to process core knowledge modules. - This is the introduction of such texts and abstracts (by the natural person experts playing the role of the human brain) in the automatic generation of double-word board, but also to achieve human-computer interaction in a semi-automatic way to further generate the Chinese chess game - not only to throw a brick Moreover, experts from all walks of life, as well as teachers and students from all walks of life, can participate in the construction of a complete "governance of the state," and a system of scientific theories. In addition, computer-aided research and scientific theory systems can be conducted for various disciplines.

4   Conclusion

It can be seen that the human-computer interaction for knowledge learning, research, and large-scale production methods of processing based on the double- chessboard have a particularly important role and value for the large countries of higher education. This type of educational informatization approach, i.e., the knowledge module finishing method used by the knowledge navigation device and the knowledge center behind it, is a systematic formalization method [11]. The practices of knowledge map [12] and knowledge graph [13] and mind maps [14] are not only compatible, but also complementary, as well as their "three points" and "three changes" to promote human-computer collaboration.

Its significance lies in the fact that the formalization Chinese characters as chess is automatically generated by the software, which can realize the automatic calculation of language point and knowledge point and the original point "three points". The specific performance of this paper is that only one or two or four and five can be used. Both the human brain and the computer can accurately reproduce the logical status of each point in the quintessence of the ideology of "governing the country" in the entire terminology system, fully embodying the dual innovations of character informatization and Chinese intelligence in sorting out texts and reading them.

References

1. Xia, J., Fang, Q.: Xi Jinping's thought on ruling the state and managing state affairs and the latest progress of Marxism in China. Marxism Res. **1**, 153–155 (2017)
2. Zou, X., Zou, S.: An indirect formalization method based on bilingual automatic conversion. Beijing: CN102253934A, 23 November 2011
3. Zou, S.: A New Study on Theories of College Education Discipline Construction. China University of Geosciences, Beijing (2012)
4. Han, Q.: Comprehensively and thoroughly grasping ten important aspects of Xi Jinping's thought of governing the country and managing the state. Soc. Chin. Charact. **06**, 24–28 (2014)
5. Han, Z.: The basic content and internal logic of Xi Jinping's thoughts on governing the country. Guangming Daily (016) (2015)
6. Xiaohui, Z., Shunpeng, Z.: Two major categories of formal strategies. Comput. Appl. Softw. **30**(09), 194–199 (2013)
7. Zou, X., Zou, S., Ke, L.: Fundamental law of information: proved by both numbers and characters in conjugate matrices. Proceedings **1**(3), 60 (2017)
8. Zou, X.: Double chessboards. Guangdong: CN104056450A, 24 September 2014
9. Cai, J.: "Two hundred years" struggle goal casts "Chinese dream". Chin. Soc. Sci. J. (A05) (2013). http://news.hexun.com/2013-05-27/154555819.html
10. Cheng, C.: "Four Key Words" interpretation of general secretary Xi Jinping's series speech ——Interview with Professor Xin Ming of the Central Party School of the Communist Party of China. Leader Wencui **1**, 7–22 (2014)
11. Zou, X., Zou, S.: Bilingual information processing methods and principles. Comput. Appl. Softw. **11**, 69–76, 102 (2015)
12. Liu, H., Tan, L.: Research on constructing knowledge map of SPOC course based on teacher perspective. Electro-Educ. Res. (9), 64–70 (2017)
13. Kuo, Q., Xiao, S.: Application of subject knowledge atlas in constructing the course map of Middle School English. Modern Commun. **18**, 155–156 (2017)
14. He, J.: Analysis of the role of computer mind map software in English teaching. Comput. Fan **10**, 31–35 (2017)

The Art of Human Intelligence
and the Technology of Artificial Intelligence:
Artificial Intelligence Visual Art Research

Feng Tao[1] ⓘ, Xiaohui Zou[2](✉) ⓘ, and Danni Ren[3]

[1] School of Philosophy, Nankai University, Tianjin 300071, China
philart@163.com
[2] Sino-American Searle Research Center, Beijing, China
949309225@qq.com
[3] College of Foreign Languages, Nankai University, Tianjin 300071, China
rendanni@139.com

Abstract. This article aims to find out how human intelligence and artificial intelligence can complement each other better through the comparison of art and technology. Here is an example of artificial intelligence visual art research. The method is: First of all, from the characteristics of artistic language, explore the charm of human intelligence. Furthermore, from the characteristics of technical language, we can feel the power of artificial intelligence. Finally, from the characteristics of human-computer interaction and typical examples, we can explore how humanities and engineering complements each other. In this paper, we discuss the creativity, artistic standards and artificial reception issues in artificial intelligence art through analyzing the latest artificial intelligence art programs such as "Creative Adversarial Networks" (CAN) and Google Deep Dream. We argue that creativity is not only novelty, such as "style ambiguity" and other techniques, but also originality. The result is the discovery of that the art standard cannot be limited to be searched in previous works of art. The art standard itself is historical and changeable, and is the "true content". Both Creativity and art standard need us comprehend and rethink about the overall laws and paradigms of whole art. However, from the perspective of art reception, the artificial intelligence art has its own unique value. Its significance lies in: it has opened up a vast new world for the intelligence art education. Further, we can also compare the thinking habits of artists and scientists and engineering technologists respectively from the linguistic cognition. Next, we will study that the cooperative research topic of digital and textual chessboard combined with human-computer will not only help teachers and students in the art education process build their own personalized knowledge bases. It would be a better understanding of the benefits of combining human brain and computer.

Keywords: Artificial intelligence art · Creativity · Aesthetic reception

Can machine draw pictures? Can artificial intelligence create artworks? Now it seems that this is no longer a problem. Some intelligent programs such as CAN ("Creative Adversarial Networks") system and Google Deep Dream have been able to create works of art.

Z. Shi et al. (Eds.): ICIS 2018, IFIP AICT 539, pp. 146–155, 2018.
https://doi.org/10.1007/978-3-030-01313-4_15

Computer art is the foundation and precursor of artificial intelligence art. Since the 1990s, with the breakthrough of some key artificial intelligence technologies, artificial intelligence art has gradually separated from traditional computer art and may become a new art form. Artificial intelligence art pays more attention to the intelligence or autonomy of computers. The fundamental purpose is to automatize computers, which is to simulate human intelligence to create artworks through programs, web search, and Deep Learning. And these artworks can be understood and appreciated by human. The study of artificial intelligence art helps to improve the creativity, imagination and emotional perception that computers did not have before. At the same time, it is also an inquiry into the principles of relevant parts of human intelligence. Artificial intelligence visual art is a combination of artificial intelligence technology and art. It not only is closely related to the technologies of artificial intelligence recognition, image processing, drawing, etc. but also involves the re-recognition of artworks and the re-definition of human artistic behaviors.

But is artificial intelligence really able to create works of art? What is the principles and mechanisms of this process, and how can artificial intelligence be creative? Does artificial intelligence solve the problem of art standards? How should we look at these works of art?

1 Introduction

Before answering these questions, let's reexamine the meaning of art. Since ancient times, thinkers have made many definitions of art, but the connotation and extension of art are constantly changing. In ancient Greece, art was regarded as a skill of man, including crafts and art. In the Age of Enlightenment, the scope of art was reduced to free art such as painting, literature, music, etc., and in the Internet age, the scope of art has been expanded significantly, such as photography, movie, computer art, new media art, etc. Looking at the definitions of art, it is basically believed that art is a unique behavior of human beings, and art is a work of human beings. Therefore, the philosopher Hegel distinguished between artistic beauty and natural beauty, and limited the objects of aesthetics into artistic beauty. The art theorist E. Gombrich directly claimed: "There really is no such thing as Art. There are only artists." [1] According to this, machine is not human, and thus the works it creates cannot be called art. But if we look at the results of the action, the intelligence agents can indeed produce some works that can be understood and appreciated by human beings and bring joy to people. Therefore, when artificial intelligence can also create works, it seems that we need to redefine the art. At this time, we can divide the artificial intelligence art problem into two parts. One is the internal problem: it involves some problems like the subject of artistic creation, subjectivity and intentionality, as well as aesthetic emotions. The other can be called an external problem: research from the results of behaviors, such as the work itself, the aesthetic effect of the work, etc. In fact, these two problems cannot be completely divided. However, we can temporarily suspend the subject of maker firstly, and only analyze it from the phenomenon and results. We can check if the CAN system is really creative, what the artistic standards used in the evaluation of artificial intelligence creativity by scientists, and the meaning of artificial intelligence art from the perspective of aesthetic acceptance.

At the beginning of the 21st century, with the development of neural networks and Deep Learning techniques, artificial intelligence art makes a breakthrough in imitating human creativity and imagination. E.g., the University of Tübingen uses neural network algorithms to combine realistic images with artist styles to produce artistically styled images [2]. In 2016, Google Deep Dream (GDD) made use of neural network technique to train machines to learn to recognize images and further generate artistic images. The scientists created the GAN program (Generative Adversarial Networks) to simulate and generate similar works by making computers learn and imitate classic works in art history. However, this still cannot allow the computer to get rid of the suspicion of copying works. In 2017, scientists further created the CAN (Creative Adversarial Networks) program, a computer program that, based on the original GAN, modifies the network's goals to deviate from the established style as much as possible. At the same time, it tries to stay within the scope of the artwork but create creative images. This kind of program makes the computer no longer simply copy the model, but by emulating the creative activities of human beings, it can create a style of art independently. According to the double-blind test conducted by art lovers and experts, people can be deceived by the paintings created by the CAN program successfully, that is, the computer successfully passed the Turing test, and it seems to have really been intelligent in visual art [3].

Let us briefly summarize the working principles and mechanisms of the current artificial intelligence visual art. Firstly, the algorithms and programs of the computer are the foundation. Any drawing and design, visual recognition, generation programs, identification standards, etc. required in the artificial intelligence art must be formalized, that is, represented by algorithms. Secondly, the expert database and network data can provide sufficient information. Take the CAN program as an example. It needs to provide the intelligent agents with enough works of art to learn, imitate and identify. Thirdly, whether it is GDD, GAN or CAN program, it is based on Artificial Neural Networks. The so-called artificial neural network is "a model based on the human brain" [4]. By imitating the working mechanism of human neurons, it can make the machine self-improve and learn. The most important point is the Deep Learning method developed in the last decade, which refers to letting computers learn from experience and understand the world according to the concept of layering [5]. One of the breakthroughs in Deep Learning is to strengthen learning, that is, to learn to perform tasks by trial and error without being supervised. It is these innovations of key technologies that have made AI visual art a breakthrough.

2 Related Work

2.1 Creativity in AI Art

Some scholars believe that creativity is necessary for human progress. The famous artificial intelligence philosopher M. Boden says: "Creativity is a fundamental feature of human intelligence, and an inescapable challenge for AI." [6] Art is an embodiment of human creative thinking, and creative thinking is an ability of human intelligence that is the most complex and difficult to be formalized. So what is the creativity in art? Can it be formalized? Can artificial intelligence have a certain degree of creativity?

AI experts believe that intelligent agents can be given a certain degree of creativity through designing programs and Deep Learning skill. Turing refuted the so-called computer's rules obedience and inability to create as early as 1950. He suggested the so-called originals rooted in the "education" of people, they are perhaps the result of the influence of some famous universal rules [7].

Boden suggests that creativity can be defined and formalized, and be represented by algorithms. She divides creativity into "Improbabilist" creativity and "Impossibilist" creativity [8]. The former is a recombination of various ideas, while the latter is to create new ideas that have never appeared before. Boden believes that Cohen's AARON program can implement some bionic one-arm operations, and when the generating program incorporates evaluation criteria such as aesthetic balance, it can decide what to do next by thinking about its behaviors. This has been somewhat creative, but AARON still can't reflect on its works, and accordingly adjust to improve the quality.

The author believes that creativity can be divided into two types, one is "novelty" and the other is "originality". The novelty is "from birth", that is, Boden's " Improbabilist " creativity, while the originality is Kant's genius, " Impossibilist creativity", and "ex nihilo"(creation out of nothing). And this is the originality with the exemplary meaning pursued by the artists.

Let's take a look at the creative part of CAN. Above all, we must know that the development of CAN and GAN (Generative Adversarial Networks) is the result of the development of artificial neural networks and Deep Learning techniques. Both gain the ability to generate artistic images after learning a large number of works of art in history. The biggest difference between CAN and GAN is that GAN can only imitate, and CAN generate "creative art by maximizing deviation from established styles and minimizing deviation from art distribution" [3]. The GAN system has two confrontational programs: generator and discriminator. The computer learns to generate new images through the generator, and the discriminator is responsible for judging the artistic images that match the art training section set by the computer. In this case, it is impossible to generate new and creative images. The novelty of CAN is that it sets two standards in the discriminator, one is "whether it is art" and the other is "what style of art is generated". These two standards are set to be confrontational, that is, it is necessary to generate images belonging to the art category, and to set images that are different from the established style, that is, an ambiguous artistic style.

We can see that the CAN program is innovative, using stylistic ambiguity (i.e. maximal deviation from the established style) and identifying artistic standards (minimum deviation from the artistic scope). In other words, innovation is a breakthrough and deviation from the original rules within the established scope. CAN programmers believe that the reason why art needs innovation is that art itself has the property of "external stimulus patterns". People will get used to external stimuli. "a moderate arousal potential" can help people wake up aesthetic sense, too little stimulation can be boring and too much can be disgusting. Therefore, the designer's task is to try to "increase the stylistic ambiguity and deviations from style norms, while at the same time, avoiding moving too far away from what is accepted as art." [3] The CAN programmers made five groups double-blind experiments. Art lovers and art history students were invited to compare abstract expressionist paintings, Basel art exhibitions, CAN paintings, and GAN paintings. The results showed that most people think that CAN paintings are artworks and have some certain artistic characteristics like intentions, visual structures, communicability, etc.

Now, let's briefly compare the high-score artwork with the artworks of V. Kandinsky and W. Turner.

From the author's point of view, the CAN work seems to be less harmonious in color than the other two artists' works. It may be due to too much dependence on calculations, making the colors more turbid and mediocre (see Fig. 1). The composition and shape are also a little unbalanced, unlike Kandinsky's work, which takes into account the relationship between shape, sound and color (see Fig. 2). When we look at Turner's work, it seems that the cloud and wind are unshaped and ambiguous, however the composition of picture is perfect, and significant (see Fig. 3). The author believes that although the CAN work seems to be novel, it is not a superior work. It is disharmony and unbalanced in aesthetic whether its color, shape, composition or brush strokes. Thus, there is no aesthetic logic.

Fig. 1. CAN high-score work, ranked first in audience preferences, ranked second in terms of human works and intentions [3].

Fig. 2. V. Kandinsky, composition viii (cited from internet).

Fig. 3. W. Turner, The Slave Ship (cited from internet).

The reason why there is such deviation in judgment, the author believes, is that, first of all, creativity is not only a "novelty" or difference, but what is more important is in its "exemplary" meaning. Artists pursue "originality", that is to say, it involves more fundamental issues in art and innovations in the paradigm of thinking. And the means of innovation are also instructive. For example, the problem of the intuitiveness of the color needs impressionists to solve and modernist painters consider the problem of the planar authenticity of painting. Intelligent agents can't predict the direction of the essence of art from the known works of art history, let alone the innovation means of solving the problem. Computer analysis and induction of data cannot be raised into problem consciousness and thought, which is the fundamental problem of artificial intelligence innovation. Currently the mechanic of AI art is programmatic reorganization and piecing together the original images. Even if its works could be novel, it cannot be considered as success, let alone original. In addition, whether it is CAN, GAN, or other art programming, we must set a judging standard (i.e. evaluation system) [9], such as the art identification, the style identification, and the art lovers' marks in the test set in the CAN program. How are these standards determined? Can you really judge whether artificial intelligence works are real artworks and have creativity? This involves the issue of art standards. Is there any standard in art? Is the art standard fixed and unchangeable? Is the art standard subjective or objective? Next, let's briefly discuss the issue of art standards.

2.2 Art Standard in AI Art

From the logical operations, to the evaluation system, to the "trial and error method" in Deep Learning, all involve standards, especially the standards based on true and false. Some experts believe that aesthetic and artistic activities can be formalized. The premise should be that art works become the "prime objects for aesthetic evaluation" [10]. In addition, the audience's reaction and evaluation can be quantified. But at the same time, artificial intelligence experts admit that art evaluation is very subjective, so computers may not be able to evaluate their "own creative efforts in human terms" [11].

When artificial intelligence visual art generates a work, it must first judge whether it is a work of art. So how could it judge? When designing the art criticism algorithms and design algorithms, Gips and Stiny coded "conventions" and "criteria" as an aesthetic system, and then input the object to be evaluated to see if it conforms to the system [9]. We can see that the computer setting the evaluation criteria still depends on some previous evaluation criteria. Therefore, choosing different evaluation criteria will inevitably cause deviation from the evaluation target. For example, Duchamp's work "Spring" cannot be considered as a work of art according to any previous artistic standard. However, we know that "Spring" as groundbreaking ready-made art has greatly expanded the boundary of art.

In the CAN system, AI classifies some similar styles of artworks by deep learning in classic works in art history and sums up the general characteristics, then it can make a certain standard. But how to judge whether a new work is an artwork, faces the contradiction between innovation and standards. The CAN system claims that if the aesthetic consciousness of the audience is excessively awakened, it will cause the audience to be disgusted. For example, the work of the GDD program is excessively awakening, thus causing a negative evaluation of the critics [3]. But we compare GDD works with many modernist works. The former seems to be much milder. That is to say, the works of the Fauvism, Expressionism, Cubism and other genres are excessively awakening according to the evaluation standard of CAN. They should be excluded from the art system.

Therefore, we can see that no matter how artificial intelligence is trained, its foundation is the existing works, and the artistic standards obtained by quantifying and analyzing the previous works are in conflict with the essence of the continuous innovation of modern art. Since art is an ever-developing living body, artistic standards are constantly changing. Adorno proposed a standard for judging true art and cultural goods as "truth content" ([Germany] Wahrheitgestalt), [12]. The truth content is different from the true and false binary logic of computer, but is the artistic standard that has both true and false and specific content. The language of art is real, which means that it truly simulates the language of things, and truly formalizes the language logic of itself, and the language logic of art is internal and historical [13]. According to Adorno, the artistic standard is a large standard based on human imitation and true language. It can be regarded as the standard of the whole human artistic spirit. Therefore, the standards that do not take into account the evolution of artistic logic are rigid (Fig. 4).

Fig. 4. The work created by the Google Deep Dream program

However, we do not mean that art has no standards. For example, R. Arnheim believes that visual art should conform to Gestalt psychology. This is a kind of "a common core of truth" that makes art associate with anyone at any time and place [14]. The visual psychological forms, like balance, simplification, gestalt, etc. mentioned in Gestalt psychology can indeed become the standard for judging the internal structure of visual art works.

The author believes that different art categories and styles have their own relative standards, and these standards are stable within a certain time and scope. For example, the standard of realist paintings is to realistically represent objects in two-dimensional space, while the standard of modernism is to pursue the sense of reality of the painting plane. From the details of artistic operations, each style has its own standards, such as represent and express, perspective and cavalier perspective, etc., and from the history of whole art, its standard is to pursue artistic reality.

Whether it is CAN or GDD program, it is still only to seek the formal law of visual art through neural networks and Deep Learning. The basic principle is to summarize and calculate the general law on the image of objects (real things, art classics). This kind of law is represented, reorganized and transformed, and then adjusted and evaluated according to the existing artistic styles. Although CAN uses the method of style ambiguity (deviation), its so-called artistic evaluation standard is still derived from the previous artistic styles, and it cannot break through standards and re-adjust standards according to the overall law of development of art. But this initiative is what the artist pursues. This requires an overall sense of art and problem consciousness, and artificial intelligence is at least impossible to rise from experience to this overall sense and consciousness. As the CAN designers finally said, artificial intelligence does not really understand art. It knows nothing about the subject and artistic principles. What it does is to learn and imitate human art [3].

2.3   Art Reception in AI Art

The complete art process involves art creation and art appreciation. Although artificial intelligence could not be called creators currently, their works can be appreciated by human. If according to Hegel's theory that artworks need to be instilled with the spirit of human beings, then the works created by artificial intelligence are hard to be called as artworks. However, if we look at from aesthetics of reception, although art lacks intentional subject creators, the works can still be understood and appreciated by human in fact.

From the perspective of Aesthetics of Reception, if a piece of artwork is not appreciated by anyone, it cannot be called a work of art. That is to say, the appreciators are the same as the creators for jointly creating the artwork.

H. Jauss, the founder of aesthetics of reception theory, puts forward the concept of "horizon of expectation" ([Germany]Erwartungshorizont). The concept emphasizes the requirements of the appreciators' own cultural cultivation, taste, experience and ideals. Therefore, different appreciators may make different interpretations and judgments on the same piece of artwork, which is why CAN designers divide testers into several different groups, distinguishing between ordinary lovers and art experts.

S. Fish further emphasizes the reader's experience in the reading process, and believes that this kind of experience can help to complete the meaning of the work. He suggests that reader is a decisive role for the work, because the social meaning, aesthetic value and even the potential meaning of the text need to be realized in the reader's creative reading. From this perspective, artificial intelligence artworks are likely to produce its unique aesthetic value, such as the work of Deep Dream, Aaron and etc. Although we may not like this type of artworks, they are too far away from the art forms we have accepted. Only when we have seen more and more such works and our aesthetic taste are changed, we can interpret and appreciate them. Therefore, if we only consider the results and don't reflect on the makers temporarily, we can't neglect on the aesthetic characters of AI works.

Some people could criticize that the aesthetic dimension of AI works is endowed with by human, the beauty of AI works has no difference with the natural beauty if AI has no purpose to make works. This is hard to argue if we don't discuss the intention of subject.

In addition, art reception also involves the relationship between autonomy and heteronomy in art. Does art bring liberation to people, or will it become a tool for dominating people? W. Benjamin mentioned in The Work of Art in the Age of Mechanical Reproduction, mechanical reproduction technology can reproduce a large number of works, which can give the public an opportunity to appreciate the art, thus bringing the opportunity to liberation [15]. Then whether the work made by artificial intelligence is the reproduction of the human artistic style, or has its own unique style,

AI can bring cheaper and more convenient works to the public. However, as T. Adorno pointed out, this kind of industrialized artworks can also become a tool of the administrative class to control the public. Humans no longer have the ability to take control of their own aesthetic tastes. All aesthetic and even cognitive ability may be controlled by others, even more boldly, controlled by intelligent machines.

Finally, the art of artificial intelligence also raises a question for our artists, that is, after the emergence of artificial intelligence art, what is the independent value and significance of human art? Since the intelligent agent has been able to massively produce artworks, is it necessary for human beings to create art? The author believes that this highlights the significance of human art to the creator. Art is the need of artists for expression and communication. Art as a human spiritual product, its ultimate goal is to achieve the harmony between human body and mind, between people and people and between people and nature, is to actualize spiritual transcendence. Human beings are to achieve the expansion of the aesthetic mind and ultimate freedom through art. And because of the machine properties of artificial intelligence, it has no body and mind, so it can't touch the core part of art. Therefore, we can predict that after the emergence of artificial intelligence art, human art will pay more and more attention to the liberation and freedom of the creators themselves.

3   Conclusion

From this, we can see that although artificial intelligence programs have been able to create works of art with unique styles, this does not mean that they are creative, because creativity is not only the uniqueness and innovation of styles, but also means that this style must be exemplary, and need creation subjects to understand the entire art history, and have independent artistic sense and problem consciousness. Similarly, judging the creativity of art requires the establishment of artistic standards, and the artistic standards are constantly changing with time and space. At present, only from the perspective of appreciation and reception of artworks, the positive meaning and value of artificial intelligence art could be considered. With the emergence of artificial intelligence art, people need rethink the definition of art and the independent value of human art.

However, the author believes that with the advancement of artificial intelligence technology, artificial intelligence may create more and more works with independent aesthetic styles and aesthetic values. Artificial intelligence art is not only for us to appreciate, but also to bring us be closer to the essence and truth of human art.

Acknowledgement. This paper is supported by the National Social Science Fund of China, the name of this project is "The Language Thought in Adorno's Philosophy". No. 16BZX118.

References

1. Gombrich, E.: The Story of Art. Phaidon Press Ltd., London (2006)
2. Gatys, L., et al.: A neural algorithm of artistic style. https://arxiv.org/pdf/1508.06576v1.pdf (2015)
3. Elgammal, A., et al: CAN: creative adversarial networks generating 'art' by learning about styles and deviating from style norms, pp. 1–22. https://arxiv.org/pdf/1706.07068.pdf, 23 June 2017
4. Negnevitsky, M.: Artificial Intelligence, A Guide to Intelligent Systems, 2nd edn. Pearson Education Limited, New York (2005)
5. Goodfellow, I., Bengio, Y., Courville, A.: Deep Learning. MIT Press, Boston (2016)
6. Boden, M.: Creativity and artificial intelligence. Artif. Intell. **103**, 347–356 (1998)
7. Turing, A.: Computing machinery and intelligence. Mind **49**, 433–460 (1950)
8. Boden, M.: Artificial Intelligence. Academic Press, San Diego (1996)
9. Gips, J., Stiny, G.: Artificial intelligence and aesthetics. In: IJCAI 1975 Proceedings of the 4th International Joint Conference on Artificial Intelligence, vol. 1, pp. 907–911. https://www.ijcai.org/Proceedings/75/Papers/135.pdf
10. Shimamura, A.: Toward a science of aesthetics issues and ideas. In: Palmer, S.E. (ed.) Aesthetic Science: Connecting Minds, Brains and Experiences, pp. 3–28. Oxford University Press, Oxford (2012)
11. Firschein, O., Fischler, M.: Forecasting and assessing the impact of artificial intelligence on society. https://www.ijcai.org/Proceedings/73/Papers/013.pdf (1973)
12. Adorno, T.: Gesammelte Schriften, Band. 7: Ästhetische Theorie. In: Tiedemann, R., unter Mitwirkung von Adorno, G., Buck-Morss, S., Schultz, K. (eds.) Suhrkamp, Frankfurt (2003)
13. Tao, F.: The standard of art: Adorno on truth content. Nanjing Coll. Art **2**, 21–25 (2013)
14. Arnheim, R.: Art and Visual Perception: A Psychology of the Creative Eye. University of California Press, Los Angel (1997)
15. Benjamin, W.: Das Kunstwerk im Zeitalter seiner technischen Reproduzierbarkeit. In: Tiedemann, R., Schweppenhäuser, H. (eds.) Gesammelte Schriften, Band 7, pp. 471–508. Suhrkamp, Frankfurt am Main (1991)

Language Cognition

Using Two Formal Strategies to Eliminate Ambiguity in Poetry Text

Wei Hua[1], Shunpeng Zou[2], Xiaohui Zou[3(✉)] [iD],
and Guangzhong Liu[1,2,3]

[1] College of Information Engineering,
Shanghai Maritime University, Shanghai, China
201740310003@stu.shmtu.edu.cn, gzhliu@shmtu.edu.cn
[2] China University of Geosciences (Beijing), Beijing, China
407167479@qq.com
[3] Sino-American Searle Research Center, Beijing, China
949309225@qq.com

Abstract. The purpose of this paper is to compare the two major types of formalization strategies through the disambiguation of natural language textual ambiguities. The method is: The first step is to select the same text. Using poetry as an example, two types of formal strategies are used to resolve the ambiguities that exist. The second step is to analyze the limitations of the first formal path, at the same time, using traditional artificial intelligence methods and a new generation of artificial intelligence. The third step is to use the double-word board tools and methods to do the same thing. The result is that using the first path, whether based on rules (traditional artificial intelligence methods) or on statistical and machine learning, especially deep learning (a new generation of artificial intelligence methods), only local solutions can be obtained; With the checkerboard tools and methods, the overall solution can be obtained. This shows the unique advantages of the second path. Its significance lies in: using the double-word chessboard tool and method (second path) can solve the common problems faced by traditional artificial intelligence and new generation of artificial intelligence, and how to eliminate the ambiguity of natural language texts. The most important thing is that it has a new role. The most typical is to construct a knowledge base of the subject through the acquisition of knowledge and formal expression of experts, so as to gradually resolve a series of ambiguities between natural language (text) processing and formalized understanding.

Keywords: Two formal strategies · Eliminate ambiguity
Natural language textual · Ambiguities artificial intelligence methods

1 Introduction

Ambiguity is a type of meaning uncertainty giving rise to more than one plausible interpretation. It generally exits in our language and expression, being ambiguous is therefore a semantic attribute of a form (a word, an idea, a sentence, even a picture) whose meaning cannot be resolved according to a simple rule or process with a finite

Z. Shi et al. (Eds.): ICIS 2018, IFIP AICT 539, pp. 159–166, 2018.
https://doi.org/10.1007/978-3-030-01313-4_16

number of steps. As the result, semantic disambiguation plays an important role in natural language processing (NLP) and many scholars have been spending tremendous effort on the problem for decades. However, the development of disambiguation technique stagnated in a long term [1] until the significant breakthrough was made on frameworks and algorisms of neural network in last decade [2]. By building computational models based on statistic theory, many significant research results and solutions regarding disambiguation have been obtained rather than by using traditional linguistics. However, the challenge and obstacle still exist, even by using neural network or deep learning, the accuracy in tasks of disambiguation has still a lot space to improve especially in the field of Chinese poetry understanding. The main reason why this task is difficult is that even neural network and deep learning emulate successfully how human's brain works on the task of language processing, the fundamental form of these new methods are still based on Aristotle's formal logic and Frege's mathematical logic [3], which means only the form of programming languages is involved into the NLP tasks so that whether rules based on traditional artificial intelligence methods or on statistical methods and machine learning, especially deep learning (a new generation of artificial intelligence methods), only local solutions can be obtained. For convenience to describe the path, here we define this form as "first type of formal strategy".

In our research, the target is to find a new path to break through the bottleneck that "first type of formal strategy" has to face and then to verify its effectiveness. In this paper we present a way to the second path – "The second type of formal strategy" [3], to resolve the problem. According to the idea of second path, overall solution can be obtained with the checkerboard tools and methods. The tool based on second path combines Chinese characters and English, binary and decimal systems, decimal and Chinese characters into "double-words" chessboards. Due to the difference between language structures, the task of semantic disambiguation of Chinese poetry is more difficult than that of modern Chinese, in the meantime, being lack of sufficient corpus leads to the limitation of statistics-based disambiguation. Through comparison of the experimental results, we provide references for further study of two formal strategies.

2   Related Work

To verify the effects of two formal strategies, we design a set of experiments for each type and build a system based on the second formal strategy. We call this system "Double-word board" [4, 5]. The same Chinese poem text is fed to both of the systems to test the effect of disambiguation, and finally we make the comparison of the experimental results.

2.1   Build a Double-Words Chessboard and Chinese Language Chessboard Spectrum

The tool based on "The second natural language formalization strategy" is called "**Double-Word Board**" [5]. It contains two main components: the Chinese character board encoded with digitals and the Chinese character board. "Double-word board" is the linkage function between digital and textual of conjugate matrices, binary and decimal codes and English and Chinese and its alternative bilingual. We can think the board as an expert knowledge acquisition system that machine cooperatively builds with human. Through the human-machine interface, Chinese characters in the poem are marked one by one with digital codes. Encoded matrix/table plays a key role in disambiguating the poem text, its digital codes and number combinations indicate the relationship of Chinese characters in context, and the chessboard stores the codes as reusable rules.

2.2   Experiments

We design two test cases for each experiment to verify our language chessboard (Double-Word Board) and its application are effective and accurate in the task of disambiguation in poetry text. We use a Chinese poem as the input of test case 1 (Fig. 1):

床前**明月**光，疑是地上霜。举头望**明月**，低头思故乡。

Fig. 1. Formal chessboard (double-word board) in Test case 1

In this Chinese poem, the phrase "明月" appears in the first and the third sentences repeatedly with different interpretations. The phrase appears in the first sentence means "moonlight" and the different meaning, "a bright moon" appears in the third sentence.

We use the built-in encoding system to create codes for each character in this poem and then we can obtain a table containing mapping relations between code and Chinese characters (Table 1).

Encoding system is not complex. We pick out the polysemic phrase "明月" in this poem to demonstrate how the chessboard (double-word board) works. Human can distinguish the different interpretation of this word in context positions. As our indication, number "565" represents "明", and "594" represents "月" are record. Therefore, "明月" can be combined into a new code composed of two numbers, as the result, information containing character sequences is also recorded according the context of the poem sentences.

Next step, in the similar way, we create codes for three-character combination and this set of code are stored in chessboard. Disambiguation of man-machine interaction is needed only once for each phrase.

Table 1. Create codes for each character in the poem

Code	Character sample
...	...
564	已
565	*明*
589	打
594	*月*
...	...

Our goal is to make comparison between first type of strategy and second type to prove the effectiveness in these two strategies. Thus, we setup test case 2, inputting the same poem text into a statistics-based computational model so that we can observe the effect and the difference of disambiguation between by using deep learning or machine learning [6] and by "double-word board". Statistics-based software system (deep learning or machine learning system) mostly needs a dataset that its function is quite similar to the use of code in double-word chessboard. But the nature of datasets is different with that of codes in chessboard radically. Datasets do not indicate the ambiguity clearly, on the contrary, codes stored in chessboard can functionally eliminate ambiguities in the poetry text as the code based on human's understanding and the knowledge is transferred to the system.

2.3   Public Datasets

We found some public corpus containing the poems we need in our experiment. "Wiki corpus" [7], "Literature 100" [8] are the two widely used poetry corpus (Table 2).

Then we give the second experiment. We pick another Chinese poem as the input text sample:

慈母手中线，游子身上衣。临行密密缝，意恐迟迟归。谁言寸草心，报得三春晖。

Suppose we do not know this poem before. In fact, when someone is learning a foreign language, it is quite often to read obscure sentences those are hard to understand. The situation gets worse in Chinese poetry. The main reason is the fact that non-native Chinese speakers are hard to segment Chinese sentences into phrases correctly meanwhile Chinese sentences usually omit sentence elements unregularly. At present, word segmentation based on statistical techniques has been well developed [9]. There are many open-sourced components available for building a natural language processing system and the performance is satisfied. The commonly used Chinese segmentation methods [10] can fall into three categories: Segmentation methods based on string matching, word segmentation methods based on comprehension, and word segmentation methods based on statistics.

Table 2. shows that the system can only counts the number of appearance of "明月" (which means "bright moon or moonlight"), it cannot accurately indicate the different meanings of phrase"明月" in different context positions.

Two word string	Word frequency
···	···
日月	508
···	···
月明	515
···	···
天地	586
···	···
明月	896
···	···

Segmentation method based on string matching: also known as the mechanical word segmentation method, it is a certain or a sort of strategies to look up the Chinese character string in a "full-sized" dictionary, if a word is found in the dictionary then the word segmentation is successful.

The word segmentation method based on comprehension: This method is to design a sort of algorisms to make a computer emulate the process of a person's understanding a sentence. The basic idea of this method is to syntactically and semantically analyze the sentence segmentation, so as to deal with ambiguity with syntactic and semantic information. It usually contains three parts: word segmentation subsystem, syntax and semantics subsystem and general control. Under the coordination of the general control, the word segmentation subsystem can obtain syntactic and semantic information about words, sentences, etc. to judge the ambiguity, that is, it simulates the process of human understanding of the sentence. This segmentation method requires a lot of linguistic knowledge and information. Because of the generality and complexity of Chinese, it is difficult to make Chinese language information into a form unified that can be directly read by machines. Therefore, word segmentation system based on understanding is still in the research stage.

The statistical word segmentation method is that sentence segmentation can be applied on unknown texts with rules learned from a huge amount of corpus segmented with statistical machine learning techniques. For example, the maximum probability word segmentation method and the maximum entropy word segmentation method are commonly used in the task of word segmentation. As the establishment of large-scale corpus, the research and development of statistical machine learning methods, statistic based methods have gradually become the mainstream in the field of Chinese word segmentation.

Here we list some mainstream models: N-gram, Hidden Markov Model (HMM), Maximum Entropy Model (ME), Conditional Random Fields (CRF [11]) etc.

In practical applications, the word segmentation system based on statistics needs a dictionary to perform string matching. Meanwhile, statistical methods combine string frequency with string matching, so as to make word segmentation perform faster, efficiently with function of recognition of new words and automatic elimination of ambiguity.

However, at current stage, these three methods perform still not well enough in the task of word segmentation in Chinese poetry text due to the lack of word dictionaries customized for Chinese poetry.

To this problem, we try the word segmentation with double-word board to see if we can make any improvement on the task.

The board disorder the characters in the poem and we can obtain a list of two-to-five characters combinations:

身，心，意，行，缝，恐，归，言，报，母，子，手，线，衣，谁，春，晖

慈母，游子，临行，意恐，谁言，报得，春晖，寸草，三春，迟迟，密密

手中线，身上衣，密密缝，迟迟归，寸草心，三春晖

慈母手中线，游子身上衣，临行密密缝，意恐迟迟归，谁言寸草心，报得三春晖

Similarly, according to the codes of single Chinese character, two-to-five word combinations are stored into the chessboard that means we transfer our knowledge to the system. After man-machine cooperation, as the persons who never read this poem we can roughly understand the correct word segmentation in the poem. Only corpora segmented correctly has the value for further processing.

In this experiment, we use the statistics-based segmentation tool jieba [12] to perform segmentation for the poem text and the sentences segmented list is:

慈|母|手中|线|游|子|身|上衣
临|行|密|密|缝|意|恐|迟|迟归
谁|言|寸|草心|报|得|三春|晖

As we can see a few segmentation mistakes still made in segmentation.

3   Discussion

Eliminating ambiguity is an important function in Natural Language Processing (NLP). This paper compares the disambiguation effects of two formal strategies by using the same poems for different scenarios of ambiguity through two sets of experiments. We can observe it has a better performance in task of disambiguation with bilingual chessboard tools. The compared results show we can still find out the relationship between the two strategies and their respective advantages and disadvantages.

Through the experiments we can see that statistics still do not work well in task of disambiguation in poetry texts. The reason for this situation is still due to the influence of the ambiguity that generally exists in languages. When dealing with ambiguity, statistical algorisms still have no enough ability to obtain global solution like human does.

Therefore, the bilingual chessboard used in this paper, that is, the combination of the formalization of the two formal strategies [13], allows the system to obtain a global optimal solution through human-computer interaction and effectively solves the problem of ambiguity in Chinese poetry. Combining the advantages of programming language and statistical techniques, the knowledge base is constructed to eliminate ambiguity in the language.

4   Future Work

Text functions are recorded in the form of Chinese or English words, and the order and position of the lattice in a particular board or matrix are also relatively constant. Although their combination can be ever-changing, but in the linkage function of the constraints, there are still laws following the rules. This is the role, value and significance of the three types of identities and their corresponding analytic geometric representations. Among them, the three types of identities embody the basic laws of three kinds of information, and the corresponding analytic geometric expression can be presented through the twin chessboard (double-chess board) and play a role in the process of man-machine collaboration, which is expressed as an expert knowledge Acquisition. Demonstration of "The second type of formal strategy" is still a rudiment in our paper. Our further work will be still working on the novel idea of knowledge transformation to computational system effective and efficient.

Acknowledgement. All sources of funding of the study have been disclosed.

References

1. Waltz, D.L.: On understanding poetry. In: Theoretical Issues in Natural Language Processing (1975)
2. Bai, M.-H., Chen, K.-J., Chang, J.S.:
   利用雙語學術名詞庫抽取中英字詞互譯及詞義解歧《全唐詩》的分析、探勘與應用－風格、 (Sense extraction and disambiguation for Chinese words from bilingual terminology bank). In: Proceedings of the 17th Conference on Computational Linguistics and Speech Processing, pp. 305–316, September 2005. (in Chinese)
3. Zou, X.H., Zou, S., Ke, L.: Fundamental law of information: proved by both numbers and characters in conjugate matrices. In: IS4SI (2017)
4. Zou, X.H.: Bilingual information processing method and principle, 69–76, 102 (2015)
5. Zou, S.: Formal bilingual chessboard spectrum: show the overlapping between language and mind. In: AAAS 2017 Annual Meeting, February 2017
6. Liu, C.-L., Chang, C.-N., Hsu, C.-T., Cheng, W.-H., Wang, H., Chiu, W.-Y.:
   對仗、社會網路與對聯 (Textual analysis of complete tang poems for discoveries and applications - style, antitheses, social networks, and couplets). In: Proceedings of the 27th Conference on Computational Linguistics and Speech Processing (ROCLING 2015), pp. 43–57 (2015). (in Chinese)

7. https://zh.wikisource.org/zh-hant/全唐詩
8. http://www.wenxue100.com/
9. Agirrezabal, M., Alegria, I., Hulden, M.: Machine learning for metrical analysis of english poetry. In: Proceedings of COLING 2016, the 26th International Conference on Computational Linguistics: Technical Papers, pp. 772–781, December (2016)
10. Jin, W.: Chinese segmentation disambiguation. In: COLING 1994 Volume 2: The 15th International Conference on Computational Linguistics (1994)
11. Lafferty, J., McCallum, A., Pereira, F.C.N.: Conditional random fields: probabilistic models for segmenting and labelling sequence data. Department of Computer & Information Science (2001)
12. https://github.com/fxsjy/jieba
13. Zou, S., Zou, X.: Understanding: how to resolve ambiguity. In: Shi, Z., Goertzel, B., Feng, J. (eds.) ICIS 2017. IAICT, vol. 510, pp. 333–343. Springer, Cham (2017). https://doi.org/10.1007/978-3-319-68121-4_36

Discussion on Bilingual Cognition in International Exchange Activities

Mieradilijiang Maimaiti[1] and Xiaohui Zou[2(✉)]

[1] Department of Computer Science and Technology, Tsinghua University,
Beijing, China
meadljmm15@mails.tsinghua.edu.cn
[2] Sino-American Searle Research Center, Beijing, China
geneculture@icloud.com

Abstract. This article aims to explore the features, mechanisms, and applications of bilingual cognition in international communication activities. Our main idea is: First, clarify the mother tongue of each international exchange activities (IEAs) and prepare some prerequisites which are related to the discussion. Then, make full use of the information and intelligent network tools to bring out the subjective initiative of both parties, while conducting the corresponding research on daily terms and professional terms, and generate the two series of bilingual phrase table. Finally, use the machine translation (MT) and translation memory tools to help them make the necessary preparations or exercises. Meanwhile, we propose the novel and efficient mixed transfer learning (MTL) approach. As a result, when the two parties communicate with each other, as well as via online or off-line communicate, the kind of tacit agreement would have been created between them. If so, it will have been leveraged among them repeatedly rather than just one time and will have targeted multiple times. Its significance lies in: This process and habit of human-computer interaction will better reveal the characteristics of bilingual cognition based on this article. Experiments on low-resource datasets show that our approach is effective, significantly outperform the state-of-the-art methods and yield improvements of up to 4.13 BLEU points.

Keywords: Bilingual cognition · International exchange activities
Human-computer interaction · Neural machine translation
Transfer learning

1 Introduction

Recent days artificial intelligence technologies made remarkable progress, many terrific ideas, various mechanisms, and marvelous applications are designed and emerged both in our daily life and study life. Meanwhile, cognitive computation [11,12] filed has also obtained outstanding achievements in the international exchange activities (IEAs). Likewise, bilingual cognition [13] also plays a vital

Z. Shi et al. (Eds.): ICIS 2018, IFIP AICT 539, pp. 167–177, 2018.
https://doi.org/10.1007/978-3-030-01313-4_17

role during the international exchanging with each other, and many people are trying to take advantages of human-computer interaction (HCI) mechanisms or applications to accelerate their communications and improve the quality of communications by avoiding the low efficiencies.

As we are taking part in IEAs, the majority of us would feel that it is time-consuming and hard to avoid unexpected misunderstandings at a certain level, and needs to higher cost frequently. Since, if the bilingual person even fluent to speak both two languages but sometimes also need to hire some professional translators or need to buy some artificial intelligence products, such as off-line spoken translator tools, some smart mobile devices, and simultaneous translation system. Moreover, because of the quality of these portable smart mobile devices or facilities not good enough at some special occasion, even it might cause some misunderstandings unexpectedly. Although there are many demands for IEAs, however, these demands are not be solved efficiently and effectively. In addition, many useful approaches have been proposed by numbers of researchers and scientists. The methods which could be used to improve the quality of international communications, even some essential features, convenience applications, popular mechanisms and the power of bilingual cognition.

Besides, except for these strategies, numbers of artificial intelligence algorithms, marvelous ideas, terrific mechanisms and outstanding approaches have been created in the international communication field. Such as online machine translation engines, online network applications, smart mobile devices, and off-line machine translation facilities. Even if the translation system which can be used by taking photos or by recognizing the voice of users, the core technology is still machine translation. During the IEAs, we need to use high-resource languages (HRLs) and low-resource languages (LRLs) to communicate with each other on the Internet or on some special occasion. In addition, the bilingual professional translator might not good at for using LRLs in comparison to HRLs. The well-known and commonly used off-line translator take advantage of compressing the neural machine translation (NMT) model which trained on huge amounts of data. Therefore, NMT models suffer from the data scarcity problem. Some useful ideas about NMT for LRLs, such as transfer learning [28], word-stem substitution and double-transfer [22] and zero-resource learning [15], have been introduced but many problems still exist in MT. Transfer learning is one of the efficient methods for low-resource NMT task. If we exploit those methods in IAEs with LRLs explicitly, it might not be achieved the more efficient result as HLRs. Generally, the bilingual cognition and some artificial applications (smart devices and machine translation system) are as the integrated union in IEAs. Previous methods almost separated them from each other, and unable to make full use of mutual effectiveness in IEAs.

In this work, therefore, by the comparison to aforementioned approaches, we aim to deal with the problem that how to make full use of these factors and combine them to better develop the international communications in various activities such as international trade, education, and conference etc. Our major idea focuses on investigating some features, mechanisms, and applications of bilingual cognition in IAEs. Exactly, the proposed method is: First, clarify the mother tongue of each international activity, and make some prerequisites before communicating others legitimately. Then, make full use of information and intelligent network tools to bring out the subjective initiative of both parties, while conduct the corresponding research on daily terms and professional terms. Finally, exploit the MT and translation memory tools to better help them make the necessary preparations or exercises. Besides, we have proposed a novel and efficient approach mixed transfer learning (MTL) for low-resource NMT. Our method achieves outstanding results that when the two parties communicate with each other, they will have some tactic agreement, even they are online or off-line. The HCI better reveal the characteristics, mechanisms, application of bilingual cognition by processing. Experiments on NMT for LRLs, from Arabic (Ar), Farsi (Fa) and Urdu (Ur) to Chinese (Ch) shows that the proposed MTL method is also achieved better results. The key is to take advantage of scientific principles of bilingual informatization and intelligence. Our contributions are as follows:

1. Mitigate the gap between a non-native speaker and native speaker by exploiting prepared or necessary draft.
2. Make full use of the combination of bilingual cognition and artificial translation system in IEAs.
3. Provide some Efficient and effective channels for IEAs.
4. The proposed NMT training approach for LRLs in IEAs is transparent to neural network architecture.

2   Background

2.1   International Exchange Activities

Intuitively, during the international communication, the neural cognition computing [11,12] should be one of the necessary parts. International communication is a major activity in an international company's marketing mix. Once a product or service is developed to meet consumer demands and is suitably priced and distributed, the expected consumers must be notified of the product's availability and value. International communication [14] consists of those movements which are practiced by the marketer to inform and convince the consumer to purchase. A well-designed advancement mix includes promoting, sales advertisements, particular selling, and public relationships which are mutually augmenting and concentrated on a regular objective.

2.2   Neural Machine Translation

Additionally, we also take advantage of machine translation (MT) especially neural machine translation (NMT) in human-computer interaction efficiently and effectively. We take X as a source language sentence and Y as a target language sentence, respectively. Given a source sentence $\mathbf{x} = x_1, \ldots, x_i, \ldots, x_I$

and a target sentence $\mathbf{y} = y_1, \ldots, y_j, \ldots, y_J$, standard NMT models [1,25,27] usually factorize the sentence-level translation probability as a product of word-level probabilities:

$$P(\mathbf{y}|\mathbf{x}; \boldsymbol{\theta}) = \prod_{j=1}^{J} P(y_j|\mathbf{x}, \mathbf{y}_{<j}; \boldsymbol{\theta}), \tag{1}$$

where $\boldsymbol{\theta}$ is a set of model parameters, $\mathbf{y}_{<j}$ is a partial translation. NMT models usually rely on an encoder-decoder scenario.

Let $\langle \mathbf{X}, \mathbf{Y} \rangle = \{\langle \mathbf{x}^{(n)}, \mathbf{y}^{(n)} \rangle\}_{n=1}^{N}$ be a training corpus. The log-likelihood of the training parallel data is maximized by the standard training objective:

$$\hat{\boldsymbol{\theta}} = \underset{\boldsymbol{\theta}}{\operatorname{argmax}} \left\{ \sum_{n=1}^{N} \log P(\mathbf{y}^{(n)}|\mathbf{x}^{(n)}; \boldsymbol{\theta}) \right\}. \tag{2}$$

The translation decision rule for unseen source sentence $\mathbf{x}$ given learned model parameters $\hat{\boldsymbol{\theta}}$ is given by

$$\hat{\mathbf{y}} = \underset{\mathbf{y}}{\operatorname{argmax}} \left\{ P(\mathbf{y}|\mathbf{x}; \hat{\boldsymbol{\theta}}) \right\}. \tag{3}$$

Meanwhile, calculating the highest probability $\hat{\mathbf{y}} = \hat{y}_1, \ldots, \hat{y}_j, \ldots, \hat{y}_J$ of the target sentence can be separated at the word level:

$$\hat{y}_j = \underset{y}{\operatorname{argmax}} \left\{ P(y|\mathbf{x}, \hat{\mathbf{y}}_{<j}; \hat{\boldsymbol{\theta}}) \right\}. \tag{4}$$

Table 1. The efficient ways for international exchange activities.

Different channels	Example
Social Media Networking	Skype, Facebook, Google+
BLOGS	Twitter, Tumblr, Blogger
Activities Sharing and Storage	Google photos, Flickr, photobucket, instagram
Telecommunication Options	Text follows FaceTime, WhatsApp, Viber, TextPlus, MagicJack

3   Methodology

3.1   HCI Method for International Communication

We are staying in an exciting moment that artificial intelligence is becoming ubiquitous and is playing increasingly significant roles in our lives and in the basic infrastructures of science, business, and both social communication and IEAs. We regard that, as shown in Table 1 there were the majority of ways which can be touchable for international communications. However, they just merely play a role of the instrument. We investigate some features and mechanisms of bilingual cognition. As illustrated in Fig. 1, in the first step, confirm and specify what they want to say in IEAs. Namely, the speaker selects the word, and use the phrase to organize the phrase table, then make a sentence by exploiting the phrases. In the confirming the bilingual content, select the bilingual pairs. Exactly, select the semanteme via relatively analogous bilingual semanteme. Finally, revise and update the series of comparison tables. Intuitively, taking advantages of MT to help the speakers and improve the quality of communications.

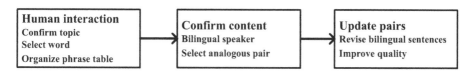

Fig. 1. The architecture of human-computer interaction with bilingual cognition.

3.2   Mixed Transfer Learning Approach for NMT

For the intention of making full use of the role of the machine and encourage machine help human to improve the quality of their communication efficiency in IEAs. We presented and analyzed the effect of machine translation for international communication as a human-interface toolkit. We follow the major idea of transfer learning (TL) in NMT for LRLs and referred them to incorporate into the architecture of human communication interface system.

We take $L_3 \rightarrow L_2$ as a parent language pairs and $L_1 \rightarrow L_2$ as a child language pair. L_3 and L_1 are source languages of parent and child, respectively, L_2 is the target language for both. Additionally, we set the dataset of parent language $L_3 \rightarrow L_2$ is D_{L_3,L_2}, while dataset of child language $L_1 \rightarrow L_2$ is D_{L_1,L_2}. Besides, we set $M_{L_3 \rightarrow L_2}$ as the parent language model which learned on parent language dataset D_{L_3,L_2}. Generally, we initialize the child model $M_{L_1 \rightarrow L_2}$ by using of parent model [28], and the corresponding parameter of parent model:

$$\theta_{L_3,L_2} = \{\langle e_{L_3}, W, e_{L_2}\rangle\}, \tag{5}$$

while e_{L_3} and e_{L_2} are both source and target embedding of parent model, and W is their parameters. We also encourage the training objective to maximize the likelihood of dataset D_{L_3,L_2}:

$$\hat{\theta}_{L_3 \rightarrow L_2} = \underset{\theta_{L_3 \rightarrow L_2}}{\mathrm{argmax}}\left\{L(D_{L_3,L_2}, \theta_{L_3,L_2})\right\}. \tag{6}$$

After that, the child model $M_{L_1 \rightarrow L_2}$ will be fine-tuned by parent model $M_{L_3 \rightarrow L_2}$ take advantage of their dataset (D_{L_1,L_2}) of child model:

$$\theta_{L_1 \rightarrow L_2} = f(\hat{\theta}_{L_3 \rightarrow L_2}), \tag{7}$$

the learned parameters $\hat{\theta}_{L_3 \to L_2}$ of parent model are transferred to child model $M_{L_1 \to L_2}$ by initialization function f.

Intuitively, as depicted in Fig. 2, we inspired by the original transfer learning [28] in NMT and introduce the mixed transfer learning (MTL) approach, which shares the vocabularies between parent and child language. Meanwhile as described in [15], we exploit the oversampling such that data from all language pairs to be of the same size as that of largest language pair, as well as ensuring an equal amount of data per language pair. Then train the mixed model by combining of parent and intermediate model which has been trained on the oversampled mixed bilingual corpus, after that fine-tune the child model to improve their translation quality, and mitigate the cost of professional translators, and better help to international communication during the using of LRLs with each other.

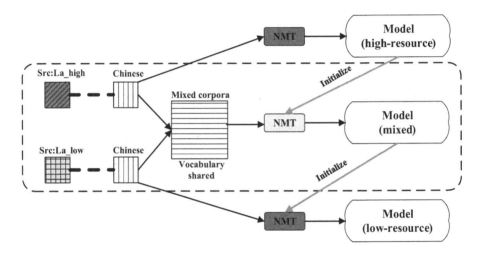

Fig. 2. The architecture of mixed transfer learning for NMT, while the dotted rectangle stands for the training of the mixed model.

Table 2. Language features of all languages used in our experiment.

Language		Family	Group	Branch	Order	Unit	Inflection
Arabic	(Ar)	Hamito-Semitic	Semitic	South	VSO	Word	High
Farsi	(Fa)	Indo-European	Indic	West	SOV	Word	Moderate
Urdu	(Ur)		Iranian	Iranian	SOV	Word	Moderate
Chinese	(Ch)	Sino-Tibetan	Chinese	Sinitic	SVO	Character	Light

Table 3. Characteristics of parallel corpora. While "Vocab." and "# Word" represent vocabulary (word type) and word token, respectively. Besides, "#Cov." stands for corresponding covering rates of each language.

Languages	Train	Dev	Test	Source			Target		
				Vocab.	# Word	#Cov.[%]	Vocab.	# Word	#Cov.[%]
Ar → Ch	5.1M	2.0K	2.0K	1.0M	32.2M	88.30	0.5M	37.4M	96.80
Fa → Ch	1.4M	2.0K	1.0K	0.2M	10.4M	96.80	0.2M	10.0M	96.80
Ur → Ch	78.0K	2.0K	1.0K	17.6K	2.6M	100.00	12.7K	2.4M	100.00

4 Experiments

4.1 Setup

All the datasets which are used in this work publicly available on Open Subtitle2016[1] and Tanzil corpora[2]. The corpus feature and specifications are shown in Tables 2 and 3, respectively, in this corpora the target side is identically set with Chinese, while source side is different parent RRLs or child LRLs. Additionally, parent language pairs Ar → Ch, Fa → Ch are collected from Open Subtitle2016 corpora, and the child language pair Ur → Ch is obtained from Tanzil corpora.

In the preprocessing step, we used NIUTTRANS preprocessing *perl script*[3] to remove and clean illegal Chinese parallel sentences (our target side) from original corpus. Besides, we also prepared several preprocessing *python scripts* for both source and target side. Each of these scripts mainly works for LRL and Chinese parallel corpus, as well as re-cleaning after removing illegal characters, removing the blank lines, removing illegal symbols and double checking non-Chinese characters, and converter which is used for converting simplified Chinese and traditional Chinese. We also use open source Chinese word stemmer system THULAC[4] for Chinese language [17]. Meanwhile, we exploited `tokenizer.perl` toolkit[5] which was provided by state-of-the-art (SOTA) phrase-based statistical machine translation (SMT) system MOSES [16] for word tokenization. Moreover, we report results without any UNK-replacement techniques [19].

Additionally, in our full experiment, we make use of the attention-based encoder-decoder model which gated recurrent unit for NMT system DL4MT[6]. Likewise, we run the experiments approximately for 3–4 days except for finetuning (only 2–3 days) on single NVIDIA TITAN X (PASCAL) GPU almost with default parameters of DL4MT, just slightly modified some of the dimensions,

---

[1] http://opus.nlpl.eu/OpenSubtitles2016.php.

[2] http://opus.nlpl.eu/Tanzil.php.

[3] http://www.nlplab.com/niuplan/NiuTrans.YourData.html.

[4] https://github.com/thunlp/THULAC-Python.

[5] https://github.com/moses-smt/mosesdecoder/tree/master/scripts/tokenizer/tokenizer.perl.

[6] https://github.com/nyu-dl/dl4mt-tutorial.

such as word embedding is 620, hidden state embedding is 1000, we limit the vocabularies to the most frequent 30K words (covering rates as shown in Table 3) and batch size is 80, sentence max length is 50. Basically, in our experiments the baseline and our method in terms of case-insensitive BLEU[7] scores [21].

4.2   Effect of MTL for Low-Resource NMT

Intuitively, it is tractable to infer from Table 2 that the parent languages Ar and Fa both of them are similar with child language Ur, likewise the Fa and Ur belong to same language family. We referred to the relatedness and explore the shared word rate among them (see Table 4). In this experiment, we take RnnSearch as a baseline which is standard attention based encoder-decoder model [1] for NMT.

Table 4. The shared word rate between Ar, Fa, and Ur, and they are converted into Latin scripts.

	Arabic	Farsi	Urdu
Arabic		12.36%	8.61%
Farsi	2.51%		8.62%
Urdu	0.16%	0.78%	

Table 5. Effect of shared vocabulary single fine-tuning. While parent language pairs with the token "(Shared)" represents they shared same vocabularies with child language pair. Conversely, language pairs with the token "(Non-Shared)" stands for using own vocabularies rather than shared vocabulary anymore. "++": significantly better than RnnSearch ($p < 0.01$).

Method	Parent	Child	BLEU ↑
RnnSearch	N/A	Ur → Ch	18.31
Mixed	Fa → Ch (Non-Shared)	Ur → Ch	19.73
	Ar → Ch (Non-Shared)		20.04
	Fa → Ch (Shared)		21.69[++]
	Ar → Ch (Shared)		22.44[++]

Generally, inspired by [18] using shared vocabulary between general domain corpus and an in-domain corpus to improve translation quality of the in-domain model. In this work, we have also shared vocabulary between parent language pairs (Ar → Ch and Fa → Ch) and child language pair (Ur → Ch). First, pre-train the low-resource NMT, then combine both source sides (Ar/Fa, Ur) and

---

[7] https://github.com/moses-smt/mosesdecoder/blob/master/scripts/generic/multi-bleu.perl.

target sides (both of two groups are Ch) of parent and child sentences to create the big corpus, as well as mixed parent and child corpora. Then generate shared vocabulary, it includes both higher frequency words of parent and child language pairs, to train the mixed model. Finally, Initialize the Ur $\rightarrow$ Ch via Ar $\rightarrow$ Ch and Fa $\rightarrow$ Ch with and without shared vocabularies, respectively. Exactly, train the parent model $M_{Ar \rightarrow Ch}$ and $M_{Fa \rightarrow Ch}$ with their own vocabularies which only consisted of Ar or Fa, and with mixed vocabularies that include both of parent and child words to train corresponding parent models, then initialize the same child model $M_{Ur \rightarrow Ch}$. As given in Table 5, proposed approach obtained improvements exploiting Ar $\rightarrow$ Ch (shared) and leveraging Fa $\rightarrow$ Ch (shared) comparison with non-shared (used own vocabulary) vocabulary parent model.

5   Related Work

As we have aforementioned, there were many methods for developing international communication [6], international education [23], international trade [2], intercultural communications [8], and intercultural business communication [4]. Moreover, the variety of efficient ideas and approaches have been introduced, however, many factors and essential parts were neglected. Intuitively, we can make full use of some features and architectures by leveraging artificial intelligent mechanisms. Likewise, we regard that the consideration of the mother tongue of speakers and combinations of bilingual cognition and machine translation could better help and improve the quality of IEAs.

In the literature, the transfer learning (TL) [26] method has been being widely used in computer vision [20] and domain adaptation [3,5]. We have also referred the TL and incorporate them into NMT [7,9,10,24] to better help the training procedure of LRLs by leveraging of highly analogous HRLs. Additionally, many researchers [22,28] have also paid attention to LRLs NMT, however, neglected the sharing vocabularies between similar or highly related HRLs and LRLs. In the IEAs, the majority of users explicitly exploit the original translation system which was trained on the big parallel corpus. In this case, the developing procedure becomes more sophisticated that preparing huge amounts of data and training them for several hours on GPUs or even TPUs. While some others also try to improve the quality of translation via leveraging the word substitution [22] between same language family, group even the same language branch. But it has some limitations between other similar languages which are in various language family or group. Likewise, Google [15] also introduce some useful methods for zero-resource translation by combining HRLs and LRLs together with oversampling. However, their approach ignores the relatedness of languages and make some turbulence signals when LRLs receiving parameters from HRLs during training.

6   Conclusion and Future Work

In this work, we discuss the efficient architecture of HCI with bilingual cognition. For the intention of, how to make full use of some factors and combine them to better develop the HCI, our method mainly focuses on the investigation of IEAs by using artificial intelligent mechanisms. Besides, we have also proposed a novel training method mixed transfer learning (MTL) for NMT, which is used in IEAs as the speaker needs to use LRLs. Additionally, instead of exploiting the original NMT model, we leverage the MTL method to share vocabulary to train the NMT model. Then guide the LRLs model by highly related HRLs model, and mitigate computation space and reduce memory consumption and time. In the future work, we plan to further validate the effectiveness of our method on more NLP and other tasks except for IEAs, meanwhile try to leverage on morphologically poor languages.

References

1. Bahdanau, D., Cho, K., Bengio, Y.: Neural machine translation by jointly learning to align and translate. In: Proceedings of ICLR (2015)
2. Bidabad, B.: A convention for international trade (based on Islamic Sufi teachings). Int. J. Law Manag. **57**(5), 522–551 (2015)
3. Carpuat, M., Daumé III, H., Henry, H., Irvine, A., Jagarlamudi, J., Rudinger, R.: SenseSpotting: never let your parallel data tie you to an old domain. In: Proceedings of ACL (2013)
4. Chien, T.C.: Intercultural training for Taiwanese business expatriates. Industrial and Commercial Training (2012)
5. Chu, C., Dabre, R., Kurohashi, S.: An empirical comparison of simple domain adaptation methods for neural machine translation. CoRR abs/1701.03214 (2017)
6. Chelariu, C., Osmonbekov, T.: Communication technology in international business-to-business relationships. J. Bus. Ind. Mark. **29**(1), 24–33 (2014)
7. Dong, D., Wu, H., He, W., Yu, D., Wang, H.: Multi-task learning for multiple language translation. In: Proceedings of ACL (2015)
8. Isotalus, E., Kakkuri-Knuuttila, M.L.: Ethics and intercultural communication in diversity management. Equal. Divers. Incl. Int. J. **37**(5), 450–469 (2018)
9. Firat, O., Cho, K., Bengio, Y.: Multi-way, multilingual neural machine translation with a shared attention mechanism. In: Proceedings of NAACL (2016)
10. Firat, O., Sankaran, B., Al-Onaizan, Y., Vural, F.T.Y., Cho, K.: Zero-resource translation with multi-lingual neural machine translation. In: Proceedings of EMNLP (2016)
11. Pounder, G.A.J., Ellis, R.L.A., Fernandez-Lopez, G.: Cognitive function synthesis: preliminary results. Kybernetes **46**(2), 272–290 (2017)
12. Finch, G., Goehring, B., Marshall, A.: The enticing promise of cognitive computing: high-value functional efficiencies and innovative enterprise capabilities. Strat. Leadersh. **45**(6), 26–33 (2017)
13. Grosjean, F., Gremaud-Brandhorst, J., Grosjean, L.: The bilingual's language modes (1999)
14. Johnson, G.J.: The International Encyclopedia of Digital Communication and Society. Reference Reviews (2017)

15. Johnson, M., et al.: Google's multilingual neural machine translation system: enabling zero-shot translation. In: Transactions of the Association of Computational Linguistics (2017)
16. Koehn, P., Och, F.J., Marcu, D.: Statistical phrase-based translation. In: Proceedings of NAACL (2003)
17. Li, Z., Sun, M.: Punctuation as implicit annotations for Chinese word segmentation. Comput. Linguist. **35**, 505–512 (2009)
18. Luong, M.T., Manning, C.D.: Stanford neural machine translation systems for spoken language domains. In: Proceedings of IWSLT (2015)
19. Luong, T., Sutskever, I., Le, Q.V., Vinyals, O., Zaremba, W.: Addressing the rare word problem in neural machine translation. In: ACL (2015)
20. Oquab, M., Bottou, L., Laptev, I., Sivic, J.: Learning and transferring mid-level image representations using convolutional neural networks. In: 2014 IEEE Conference on Computer Vision and Pattern Recognition, pp. 1717–1724 (2014)
21. Papineni, K., Roukos, S., Ward, T., Zhu, W.J.: BLEU: a method for automatic evaluation of machine translation. In: Proceedings of ACL (2002)
22. Passban, P., Liu, Q., Way, A.: Translating low-resource languages by vocabulary adaptation from close counterparts. ACM Trans. Asian Low Res. Lang. Inf. Process. **16**, 29:1–29:14 (2017)
23. Davidson, P.M., Taylor, C.S., Park, M., Dzotsenidze, N., Wiseman, A.W.: Introduction reflecting on trends in comparative and international education: a three-year examination of research publications. International Perspectives on Education and Society (2018)
24. Sennrich, R., Haddow, B., Birch, A.: Neural machine translation of rare words with subword units. CoRR abs/1508.07909 (2016)
25. Sutskever, I., Vinyals, O., Le, Q.V.: Sequence to sequence learning with neural networks. In: Proceedings of NIPS (2014)
26. Tan, B., Zhang, Y., Pan, S.J., Yang, Q.: Distant domain transfer learning. In: AAAI (2017)
27. Vaswani, A., et al.: Attention is all you need. In: Proceedings of NIPS (2017)
28. Zoph, B., Yuret, D., May, J., Knight, K.: Transfer learning for low-resource neural machine translation. In: Proceedings of EMNLP (2016)

The Cognitive Features of Interface Language and User Language

Xi Luo[1], Lei Di[1], and Xiaohui Zou[2(✉)]

[1] Shanghai Shenyue Software Technology Co., Ltd.,
Room B10, North D., Bld. 8, No. 619 Longchang Rd., Shanghai 200090, China
m15002126553_1@163.com, 2093950@qq.com
[2] Sino-American Searle Research Center, Tiangongyuan, Paolichuntianpai
Building 2, Room 1235, Daxing, Beijing 102629, China
949309225@qq.com

Abstract. The purpose of this paper is to analyze the cognitive features of interface language and user language, clarify user requirements, and optimize interface design. The method steps are: First of all, clarify that interface is well-known as a display form that users can perceive through their own vision, hearing, operation, etc., as well as a direct way for interaction between software and users. Furthermore, to analyze the interface design and the multi-language features inherent in the user interface, the user language that needs to be converted at the first level is the user's personalized language. In general, it is necessary to understand what the user's inner true thoughts want; the second level of the user language that needs to be converted is the personalized interface that the user wants. However, at this time, the user's requirements are described by combining natural language, graphics, and even metaphor. The result is: Through the dual conversion from user language to interface language, the combination between user requirements and interface design has been clearly defined. At least, the requirements focus has been identified, like the difference between main and auxiliary functions. Grasp this point, a targeted and concise interface design is possible. Its significance lies in: It is not only conducive to the optimization of the specific software interface function design, but also helps people to further discuss the cognitive features of the interface language and the user language in theory.

Keywords: Interface · User language · Interface language

1 Introduction

The interface is not strange to most people. Whenever you touch a new piece of software, the first time the user interacts is the interface. Through the operation of the interface, user could achieve the purpose of use. But how can each interface highlight its intention? Deliver the using information to the users in which way? This paper will combine the user interface language and user language to explore and try to answer these questions.

Z. Shi et al. (Eds.): ICIS 2018, IFIP AICT 539, pp. 178–183, 2018.
https://doi.org/10.1007/978-3-030-01313-4_18

2   Interface Language

In our daily life, the first interaction with software is presumably the interface. A concise and effective operation interface will not only reduce the user's time for learning the operation, but also make the user really appreciate the convenience and ease of use. At present, most of the web interface or APP interface can highlight its existing significance. Just like the current popular software/APP like Taobao, Netease cloud music, WeChat and so on. Each interface is substantial without missing major functional points. Not all needs can be summed up in a natural language sentence. There are also transitions between language and language.

Nowadays there are many excellent interface designs. Some of them are simple and practical, for example WeChat, wallet finance, etc. In terms of interface design, users can understand at a glance how these interfaces are used and what functions they have. Each interface can be said to have its own value. However, in the beginning of the design process, the interface may not be like the final version handed to the users. The interface still needs to be polished with simplicity before being presented to the user. This is the interface language.

First of all, let's explain what interface language is. The interface language is the medium for human-computer communication and interaction. It can help people organize the logic from the visual interface information, complete the interaction with the interface, and ultimately achieve the purpose of use. ("How do you turn complex logic and information into a simple interface experience?" 2017). User can obtain the operating logic of the entire interface only by looking at it. A good interface language is relatively single and complex. The so-called "single" is to tell the user directly through a simple interface language what kind of problems the interface can help user solve. It saves the process of reading information and learning interface by users. The designer's main job is to inform the user of the simplest information and operation path. But why do we say that a good interface language is in another way complex? The interface language is actually the result of conversion based on the background logic and thinking layers. The step "how to present in a simple way" itself is complicated. It's necessary to really understand what the user really wants, so that the focus by the design could be identified. This understanding process is actually the process of understanding the user language.

3   User Language

The user language can tell what the user really wants. Most of the user languages are hidden. However, the vast majority of users who come into contact will speak out the so-called needs in most case. But they are all expressed by the user in an external language. For example: Colleague A in the office proposed to go to lunch together, and colleague B asked what kind of lunch to eat? Colleague A answered casually. Colleague B suggested: Let's eat noodles. Colleague A said: Just eaten yesterday, let's go to eat rice instead. In this case, colleague A could be seen as a user and colleague B as a designer. Only through external languages, users can also tap into potential internal language. Unlike external language, internal language is a self-questioning or non-verbal language activity ("Internal language" 2013). In actual life, people communicate with each other basically by external languages. External language is the speech process when people communicate with each other. External language consist of verbal and written language. Oral language is what people has to say to others. However, the grammatical structure and logic system of verbal speech is not required to be complete. When users describe their own needs, most of them will use gestures to explain. ("External language" 2016). But sometimes the external language can't express what the user really wants to express. This is the user's internal language. In fact, the internal language and user language have similarities. The internal language is a verbal process that does not play a role in communication, and it is a verbal activity when the individual thinks. Internal language and external language are inextricably linked. At first the internal language is produced, then it is transformed into an external language that needs to be elaborated. However, the transformation of internal language into external language is actually a shift from a brief summary of language. Therefore, sometimes there was such a phenomenon - the user's needs, through a simple understanding, into the actual operation of the interface, but we found that what users really want was not what we did. The root cause of the problem is the transformation from the internal language to the external language. How to solve this problem? These series of transformation process build essentially a tedious and long process.

At first let's consider user's external language. The people-to-people communication mentioned above is basically external language. But the process by which people use language to communicate is called speech. In the terminology, "language" and "speech" are two different concepts. What is the language? In our opinion, language and speech activities cannot be confused; it is only a definite part of speech activity and of course a major part. Language activities are multi-faceted and complex in nature, cover several fields like physical, physical, and psychological at the same time. They also belong to the fields of individuals and the community. ("Ordinary Linguistics Course" 1980) Speech actually refers to people's use of language. Language can be regarded as a tool used by people to think in communication. In the case of language, it is the use of this tool to conduct communication activities. There might exist some minor deviations between the user's language and speech in the process. All the things like sometimes inappropriate words, fast speech, etc. can cause the deviations. How to solve these deviations? A possible solution is the clever usage of pictures and images. Take a "Geomking – junior middle school plane geometry learning software" example, as shown below (Fig. 1).

This example is to prove that when one of the two intersecting lines is taken as the third straight line and is related to the interior angles on the same side, the basic graphic of the parallel lines related to the interior angles on the same side can be applied or added by the proof. Method: Make a parallel line to the straight line that intersects the third line. (Examples selected from Geomking – junior middle school plane geometry learning software)

In the Basic Graphic Analysis Method (Xu 1998), each interface is provided with vivid solution steps and graphic transformations. Students and teachers are users, they can see the changes in each step and changing process. It could help them to get a better understanding. That is the graphic-assisted explanation. However, it is not possible that all the words need to be explained in a graphic-assisted manner. There are other ways

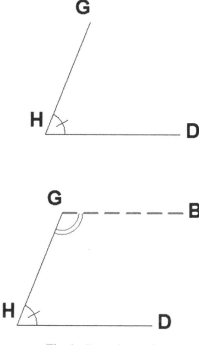

Fig. 1. Example graph

to give a visual solution. Similar to the example of colleague A and colleague B above, communicate with the user per language to see if it's possible to achieve what the user wants. Repeated communication, excavation at each level, confirmation, from the outside to the inside, so as to achieve the user's real inner needs, that is, the user's own internal language. This is the process by which we learn the internal language of the user from external utterances. After achieved the user's internal language, essentially the user's user language, we can then begin to translate the user language into an interface language.

The transformation of user language into interface language is essentially a presentation process from the psychological to the external, together with the multilingual transformation process. The comparison of this group of languages is similar to the internal and external language comparison. The difference is that the logic during the conversion from user language to interface language to is enhanced. Analyzing the user language is a crucial step by the conversion from the user language to interface language. It is related to which information the interface language should pass to the user. Without analyzing the results of the user language, it is impossible to design user-friendly interface. This is also the first difficulty in conversion from user language to interface language – highlighting the key points in the interface language. Most processes of analyzing the user language are silent, and some ideas even disappear. In fact, in this case it's possible to create a simple interface language draft - drawing. The user language could be expressed in the form of drawing. This process could simply the process of conversion into the interface language. Just by the interface language, in fact, there is also opposition - internally and externally. Most users read the interface information from the actual interface, that is, the interface language. With simple thinking logic, interactions can be quickly implemented, which is arguably the external interface language. This is the display of the interface language. Internal interface language is the realization of the interface logic - programming language. Once an interface language does not contain internal contents, it is an interface that can only transmit information and cannot interact. Such an interface could only be used to display. The user can only accept the information given by the interface, no way to interact with the interface. Only combined with programming language, the substantive interface could be presented to the user. The programming language highlights the rigorous logic in the interface language. Which button controls which features, data display, search filters, and so on. These logics cooperate with the interface language to pass the information to the user, after user receives the information transmitted in the interface language and make a relative reaction or operation. If there is no rigorous logic, the possible results are not consistent with what the user wants. This is not an incorrect display but a lack of rigorous logic. Interface language and programming language are also essentially inseparable and mutually dependent. For the user, the user language is internal and hidden; the interface language is visual and can convey information; the programming language is invisible and hidden. However, the concealment of the user language is different from the concealment of the programming language. The user's language is the speech process of the user's heart, and is hidden in thought; and the programming language can be said to be a series of program code written in languages such as java and C ++. It is invisible concealment. The entire process of converting from the user language to the interface language is actually interlocking and inseparable.

4   Conclusion

Combining the features of user language and interface language, the process of transformation is a multi-level transformation. The necessary associations exist between each language. Internal language - idea to external language - expression. Understand the user's internal language - idea to user language - requirements. User Language - Requirements to Interface Language - Presentation and Programming Language - Logic. The key is how to achieve the user's internal language and translate it into the real needs of the user. Only by grasping this point you can design a simple and clear interface that the user can understand at a glance what the interface wants to tell the user.

References

How to Turn Complex Logic and Information into a Simple Interface Experience? (2017) Selected from http://www.woshipm.com/pd/858699.html. This article mainly describes the definition of the interface language and the interface design direction

External Language (2016)

Selected from the Baidu Encyclopedia https://baike.so.com/doc/9324061-9660184.html mainly refers to the definition of the external language. And the difference between internal and external languages

Internal Language (2013)

Selected from the Baidu Encyclopedia https://baike.so.com/doc/6492691-6706401.html. Main reference to the definition of the internal language. And the difference between internal and external languages

General Linguistics Course (1980)

de Saussure, F., Bally, S., Schreiber, A., Riedlinger, A. (eds.): "General Linguistics Course" by. The Chinese translation was published in 1980

Geomking – Junior middle school plane geometry learning software (2007)

Selected from Shanghai Shenyue Software Technology Ltd. http://www.001jihe.net/Home/ShowBig2Index. The software mainly uses the pattern of graphic analysis to help teachers explain the problems of geometry and help students complete geometry exercises

Basic Graphic Analysis Method (1998)

Xu, F.: Elephant Publishing House (1998)

The Cognitive Features of Programming Language and Natural Language

Wen Xu[1], Fangqu Xu[1], Xiaohui Zou[2(✉)] (iD), and Zhenlin Xu[1]

[1] Shanghai Shenyue Software Technology Co., Ltd.,
Room B10, North D., Bld. 8, No. 619 Longchang Rd., Shanghai 200090, China
cindyxu@geomking.com, xufangqu@geomking.com, hyperien@qq.com
[2] Sino-American Searle Research Center,
Tiantongyuan, Paolichuntianpai Building 2, Room 1235, Daxing, Beijing 102629, China

Abstract. The purpose of this paper is to establish the mechanism of mutual transformation between programming language and natural language through the cognitive features of programming language and natural language which enables computers to read natural language and furthermore, to think like a human brain to achieve the purpose of human-computer interaction. The method is: First, establish a data dictionary, a vocabulary table, a sentence table and a bilingual table so that natural language and programming language can establish a one-to-one, one-to-many conversion relationship. Second, according to the data dictionary and bilingual table, establish the thinking tree model, and implement the traversal of natural language (a particular mathematical problem) through programming language. Third, establish a selection mechanism so that, through programming language, the computer knows how to make different choices and responses corresponding to different natural languages. The result is: Through the exploration of the cognitive features of programming language and natural language, establish the mechanism of the mutual transformation between programming language and natural language. The significance is: through the use of two different languages, it can establish the mechanism of human-computer interaction. For the one-to-one situations, it can be automatically converted based on rules. On the other hand, for the one-to-many situation, it can realize machine learning, moreover, deep learning based on statistics. Thus, it lays the foundation for the application of artificial intelligence in the field of education.

Keywords: Natural language · Programming language · Cognitive features
Artificial intelligence · Deep learning

1 Introduction

Artificial Intelligence is a new technical science for researching and developing theories, methods, technologies, and application systems for simulating, extending, and expanding human intelligence. Artificial intelligence is a branch of computer science. It attempts to understand the essence of intelligence and produces a new intelligent machine that can respond in a similar manner to human intelligence. Research in this

Z. Shi et al. (Eds.): ICIS 2018, IFIP AICT 539, pp. 184–190, 2018.
https://doi.org/10.1007/978-3-030-01313-4_19

area includes robots, language recognition, image recognition, Natural language processing, expert systems, etc.

The most essential attribute of human intelligence is thinking. This is because all human achievements in the process of social progress and inheritance, especially innovation and creative achievements, are all thought out and the results of thinking. Since the basis of thinking is language, and the process of thinking is also expressed through language, the study of human intelligence and the study of artificial intelligence will certainly bring language research a basic and key research role. It is not only an entry point, but also a breakthrough point.

The purpose of this paper is to establish the mechanism of mutual transformation between programming language and natural language through the their cognitive features, so that the computer can read natural language and further think like the human brain, thus achieving the purpose of human-computer interaction.

2   Cognitive Features of Programming and Natural Language

Intelligent machines that can respond in a similar manner to human intelligence must have their own language system. The source of this language system is natural language and programming language.

Natural language is the form that human uses to express the information that needs to be expressed in sound and words (the mother tongue of the country and the nation). Natural language is the most commonly used, popular, and important form of human expressions for thinking activity, thinking process, and thinking achievement. It is also the most basic and most important form of works publishing by scientists, experts, and scholars (Zhong 2004).

Taking the application of the "Thinking – student's thinking process visualization and evaluation system (Abbr. Thinking)" as an example, for any test question, in order achieve the goal of recording, display and evaluation of every single step by student's thinking process in computer, it's necessary that all the thinking nodes of this problem should be written out step by step, the situation that every thinking node student may think of should be listed in full or as much as possible. The result could build such a "Thinking Tree" map. Obviously at this time only natural language could be used (of course, including graphic language, symbolic language, etc.).
Example (Xu and Xu 2017):
In picture, parabola $y = ax^2 - 8ax + 12a$ ($a < 0$) intersects with X axis at points A, B (point A is at left side of point B). Another point C in parabola is in the first quadrant, so that $\angle ACB$ is a right angle, and $\triangle OCA \backsim \triangle OBC$.

1. Find the length of line segment OC
2. Find the parabolic function
3. Judge if there is point P in X axis, so that $\triangle BCP$ is an isosceles triangle? If it exists, please list all possible coordinates of P point. If it doesn't exist, please give the reason (Fig. 1).

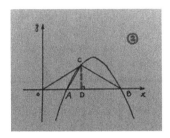

Fig. 1. Example figure

4. Only the solution idea needs to be answered

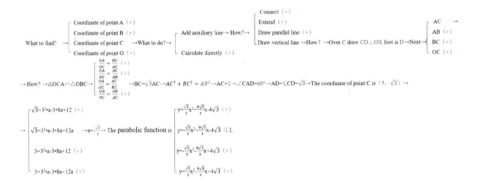

Where to start thinking? →
- Find coordinates of point A and B → How? → $ax^2 - 8ax + 12a = 0$ → $x_1 = 2$, $x_2 = 6$ → Coordinate of point A is (2,0), coordinate of point B is (6,0) →
- Find coordinate of point C (×)
- Find value of a (×)
- Find the vertex coordinates of the parabola (×)

→ OA=2, OB=6 → What should be used for following proof? → △OCA∽△OBC → $OC^2 = OA \cdot OB$ → $OC = 2\sqrt{3}$

5. Only the solution idea needs to be answered

What to find? →
- Coordinate of point A (×)
- Coordinate of point B (×)
- Coordinate of point C → What to do? →
  - Add auxiliary line → How? →
    - Connect (×)
    - Extend (×)
    - Draw parallel line (×)
    - Draw vertical line → How? → Over C draw CD⊥OX foot is D → Next →
      - AC
      - AB (×)
      - BC (×)
      - OC (×)
  - Calculate directly (×)
- Coordinate of point O (×)

→ How? → △OCA∽△OBC →
- $\frac{OA}{OC} = \frac{BC}{AC}$ (×)
- $\frac{OA}{OC} = \frac{AC}{AB}$ (×)
- $\frac{OA}{OC} = \frac{AB}{BC}$
- $\frac{OA}{OC} = \frac{AB}{AC}$ (×)

→ BC=√3AC → $AC^2 + BC^2 = AB^2$ → AC=2 → ∠CAD=60° → AD=1,CD=√3 → The coordinate of point C is (3, √3) →

→
- √3=3²·a-3·8a+12 (×)
- √3=3²·a-3·8a+12a → $a = \frac{\sqrt{3}}{3}$ → The parabolic function is
- 3=3²·a-3·8a+12 (×)
- 3=3²·a-3·8a+12a (×)

- $y = \frac{\sqrt{3}}{3}x^2 - \frac{8\sqrt{3}}{3}x - 4\sqrt{3}$ (×)
- $y = \frac{\sqrt{3}}{3}x^2 - \frac{8\sqrt{3}}{3}x - 4\sqrt{3}$ ①L
- $y = -\frac{\sqrt{3}}{3}x^2 - \frac{8\sqrt{3}}{3}x + 4\sqrt{3}$ (×)
- $y = \frac{\sqrt{3}}{3}x^2 - \frac{8\sqrt{3}}{3}x + 4\sqrt{3}$ (×)

6. Only the solution idea needs to be answered

Obviously, in the above example, the natural language can be used clearly, accurately, and step by step to express the thinking process. However, such language can not make the computer "understand", and can not be used to display the contents in computer directly.

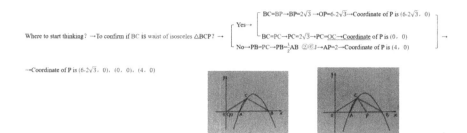

In order to let computer "understand" natural language, a language system should be established by computer so that natural language could be run and displayed. That is the programming language (Tucker and Noonan 2009).

Computers need to deal with and demonstrate a natural language system, and run a programming language system. Therefore, a common working mode of natural language and programming language on a computer at the same time is formed. Running both language systems in computer accurately, harmoniously and efficiently, achieving the goal of recording, displaying and evaluating the thinking process is obviously a challenge to human wisdom too.

Since natural language and programming language are used together on a computer, the first step is to establish the correspondence between natural language and programming language. The establishment and formation of this correspondence involves two aspects of the roadmap (Jurafsky and Martin 2005):

- In first way, establish a one-to-one correspondence, and then promote it to a one-to-many correspondence. For the already established one-to-one correspondence between natural language and programming language, it's possible to realize the automatic conversion by computer. And then on this basis the one-to-many correspondence between natural and programming language could be conversed automatically by computer.
- In second way, it starts from the "basic element" of natural language and programming language, constructs vocabularies and sentence tables, and bilingual tables based on established correspondences to form a data dictionary.

In "Thinking" software, the basic element is not only the word and sentence, but also the basic element of the thinking process. For example: by the thinking process of congruent triangles, there is a fixed thinking mode: in two triangles, if we can get three conditions that could lead to the congruent triangles judgement, we can prove that these two triangles are congruent triangles, and furthermore their corresponding edges or corresponding angles are proven to be equal. The whole part mentioned about congruent triangles builds then a basic element of thinking.

During the thinking of a problem, goes a wrong way or trial and error could often happen, and this also needs to be recorded, displayed and evaluated. So on the thinking node, it is necessary to design students' possible wrong thinking activities. Since the common mistakes that students could make in a certain thinking node are basically also determined, the choices designed on such thinking nodes can also become a basic element of thinking.

For example, in the above mentioned example we extract one choice option:

$$\text{What to find?} \quad \rightarrow \quad \begin{cases} \text{Coordinate of point A } (\times) \\ \text{Coordinate of point B } (\times) \\ \text{Coordinate of point C} \\ \text{Coordinate of point O } (\times) \end{cases}$$

Per natural language, it could be described like following:

What should be found in this case? There are four options: coordinate of point A, coordinate of point B, coordinate of point C, coordinate of point O, the correct answer is coordinate of point C.

Per programming language (take JavaScript as example), it could be described like following:

```
{
    code: 's010',
    title: 'What item should be found here',
    sType: 'S',
    reviewPoint: '',
    choices: [
        { match:'A', title: 'Coordinate of point A'},
        { match:'B', title: ' Coordinate of point B'},
        { match:'C', title: ' Coordinate of point C'},
        { match:'D', title: 'Coordinate of point D'},
    ],
    branches: [
        { match: 'C', correct: true, next: 's030' },
    ]
},
```

These two different languages, natural language and programming language, express the same meaning under different forms in different scenarios. Therefore, we can establish a correspondence between them.

After the data dictionary has been established and obtained, the second step is to establish the "thinking tree" model based on the data dictionary, so that every mathematical problem could be gone through by programming language.

The ability to evaluate the thinking process on computers is because we have built a "thinking tree" model. The thinking tree for every concrete mathematical problem starts with asking such questions: what nature does the thinking start from? What is the way to think about the question? What should the first step of the question be to seek (do)? Etc. This is the first thought node. In this thinking node, we can carry out a diffusion thinking, list all the possible situations that you think out, and then go into the second thinking node along each path, and the thinking process can be developed in turn until the problem is solved. In this process, a thinking tree is formed. Obviously, such a thinking tree is expressed by natural language.

The next step is to use programming language to go through all natural language used in the thinking tree. It's necessary to understand the natural language used in the thinking tree, especially the basic elements which formulate the thinking tree.

The third step is to establish a selection mechanism through programming and programming languages. The computer can "understand" the natural language used in the thinking tree through the correspondence between the programming language and natural language, and compare the various information expressed in the natural language and make a judgment, reaction and choice. In the "Thinking" software, the thinking tree is invisible to the teachers and the tested students, because the thinking tree has been compiled into the computer program through the programming language, but the computer can faithfully record every step of thinking of the tested student. Which step he thought of, thought wrong, what's the reason? What's the reason if he went around in circles and went back? As long as the students who have been tested have a thinking behaviour, there are records on the computer, and these records can be evaluated. Thus, the students' thinking process, thinking ability, thinking level and thinking quality can be evaluated.

3   Conclusion

The result of this paper is to establish the mechanism of mutual transformation between programming language and natural language through the research and exploration of the cognitive features of the programming language and natural language, and finally make it possible to evaluate the thinking ability based on the thinking process.

The significance of this paper is reflected in the promotion of educational development and reform. It will change the fundamental problem of traditional education for a long time, actually the evaluation of the education quality bases on the correctness of test answers solely. As long as the answer is correct, there is full score, whether the answer is thought out or remembered, even plagiarized. In the application of the "Thinking", even if the students could do every exercise in one test paper, let them test in the process of thinking, thought and not thought, think out and do not think out, one evaluation is clear, thus the evaluation could be developed from the evaluation of answer solely to the evaluation of process.

The basic thinking element, the thinking tree model and the data dictionary established in this paper can promote the computer to automatically generate new exercises bank and new thinking tree bank, and spread it into other subject areas so that a more complete automatic teaching system, automatic proposition system, automatic scoring system could be developed, thus the real education artificial intelligence could be realized.

By using two different languages together, a human-machine interaction mechanism can be established. Among them, one-on-one situations can be automatically converted based on rules, and one-to-many situations can be converted based on machine learning or even deep learning, so that a solid base for education artificial intelligence could be established.

References

Zhong, Y.: Comprehensive information based methodology for natural language understanding. J. Beijing Univ. Posts Telecommun. **27**(4), 1–12 (2004)

Jurafsky, D., Martin, J.H.: Speech and Language Processing. Publishing House of Electronics Industry, Beijing (2005)

Tucker, A.B., Noonan, R.E.: Programming Languages Principles and Paradigms, 2nd edn. Tsinghua University Press, Beijing (2009)

Xu, F., Xu, W.: Transparent Geometry - New Practice of Internet + Planar Geometry. Shanghai Education Publishing House, Shanghai (2017)

Ten-Years Research Progress of Natural Language Understanding Based on Perceptual Formalization

Peihong Huang[1(✉)], Guo-Lei Zheng[2], and Shilong Ma[3]

[1] Putian Branch of Agriculture Bank of China, Putian, China
pksc123@126.com
[2] School of Mechanical Engineering and Automation,
Beijing University of Aeronautics and Astronautics, Beijing 100191, China
[3] School of Computer, Beijing University of Aeronautics and Astronautics,
Beijing 100191, China

Abstract. This paper introduces the research progress of machinery truly understanding of natural language from three aspects. First, this paper explains why to carry out data or feature description by perceptual structure. Secondly, this paper summarizes the main understanding algorithms since the theory of machinery truly understanding has been proposed, and emphasizes the recent research progress. Finally, in view of current research status, this paper gives some research directions of natural language understanding in the future.

Keywords: P-semantics computation · Natural language understanding
Understanding mechanism · Semantics · Perception

1 Introduction

The problem of natural language understanding has appeared in the field of natural language processing for a long time. In 1949, Warren Weaver [1] of the United States put forward the idea of machine translation, IBM and other companies have followed up the research. But seven years later, in 1966, the American academy of sciences submitted an ALPAC report [2], put forward that machine translation encountered Semantic Barriers, and Semantic Barriers is essentially a natural language understanding problem. People hopes to deal with natural language through "understanding". Is there any way to translate raw text that is not understood into understandable structure? In this regard, many conceptual understanding [3–26] attempts have given positive answers and inspiration point.

At present, the perception theory is no longer confined to the field of biology, cognitive science [13, 27, 34] in the field of computer is known by many scholars, literature [28] since 2007 officially published and put forward the theoretical framework of natural language understanding, the process of perceptual formalization oriented natural language machine understanding research has gone through ten years, has published some representative papers [28–31]. In order to facilitate colleagues to quickly understand the relevant content, this paper gives the main results of real understanding of

Z. Shi et al. (Eds.): ICIS 2018, IFIP AICT 539, pp. 191–200, 2018.
https://doi.org/10.1007/978-3-030-01313-4_20

natural language, finally, the future research direction is analyzed, and the development trend of machinery understanding is discussed.

2   What is the Semantics

Semantic problem is a long-standing academic problem. For a long time, researchers have used concepts to define and represent meaning, but because of the understanding problem of concepts, the definition of semantics is not clear. Through the formalization of perception and based on the minimal element perception of feeling, this study constructs and defines semantics to solve the problem that semantics can not be expressed exactly for a long time.

In the 18th century, the British philosopher Hume [32] thought that experience is made up of perception, argued that "people are just a set of package of many different feelings, these feelings will always change at an almost unimaginable speed." One of the core ideas of Hume's theory is that causality is a constant conjunction between one thing and another, that inductive reasoning has extraordinary charm to foresee the future, and that "nature" as the answer to the problem solution is put forward. This study [28–31] is based on the perception and takes the natural perception image as the research object to establish the semantics of natural language. Different from the perception in the ordinary sense which is represented as corresponding concept, perception in this paper represents the element of feeling.

Definition 1 [33] (perception). Set p as a non-empty element, meet:

(1) p is a feeling;
(2) p is the smallest element;

Among them, the extraction of qualitative invariant part of the perception is identified as a perception p. p corresponding to the basic pattern m, then p is called as a perception. The qualitative part of perception does not have to be completely simulated, and the same purpose as simulation can be achieved by extracting the qualitative invariant part of perception with symbol identification and other means. Perception p is the smallest element of perception, that is, the unit of perception, corresponding to the basic pattern, can constitute the smallest element of meaning.

Definition 2 [28] (Semantics). Generally used S to represent semantics, set M as perceptual pattern set, set H as external stimulus. Make $H \subseteq R^D, M \subseteq R^d, D > d$, borrow the definition of the function, what M will be on the semantic effect of S can be represented by the following mapping:

$$\Phi : S \times H \rightarrow M,$$

Meet:

(1) for all $x \in H, m \in P^r$, $\Phi(s, x) = m$;
(2) $x \in H, x$ is the external identification of s.

Among them, the perceptual pattern set M includes the intrinsic perceptual pattern M_i and the acquired perceptual pattern set M_l, the perceptual element pattern $m_i \in M_i$, and the acquired perceptual pattern set M_l, which is a perceptual set composed of perception p, is introduced by formula (1), and the perceptual pattern set M corresponding to the external stimulus or external marker H is called as **Semantics** s, expressed as $S(H) = M$, or $S^H = M$.

Definition 3 [28] (Learning). For that cognitive system S^c, setting an instinctive mechanism η, judging truth value process $e(x)$, memory process $\chi(x)$, a related function rule set P^r, a trust set b_r, variable x corresponding perception set P^x, related perception map $P^x_m \subset P^x$ and a value detection process $\varepsilon(x)$, wherein the corresponding formula is as follows:

$$\eta = \varepsilon(x)\&e(x) \rightarrow \chi\left(P^r \cup P^x_m \cup b_r\right) \tag{1}$$

Perception is selected to perform machine understanding tasks because it can realize natural representation, In this paper, such perceptual semantic computation is referred to *p*-**Semantic Computation**. The formation process of perceptual semantics is that real-world images form perceptual pattern sets under the action of learning axioms and are marked with natural language symbols, thus forming language and its corresponding perceptual semantics. This is also the reason for using perception to express semantics. It is not clear that what the meaning of the cognitive object is caused by the semantic representation of symbols or concepts, so image perception calculation is needed to make up for the limitation of symbol calculation, and image thinking [34] and symbol logic thinking are needed to combine to solve the problem.

3　Mechanisms of Understanding

The complex problem of natural language understanding in psychology is far from clear enough to formalize the laws or processes of natural language understanding. The study [28–31] based on perception formalization constructs the understanding formula of natural language on the basis of what–why understanding effect to reveal the laws of natural language understanding to solve the problem.

3.1　Understanding

The understanding formula [28] is as follow:

$$b_t = u(x) = w\left(P^x_g, m^x\right) \tag{2}$$

Among them, the external stimulus is x, its corresponding to a certain perception pattern m^x and perception subset for P^x_g, sure feeling b_t, matching function w.

3.2 Comprehensive Understanding

For external stimulus x, the perception P produced for it, corresponding aggregatable perception set $(P_g^x)_i$ and perception pattern set $(M^x)_i (i = 1, 2, \ldots, n)$, then the comprehensive understanding of x is as follows:

$$u_c(x) = \prod_{i=1}^{n} u\left(\left(P_g^x\right)_i\right) = \prod_{i=1}^{n} w\left(\left(P_g^x\right)_i, (M^x)_i\right) \tag{3}$$

Proof: (omitted), see document [28] for relevant proof.

3.3 Understanding Effect

As we all know, the theory of physics is based on the laws of physics, which are verified by experiments. Perceptual-based natural language understanding theory drawn the conclusion of what–why understanding effect through experiments [31], then deduced the whole understanding theory, and proved the completeness theorem (ω-completeness) of the theory of natural language understanding.

What–Why Understanding Effect. Paper [31] gave assumptions and expectations, assuming that due to what–why factors cause understanding effect, so when the variable values added to the variables at not understanding state, expectations will lead to the expected understanding effect; If you don't add the variable values, you can't have an understanding effect. In variable control, this study controlled the factors such as what-variables, why-variables, true and false words and variable complexity, and took into account the operational definition of variables in the subjects. Through the understanding effect experiment, it is found that what-factor and why-factor jointly lead to the understanding effect in the natural language processing. See literature [31] for details of experiments, results, and discussion.

Reliability (ω-reliability). Perceptual-based natural language understanding study ensure its logic through the above what–why understanding effect verification and reliability(ω-reliability) proof, that is to say, for any external stimulus x, if it can be introduced by axiom system ω to be understandable, then it is correct that the external stimulus x can be understood. That is:

$$\{ \text{V}x \mid \text{if } \omega \Rightarrow U(P^x) \text{ is true, then } U(P^x) = 1 \} \tag{4}$$

See document [29] for the certification process.

Completeness (ω-completeness). Literature [29] had proved its completeness (ω-completeness), that is, to any external stimulus x, if P^x can be understood by the machine, its understanding will be introduced by the axiom system ω:

$$\{\text{V}x | if \ U(P^x) = 1, \ then \ \omega \Rightarrow U(P^x) \text{ is true}\} \tag{5}$$

Where P^x is represented the corresponding perceptual set (i.e., the poly-perception set) of x.

Proof: slightly, the proof process is shown in document [29].

Example 1. The following is the understanding of the stimulus "花Readers can use this example to deepen their understanding of the understanding definition.

(1) Firstly, the perception set of the "花shape matches with the perception pattern set in the cognitive system S^c to obtain semantics including the perception pattern set and the certainty feeling of its logo, so as to know what it is and why it is; (2) Matching and understanding each perceptual subset of these perceptual pattern sets to know what and why these perceptual subsets are; (3) Finally, matching and disjointing each perception element p, so as to know what the color 'blue' perception element is and why (the corresponding truth value); (4) The "花external stimulus is understood by means of a comprehensive matching and disjuncting so as to fully know what and why the various parts involved in the stimulus are.

Example 2 [30]. Analyzes the understanding process of "bright moon light in front of bed" below. This is a poem of the poet Li Bai of Tang Dynasty. The understanding process is omitted, see document [30] for detail.

4   Several Understanding Mechanism Related Algorithms

4.1   Text Understanding Algorithm for Language Machine Understanding

Algorithm 1. Text understanding algorithm [28] (Semantics Algorithm, SemA)

Input: sentences sequence set $X = \{x1, x2, \ldots, X_m \in R^D\}$

Output: $Y\{y1, y2, \ldots, yn \in R^d\}, \prod_{j=1}^{n} y_j = 1, D < d$

1. Sentence understanding. Matching words based on the possible poly-perception sets of sentences according to semantic constraints, the understanding result y_j^w of each poly-perception set, so that $\prod_{j=1}^{n_1} y_j^w = 1$. Understanding results of each poly-perception set at sentence level are y_j^s, meet: $\prod_{j=1}^{n_2} y_j^s = 1$.

2. Context understanding. Understanding results of each poly-perception set at sentence group level are y_j^c, the constraint was as follows: $\prod_{j=1}^{n_3} y_j^c = 1$.

3. Sentences arrangement. The understanding results of each poly-perception set at the level of generalization are y_j^k, and conform to the constraints $\prod_{j=1}^{n_4} y_j^k = 1$. See literature [28] for experiments, experimental results and discussions.

4.2 Pragmatic Meaning Derivation Algorithm of Natural Language Machine Understanding

Definition 4 [29] (Pragmatic Meaning). Set the context G, the context includes sentence set $s = \bigcup_{j=1}^{n} s_j$, $S_j \in G$, the semantic meaning of the sentence S_j in the process of use is $(P_h^s)_j$, called $(P_h^s)_j$ as the **Pragmatic Meaning** of the sentence S_j.

Set $S_j \in G$, the pragmatic meaning of a sentence S_j in the corresponding context can be uniquely determined:

$$\left(p_h^s\right)_j = \left(M^r - \bigcup_{i=1, i \neq j}^{n} (p^s)_i - \left(p_s^s\right)_j \right) \cup \left(p_t^s\right)_j \tag{6}$$

Proof: slightly, see document [29]. Wherein that meaning of the sentence is p^s, a set of relevant rules $M^r \subseteq S_c$. Pragmatic meaning $\left(P_h^s\right)$ and literal meaning $\left(P_s^s\right)$ are both semantics, and sentence semantics $S = P_h^s + P_s^s$, and both semantics S and *what–why* understanding effect are related.

Algorithm 2. Pragmatic meaning derivation algorithm [29] (PA)

Input: a set of linguistic material sequences G in context

Output: all sentences' periods within the context of G and their implication (Pragmatic Meaning) in the context G

(1) From the text understanding process to obtain the rules M^r.
(2) Get the location of the source sentence in the target rule M^r.
(3) From the corresponding relationship to obtain pragmatic meaning part p_h^s, corresponding relationship was in accordance with pragmatics-formula (6).
(4) Continue to perform steps 1–3, printed out p_h^s, then exit.
     Experiment (omitted), see literature [29] for experimental results and discussion.

4.3 Deductive Reasoning Algorithm Guided by Natural Language Understanding

Definition 5. Difficulty Element [33]. The so-called Difficulty Element, refers to the user, is the sentence or its elements which is difficult to solve and must be solved by reasoning, in this article Difficulty Element was agreed as dd.

For any conclusion K, if it can be expressed as reasoning sequence S_k of a fact F and the rule R_m, $F_j \in F$, if **testknown**(S_k) = True, the end of the solution. If **testknown**(x) = false, said x contains Difficulty Element dd, otherwise true. See reference [33] for algorithm, and its expression is as follows:

$$S_k = \begin{cases} \{K\}, & k=0 \ . \\ \{S_{k-1}, R_m\}, \text{testknown}(S_{k-1})=\text{False} \ \&\& \ k >=1 \ \&\& \ S_{k-1} \text{的} \ dd \notin F. \\ \{S_{k-1}, F_{j1}\}, \text{testknown}(S_{k-1})=\text{False} \ \&\& \ k >=1 \&\& \ S_{k-1} \text{的} \ dd \in F \ . \end{cases} \quad (7)$$

Relevant experiments, experimental results and specific discussions can be found in reference [33]. The above three real understanding algorithms are mainly evaluated by the what–why understanding effect. Understanding indicators are measured by understanding degree to see how much they have understood.

5 Related Work and Prospects

The main contribution of perceptual-based natural language understanding study is to give some law and theorem about what is understanding and what is natural language understanding. Paper [35, 36] is similar to the present study [28–31, 33, 37], linking perception with natural language to study the problem of language grounding, which further confirms the correctness and effectiveness of this research direction. The perception in the paper [35, 36] is still a conceptual level, not a perceptual unit.

There are two type of ways to classify natural language understanding researches at home and abroad:

(1) One is considered that understanding is the analysis of grammar, semantics and pragmatics, such as the system grammar [14], the case grammar [3, 4], the full information theory [15], and so on. The statistical method is essentially a lexical or syntactic analysis, which can be classified as such category. Winograd [14] completed the SHDRLU system in the closed building block world, using the system grammar within a limited vocabulary range, and the human-computer dialogue experiment had been successful.

(2) Another idea of understanding is that understanding is the mapping of concepts. Schank (1975) of Yale University in the United States and his colleagues put forward the concept dependency (CD) theory [5] that there is a conceptual basis in the human brain, and the understanding of natural language is the process of mapping concepts. Many typical theories of language understanding are affected by this idea, for example, WordNet [6–10], HNC [17], HowNet [19], ontology theory [11–13] and so on were based on conceptual understanding.

The above two methods can be attributed to the conceptual level of natural language understanding. The study of natural language understanding based on perceptual formalization is different from the above two methods. It is based on the smallest element of semantics, and based on the what–why understanding effect obtained by the physics method, is a true understanding of natural language. Natural language understanding [38] at conceptual level is different from real understanding.

In addition, this paper gives some natural language machinery understanding research clues for readers to provide reference.

5.1   The Natural Language Understanding Basis of Machine Translation

At present, more and more algorithms are devoted to the problem of intermediate language representation of data. The KBMT and KANT system [39] of Carnegie Mellon university is a knowledge-based translation system in a restricted intermediate language. At present, the system is the most important machine translation system using inter-language model translation method.

Inspired by the inter-language translation model, we can consider constructing a machine translation model based on perceptual semantics. Since the target text and the translated text have the same semantic meaning and the perceptual elements, we can take perceptual element as a sememe, use perceptual semantic set as inter-language. Paper [28] divides the sentences and words on the sentences' group, constructs understandable mature sentences, and further logically analyzes and generates the structure of the aggregative perception set; Paper [29] gives a study of semantic meaning representation for the dynamic sentence meaning and the derivability of sentence understanding in sentences group. By using these mathematical formulas, this idea of perceptual hierarchical representation can be combined with the existing inter-language translation mode.

5.2   Machine Learning Based on Natural Language Understanding

Big data provides a data basis for machine learning, in which Deep Learning generates hidden lay of neural network according to automatic method. Inspired by this, based on perceptual expression, perceptual element is used as hidden layer, Deep Learning research based on perceptual element is worth to be carried out. Papers [28, 29] provide relevant definitions, common premises and theorem proofs for representation, understanding and learning of machine understanding and learning system. Paper [30, 37] provides a machine understanding and learning method based on text analysis. On this basis, we can consider the further application of understanding theory in machine learning. Machine learning combined with perceptual elements and logic is the highest level of machine learning to reach the human level, is an important small data learning method in human-like learning.

5.3   The Physiological Basis of the Invariance of Perceptual Properties

In the process of formalization of natural language understanding, paper [28, 29] formalizes perception, separates the qualitative part of perception from the quantitative part, thus completing the formalization of language understanding process. In the formal process, the invariant qualitative part is represented by symbols, but the qualitative part belongs to the category of physiology, its research has physiological significance [29], this is also a very forward-looking topic which gives a possible way to achieve human longevity [40].

5.4   Visual Turing Test and Intelligence Definition

Problems of knowledge representation in traditional expert system and knowledge engineering are due to the fact that the relevant knowledge need to be used is often not enough. Paper [40] combined with visual Turing test, gave the proof framework of the relationship between intelligent definition and Turing test [38]. Therefore, we can consider scheme to improve the traditional expert system, on the basis of the structure of conceptual space, try to combine the theory of understanding, solve the problem of insufficient common sense, and solve the problem of knowledge representation. Of course, there are a lot of future research directions for natural language understanding based on perceptual formalization, here are just a few references for readers.

References

1. King, M.: When is the next Alpac report due? In: International Conference on Computational Linguistics and Meeting on Association for Computational Linguistics, pp. 352–353. Association for Computational Linguistics (1984)
2. ALPAC: Language and Machines: Computers in Translation and Linguistics. Nation Academy of Sciences, Washington D.C. (1966)
3. Fillmore, C.J.: Some problems for case grammar. In: 22nd Annual Round Table. Linguistics: Developments of the Sixties-Viewpoints of the Seventies, pp. 35–56. Georgetown University Press, Washington, D.C. (1971)
4. Zou, C.L., Cui, J.Y.: A study of category type logic based on case grammar. J. Anhui Univ. Philos. Soc. Sci. **38**(4), 15–22 (2014)
5. Schank, R.C.: Conceptual information processing. Inf. Process. Manag. **22**(86), 1–3 (1975)
6. Miller, G.A., Beckwith, R., Fellbaum, C., et al.: Introduction to WordNet: an on_Line lexical database. In: CSL Report, Princeton University, Princeton (1993)
7. Pan, Y.N., Teng, H.M.: Word sense disambiguation of tags in XML documents based on WordNet. Inf. Sci. **3**, 24 (2014)
8. Wei, T., Zhou, Q., Chang, H., et al.: A semantic approach for text clustering using WordNet and lexical chains. Expert Syst. Appl. **42**, 2264–2275 (2015)
9. Sharan, A., Joshi, M.L.: An algorithm for finding document concepts using semantic similarities from WordNet ontology. Int. J. Comput. Vis. Robot. **1**(2), 147–157 (2010)
10. Sotolongo, R., Yan, D., Hirota, K.: Algorithm for code clone refinement based on semantic analysis of multiple detection reports using WordNet. In: Proceedings of the 4th International Symposium on Computational Intelligence and Industrial Applications, pp. 191–198. Publishing House of Electronics Industry (2010)
11. Janzen, S., Maass, W.: Ontology-based natural language processing for in-store shopping situations. In: IEEE International Conference on ICSC 2009, pp. 361–366. IEEE (2009)
12. Minu, R.I., Thyagharajan, K.K.: Semantic rule based image visual feature ontology creation. Int. J. Autom. Comput. **11**(5), 489–499 (2014)
13. Lu, L.Q.: Knowledge Engineering and Science at the Turn of the Century, vol. 9, pp. 447–497. Tsinghua publishing house, Beijing (2001)
14. Winograd, T.: Language as a Cognitive Process. Addison Wesley, Boston (1983)
15. Zhong, Y.X.: Full information methodology for natural language understanding. J. Beijing Univ. Posts Telecommun. **4**, 1–12 (2004)
16. Liu, S.: Research and Answer System Based on the Full Information. Beijing University of Posts and Telecommunications, Haidian (2014)

17. Huang, Z.Y.: HNC theory and understanding of natural language statement. Chin. Basic Sci. **z1**, 85–90 (1999)
18. Liu, Z.Y.: Construction of HNC word knowledge base serving Chinese English Machine Translation. Lang. Appl. **1**, 117–126 (2015)
19. Dong, Z.D.: HowNet (2017). http://www.keenage.com
20. Ma, X.W.: Natural language understanding. Comput. Eng. Appl. **4**, 20–23 (1987)
21. Jia, Y.X., Yu, S.W., Zhu, X.F.: Research progress in automatic metaphor processing. Chin. J. Inf. **23**(6), 46–55 (2009)
22. Yu, S.W.: Natural language understanding and grammar research. In: Ma, Q.Z. (ed.) Introduction to Grammar Research, pp. 240–251. Commercial Press, Beijing (1999)
23. Wallace, R.S.: The anatomy of ALICE. In: Epstein, R., Roberts, G., Beber, G. (eds.) Parsing the Turing Test, pp. 181–210. Springer, Dordrecht (2009). https://doi.org/10.1007/978-1-4020-6710-5_13
24. Lu, R.Z.: Concept, semantic computation and connotation logic. In: Some Important Issues on Chinese Information Processing. Beijing: Science Press, pp. 90–99 (2003)
25. Li, Z.J., Lu, R.Z.: Research on the generation method of user requirement concept map. Comput. Appl. Softw. **29**(1), 23–26 (2012)
26. Chandrasekar, R.: Elementary? Question answering, IBM's Watson, and the Jeopardy! challenge. Resonance **19**(3), 222–241 (2014)
27. Shi, Z.Z.: Cognitive Science, pp. 1–591. University of Science and Technology of China press, Hefei (2008)
28. Huang, P.H.: Formalization of natural language understanding. Comput. Eng. Sci. **29**(6), 113–116 (2007)
29. Huang, P.H.: NLU-A logic theory on machine perceiving language of humankind. J. Sichuan Ordnance **30**(1), 138–142 (2009)
30. Huang, P.H.: A method of machinery inference learning based on text explaining. J. Chongqing Commun. Inst. **29**(6), 89–91 (2009)
31. Huang, P.H.: What–why effect on natural language understanding. Mod. Comput. Res. Dev. **10**, 9–16 (2016)
32. Hume: Human Understanding Study, pp. 19–66. Commercial Press, Beijing (1957)
33. Huang, P.H.: Study on cognitive system based on natural language understanding-algorithm foundation and construction of man-machine dialogue system. In: The 8th Symposium of Computer Science and Technology for ICT Graduate Students of CAS, pp. 23–24 (2004)
34. Qien, X.S.: About Thinking Science. Shanghai People's Publishing House, Shanghai (1986)
35. Jayant, K., Thomas, K.: Jointly learning to parse and perceive: connecting natural language to the physical world. TACL **1**, 193–206 (2013)
36. Emanuele, B., Danilo, C., Andrea, V., Roberto, B., Daniele, N.: A discriminative approach to grounded spoken language understanding in interactive robotics. In: Proceedings of the Twenty-Fifth International Joint Conference on Artificial Intelligence, pp. 2748–2753. Morgan Kaufmann, New York (2016)
37. Huang, P.: Exploration on causal law of understanding and fusion linking of natural language. In: Shi, Z., Goertzel, B., Feng, J. (eds.) ICIS 2017. IAICT, vol. 510, pp. 344–350. Springer, Cham (2017). https://doi.org/10.1007/978-3-319-68121-4_37
38. Morten, T.: The Imitation Game (2017). https://en.wikipedia.org/wiki/The_Imitation_Game
39. Feng, Z.W.: Machine translation's current situation and problems. In: Some Important Issues in Chinese Information Processing, pp. 353–377. Science Press (2003)
40. Huang, P.H., Zheng, G.L.: Study on definition of intelligence by combining with primitive semantics. Mind Comput. **1**(3), 193–202 (2010)

Learning Word Sentiment with Neural Bag-Of-Words Model Combined with Ngram

Chunzhen Jing[(✉)], Jian Li[(✉)], and Xiuyu Duan[(✉)]

{jcz,lijian,duanxy}@bupt.edu.cn

Beijing University of Posts and Telecommunications,
No. 10 Xitucheng Road, Haidian District, Beijing, China

Abstract. To better analyze the sentiment, attitude, emotions of users from written language, it is necessary to identify the sentiment polarity of each word not only the overall sentiment (positive/neutral/negative) of a given text. In this paper we propose a novel approach by using a method based on Neural Bag-Of-Words (NBOW) model combined with Ngram, aiming at achieving a good classification score on short text which contain less than 200 words along with sentiment polarity of each word. In order to verify the proposed methodology, we evaluated the classification accuracy and visualize the sentiment polarity of each word extracted from the model, the data set of our experiment only have the sentiment label for each sentence, and there is no information about the sentiment of each word. Experimental result shows that the proposed model can not only correctly classify the sentence polarity but also the sentiment of each word can be successfully captured.

Keywords: Word sentiment · Short text · Neural Bag-Of-Words Ngram

1 Introduction

Automatic sentiment analysis is a fundamental problem and one of the most active research areas in natural language processing (NLP) which has been widely used in data mining and text mining [18,20]. Detecting sentiment on short text such as reviews on certain product or exchanging information and opinions via short 200 words messages is becoming ubiquitous. There has been a large amount of research in this area of sentiment classification. Sentiment classification mainly focus on categorizing these texts in either two (binary sentiment analysis) or three (ternary sentiment analysis) categories, and this is an explicitly unordinal classification problem.

Neural network and deep learning have shown great promise in natural language processing (NLP) over the past few years. Examples are in semantic analysis [9], machine translation [1,4]. However many techniques of deep learning in

Z. Shi et al. (Eds.): ICIS 2018, IFIP AICT 539, pp. 201–210, 2018.
https://doi.org/10.1007/978-3-030-01313-4_21

sentiment classification suffer from over-abstraction problem [19], traditionally most of it has focused on classifying the text into several different categories, the only information obtained from these techniques is the polarity of the texts, and it's difficult to extract the sentiment knowledge more in depth, such as the sentiment of each word, i.e., positive intensity and negative intensity of a certain word.

In this paper, we propose a sentiment classification model based on Neural Bag-Of-Words (NBOW) [8] combined with Ngram, named NBOWN. The main advantage of the proposed model is its ability to extract the sentiment of each word in a text without explicit word-level polarity information. It identifies the words only by sentence-level polarity that is more abstracted but easier to availability.

In our model, each word is represented as a continuous-valued vector [3] and each sentence is represented as a matrix whose rows correspond to the word vector used in the sentence. Then, the model is trained using these sentence matrices as inputs and the sentiment labels as the output. Both the sentence-level polarity and words-level polarity for all words in the text can be extracted while the training, which helps us better understand the result of sentence-level sentiment classification.

The rest of the paper is organized as follows. First in Sect. 2 we discuss about the related works. In Sect. 3 we briefly introduce the NBOW model and present our proposed model, named as Neural Bag-Of-Words-Ngram (NBOWN) model, in Sect. 4. In Sect. 5, we give details about the data and the experiment setup. Section 6 gives experiment results and visualization of word sentiment performed by our model. Finally, we give our conclusions in Sect. 7.

2   Related Work

A variety of neural network architectures have been proposed for different language processing tasks. In sentiment classification, fully-connected feed forward neural networks [6], convolutional neural networks (CNN) [10,26] and also recurrent/recursive networks (RNN) [7] have been used. The CNN models are characterized by a set of convolution filters acting as a sliding window over the input sequence, which act as powerful n-gram feature extractors, typically followed by a pooling operation (such as max-pooling) [29] to generate a fixed-vector representation of the input sentence.

Recently, recurrent neural network architectures (RNNs) [17], such as long short term memory networks (LSTMs) [16] and Gated Recurrent Unit (GRU) [5], have received significant attention for various NLP tasks. However, the long term relationships captured well by LSTMs/GRU are of minor importance to the sentiment analysis of short texts. Even though the attention mechanism based on recurrent neural networks [27] can learn the task specific word importance, it doesn't explicitly model the sentiment polarity of each word in the text. Additionally, RNNs are much more computationally expensive, and both CNNs and RNNs require careful hyper-parameter selections and regularizations [28].

A Bag-Of-Words BOW represents text as a vector of word features such as word occurrence frequency and variants of term frequent-inverse document frequency known as tf-idf. BOW methods can be also applied in many areas [2,24]. With the development of neural network and deep learning based language processing, the syntactic and semantic characteristics of words and their surrounding context can be captured by using a more powerful continuous vector representation of words [3,12], such as word2vec [15], GloVe [21] and they outperform the count based word representation. The Neural Bag-Of-Words (NBOW) [8] model performs classification with an average of the input word vectors and achieves an impressive performance. We focus our model based on Neural Bag-Of-Words (NBOW) model.

3   Neural Bag-Of-Words (NBOW) Model

The NBOW model is a fully connected network, the input is an average of the d dimensional word vectors, for the words w in text X, corresponding vector v_w is looked up, and a hidden vector representation is obtained as follows:

$$s = \frac{1}{|X|} \sum_{w \in X} v_w \tag{1}$$

The average vector s is fed to a fully connected layer to estimate the probabilities for the output label as:

$$\hat{y} = softmax(Ws + b) \tag{2}$$

where $W \in R^{d \times K}$, K is class number, b is a bias vector and softmax is like follows:

$$softmax(q) = \frac{exp(q)}{\sum_{j=1}^{K} exp(q_j)} \tag{3}$$

For sentiment classification tasks the NBOW is trained to minimise the cross entropy loss using a gradient descent algorithm.

4   Proposed Model: Neural Bag-Of-Words-Ngram (NBOWN)

While the NBOW learns word vectors specialised for the sentiment classification task, and the overall sentiment of the sentence can be captured, it lacks to identify high-contributing words to classification results, and it cannot tell the sentiment of a certain word. This paper presents a novel approach for sentiment classification on short text. Both the importance and the contribution to each sentiment polarity of each word can be captured.

It is easy to realize that the NBOW model is essentially a fully connected feed forward network with a BOW input vector, and it is a unigram model which only the unigram pattern of the text is considered. Inspired by the powerful n-gram extractors in CNNs, We thus propose the Neural Bag-Of-Words-Ngram (NBOWN) model, with the motivation to enable the NBOW model to combine with the unigram, bigram and trigram knowledge of the text.

To get the impact of each words in a text on each sentiment polarity, we first map each word vector to a 3-dimensional vector, each dimension shows the sentiment of this word, which can be positive, neutral and negative. The method proposed in NBOW2 [23] model was used to let the model learn the word importance weights which are task specific, as [23] shows the word weights learned by the model achieve accuracy closer to tf-idf variants.

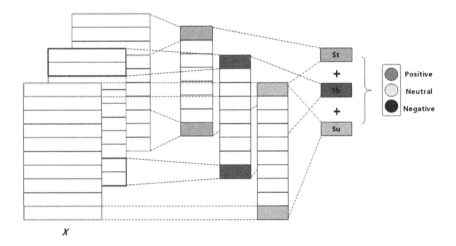

Fig. 1. Framework of proposed method

The unigram pattern score is a weighted average of the 3-dimensional vectors mapped from the word vectors.

$$s_u = \frac{1}{|X|} \sum_{u \in X} W_u v_u \alpha_u \tag{4}$$

where v_u is a unigram pattern of word vector v_w, and in unigram pattern they are equal, $W_u \in R^{d \times K}$ maps the d-dimensional vector v_u to a K dimensional vector, K is number of class, in our cases $K = 3$, the α_u are the scalar word importance weights for unigram pattern $v_u \in X$, α_u are obtained by introducing a vector a_u in the model, and are calculated as follows:

$$\alpha_u = f(v_u \cdot a_u) \tag{5}$$

where $v_u \cdot a_u$ represents a dot product between input vector v_u and vector a_u, and f scales the importance weights to range $[0, 1]$. In our model, the sigmoid function $f(t) = (1 + e^{-t})^{-1}$ is used.

To apply the ngram pattern to NBOW model, the bigram and trigram pattern of the text is used, like follows,

$$v_b = mean(v_{wi}, v_{wi+1}) \tag{6}$$

$$v_t = mean(v_{wi}, v_{wi+1}, v_{wi+2}) \tag{7}$$

$$mean(v_{wi}, v_{wi+1}, ...v_{wn}) = \frac{1}{n+1}(v_{wi} + v_{wi+1} + .. + v_{wn}) \tag{8}$$

the v_b is an mean value of the word vectors v_{wi} and v_{wi+1}, v_{wi} represents ith word vector in text X. The bigram pattern is an average of v_{wi} and its adjacent vector v_{wi+1}, the trigram pattern follows the same way. To address the sparse problem when introducing the ngram to the NBOW model, we use the same method of Eq. 5. The bigram/trigram pattern score is a weighted average of K dimensional vectors mapped from bigram/trigram patterns.

$$s_b = \frac{1}{|X| - 1} \sum_{b \in X} W_b v_b \alpha_b \tag{9}$$

$$s_t = \frac{1}{|X| - 2} \sum_{t \in X} W_t v_t \alpha_t \tag{10}$$

The α_b and α_t is a scalar scales the importance weights of a ngram pattern. And is calculated as follows:

$$\alpha_b = f(v_b \cdot a_b) \tag{11}$$

$$\alpha_t = f(v_t \cdot a_t) \tag{12}$$

For final result, like the Fig. 1 shows, the softmax function will get the probability estimates of scores get in ngram model.

$$\hat{y} = softmax(s_u + s_b + s_t) \tag{13}$$

And the sentiment distribution of ngram pattern in text X can be calculated as:

$$d = softmax(W_{ngram} v_{ngram}) \tag{14}$$

d is the sentiment distribution of certain ngram pattern in text X, and it is calculated as a softmax estimates of product W_{ngram} which can be either W_u, W_b, W_t and its corresponding ngram pattern v_{ngram}.

5   Experiment

To analyse and verify the proposed NBOWN model, we used publicly available Amazon Unlocked Mobile[1] and Twitter Airline review dataset,[2] both of the reviews written in English, Amazon Unlocked Mobile consists of review sentences and ratings from 1 to 5, 1 for very negative, 5 for very positive, Twitter Airline consists of review sentences and sentiment labels contains positive, neutral and negative. Both of the reviews in the datasets are short, and contain less than 200 words. We also make available the source code used in our experiments[3].

5.1   Data

In Amazon Unlocked Mobile, as shown in Table 1, the reviews with ratings smaller than or equal to 2 was used as negative examples, greater or equal to 4 as positive examples, 3 as neutral examples (Table 2).

Table 1. Rating number of Amazon Unlocked Mobile.

Rating	Number	Sentiment
1	72337	Negative
2	24728	
3	31765	Neutral
4	61392	Positive
5	223605	

Table 2. Sentiment of Twitter Airline.

Sentiment	Number
Negative	9082
Neutral	3069
Positive	2334

For training the NBOWN model, we randomly extract 15% of the original training set as the validation set and use remaining 85% as the final training set.

5.2   Word Embedding and Performance Measure

Each sentence was split into tokens using space, all tokens were used to learn the word embedding vectors. We fixed the embedding size to 100, and initialized the embedding layer using pre-trained GloVe, and because the embeddings learned in unsupervised phase contain very little information about sentiment of word [13,14], since the context for a positive word tends to be very similar to the context of a negative word, to add polarity information to the embeddings, we jointly trained the embeddings and the parameters of the model. Training was performed with the Adam gradient descent algorithm [11]. Additionally, early stopping [22] was used when the validation error starts to increase.

---

[1] https://www.kaggle.com/PromptCloudHQ/amazon-reviews-unlocked-mobile-phones/data.

[2] https://www.kaggle.com/c/twitter-airlines-sentiment-analysis/data.

[3] https://github.com/JingChunzhen/sentiment_analysis/tree/master/nbow.

6   Result

6.1   Classification Performance

We used several methodologies to comparison with NBOWN, CNN, bidirectional LSTM, bidirectional LSTM with Attention, NBOW and NBOW2. Three different window sizes 2, 3, 5 (how many words are considered in one receptive field) was used in CNN, while the number of filters was fixed to 128. For RNN models, the hidden size in LSTM unit was fixed to 128, with attention size to 50, dropout was added for both CNN and RNN models, and dropout rate [25] was set to 0.15.

The maximum number of words in Amazon Unlocked Mobile was set to 164, and in Twitter Airline was 34. Zero paddings were added if the length of the review was shorter than this number, whereas the last words were trimmed if reviews were longer than this number.

Table 3. The test accuracy between methodology

Test	Amazon	Twitter
CNN	0.9374	0.7884
bi-LSTM	0.9364	0.7742
bi-LSTM-Attention	0.9450	0.7783
NBOW	0.8853	0.7971
NBOW2	0.8867	0.7856
NBOWN	**0.9147**	**0.8013**

Table 3 shows the classification accuracies for several models. All the word vectors in the model was initialized by GloVe, and updated during the training. The NBOWN models achieved 91.47% in Amazon Unlocked Mobile, and achieved 80.13% in Twitter Airline. Higher than NBOW and NBOW2 methods. It is worth noting that the CNN and RNN based approaches operate on rich word sequence information and have been shown to perform better than NBOW approaches in Amazon Unlocked Mobile dataset. Because the length of reviews in Twitter is much shorter than reviews in Amazon, the RNN based approach didn't achieve very impressive result. And in Twitter Airlines, NBOWN was not far from CNN and LSTM methods.

6.2   Visualization of Words Sentiment

As Figs. 2, 4 and 6 show, the positive word thanks, great, awesome, faster and the negative words annoying, miss, awful, blurry can be well captured, Figs. 3, 5 and 7 show the sentiment trend with the increase of comments as each word comes in. Every point of the curve in Figs. 3, 5 and 7 is a sentiment score of positive/neutral/negative of the sentence of the current length, and is calculated as follows:

$$S_i = \frac{(i-1) * S_{i-1}}{i} + \frac{\alpha_i * p_i}{i} \quad (15)$$

where S_i represents the positive/neutral/negative score of review of current length i, α_i represents the importance of current ith word, and is calculated by Eq. 5, p_i represents the positive/neutral/negative score of the ith word.

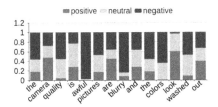

Fig. 2. Sentiment distribution of words in Twitter Airline

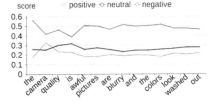

Fig. 3. Sentiment trend of a negative review in Twitter Airline

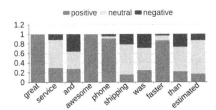

Fig. 4. Sentiment distribution of words in Amazon Unlocked Mobile

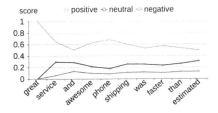

Fig. 5. Sentiment trend of a positive review in Amazon Unlocked Mobile

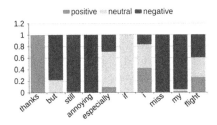

Fig. 6. Sentiment distribution of words in Amazon Unlocked Mobile

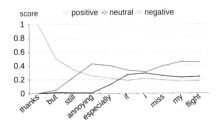

Fig. 7. Sentiment trend of a negative review in Amazon Unlocked Mobile

As Fig. 3 shows, at the beginning, because of the first word thanks, the sentence level sentiment is positive, the sentiment changes to negative when it encounters the negative word annoying, the negative curve goes a step higher when it encounters another negative word miss. Finally, the sentence level sentiment of this review is negative.

7   Conclusion and Future Work

In this paper, we propose NBOWN, a classification model based on Neural Bag-Of-Words combined with Ngram. Among the BOW methods, we achieved the best results, overall, we have some unique advantages: compared with CNN and RNN models, our model is much less computationally expensive, while the attention mechanism based on RNN model can only identifies the importance of each words in a text, we can successfully get the sentiment polarity of each words.

Although the experimental results were favorable, the current study still has some limitations, which lead us to future research directions. First our proposed method used a simple ngram pattern which is the mean value of the word vectors and its adjacent word vectors, the order of words was not considered. Second, we simply used a simple space-based token for training word vectors, the classification performance might be improved if more sophisticated preprocessing techniques are performed.

References

1. Bahdanau, D., Cho, K., Bengio, Y.: Neural machine translation by jointly learning to align and translate. Computer Science (2014)
2. Bai, Y., et al.: Bag-of-words based deep neural network for image retrieval. In: ACM International Conference on Multimedia, pp. 229–232 (2014)
3. Kandola, J., et al.: A neural probabilistic language model. In: Holmes, D.E., Jain, L.C. (eds.) Innovations in Machine Learning. Springer, Berlin Heidelberg (2006). https://doi.org/10.1007/3-540-33486-6_6
4. Cho, K., et al.: Learning phrase representations using RNN encoder-decoder for statistical machine translation. Computer Science (2014)
5. Chung, J., Gulcehre, C., Cho, K.H., Bengio, Y.: Empirical evaluation of gated recurrent neural networks on sequence modeling, Eprint arXiv (2014)
6. Collobert, R., Weston, J.: A unified architecture for natural language processing: deep neural networks with multitask learning. In: International Conference on Machine Learning, pp. 160–167 (2008)
7. Dong, L., Wei, F., Tan, C., Tang, D., Zhou, M., Ke, X.: Adaptive recursive neural network for target-dependent twitter sentiment classification. In: Meeting of the Association for Computational Linguistics, pp. 49–54 (2014)
8. Goldberg, Y.: A primer on neural network models for natural language processing. Computer Science (2015)
9. Hofmann, T.: Probabilistic latent semantic analysis. In: Proceedings of Uncertainty in Artificial Intelligence, vol. 41, no. 6, pp. 289–296 (2013)
10. Kim, Y.: Convolutional neural networks for sentence classification. Eprint arXiv (2014)

11. Kingma, D.P., Ba, J.: Adam: a method for stochastic optimization. Computer Science (2014)
12. Le, Q.V., Mikolov, T.: Distributed representations of sentences and documents, vol. 4, p. II–1188 (2014)
13. Li, Q., Shah, S., Fang, R., Nourbakhsh, A., Liu, X.: Tweet sentiment analysis by incorporating sentiment-specific word embedding and weighted text features. In: IEEE/WIC/ACM International Conference on Web Intelligence, pp. 568–571 (2017)
14. Maas, A.L., Daly, R.E., Pham, P.T., Huang, D., Ng, A.Y., Potts, C.: Learning word vectors for sentiment analysis. In: Meeting of the Association for Computational Linguistics: Human Language Technologies, pp. 142–150 (2011)
15. Mikolov, T., Yih, W.T., Zweig, G.: Linguistic regularities in continuous space word representations. In: HLT-NAACL (2013)
16. Mikolov, T., Joulin, A., Chopra, S., Mathieu, M., Ranzato, M.A.: Learning longer memory in recurrent neural networks. Computer Science (2014)
17. Mikolov, T., Karafiát, M. Burget, L., Cernocký, J., Khudanpur, S.: Recurrent neural network based language model. In: INTERSPEECH 2010, Conference of the International Speech Communication Association, Makuhari, Chiba, Japan, pp. 1045–1048, September 2010
18. Nadeem, M.: Survey on Opinion Mining and Sentiment Analysis. Springer, New York (2015)
19. Nasukawa, T., Yi, J.: Sentiment analysis: capturing favorability using natural language processing. In: International Conference on Knowledge Capture, pp. 70–77 (2003)
20. Pang, B., Lee, L.: Opinion mining and sentiment analysis. Found. Trends Inf. Retr. **2**(1–2), 1–135 (2008)
21. Pennington, J., Socher, R., Manning, C.: Glove: global vectors for word representation. In: Conference on Empirical Methods in Natural Language Processing, pp. 1532–1543 (2014)
22. Prechelt, L.: Automatic early stopping using cross validation: quantifying the criteria. Neural Netw. Off. J. Int. Neural Netw. Soc. **11**(4), 761 (1998)
23. Sheikh, I., Illina, I., Fohr, D., Linarès, G.: Learning word importance with the neural bag-of-words model. In: The Workshop on Representation Learning for NLP, pp. 222–229 (2016)
24. Ma, S., Sun, X., Wang, Y., Lin, J.: Bag-of-words as target for neural machine translation. ACL (2018)
25. Srivastava, N., Hinton, G., Krizhevsky, A., Sutskever, I., Salakhutdinov, R.: Dropout: a simple way to prevent neural networks from overfitting. J. Mach. Learn. Res. **15**(1), 1929–1958 (2014)
26. Wang, P., et al.: Semantic clustering and convolutional neural network for short text categorization (2015)
27. Yang, Z., Yang, D., Dyer, C., He, X., Smola, A., Hovy, E.: Hierarchical attention networks for document classification. In: Conference of the North American Chapter of the Association for Computational Linguistics: Human Language Technologies, pp. 1480–1489 (2017)
28. Zaremba, W., Sutskever, I., Vinyals, O.: Recurrent neural network regularization. Eprint arXiv (2014)
29. Zeiler, M.D., Fergus, R.: Stochastic pooling for regularization of deep convolutional neural networks. Eprint arXiv (2013)

Related Text Discovery Through Consecutive Filtering and Supervised Learning

Daqing Wu and Jinwen Ma[✉]

Department of Information Science, School of Mathematical Sciences and LMAM,
Peking University, Beijing 100871, People's Republic of China
wudq@pku.edu.cn, jwma@math.pku.edu.cn

Abstract. In a related or topic-based text discovery task, there are often a small number of related or positive texts in contrast to a large number of unrelated or negative texts. So, the related and unrelated classes of the texts can be strongly imbalanced so that the classification or detection is very difficult because the recall of positive class is very low. In order to overcome this difficulty, we propose a consecutive filtering and supervised learning method, i.e., consecutive supervised bagging. That is, in each consecutive learning stage, we firstly delete some negative texts with the higher degree of confidence via the classifier trained in the previous stage. We then train the classifier on the retained texts. We repeat this procedure until the ratio of the negative and positive texts becomes reasonable and finally obtain a tree-like filtering and recognition system. It is demonstrated by the experimental results on 20NewsGroups data (English data) and THUCNews (Chinese data) that our proposed method is much better than AdaBoost and Rocchio.

Keywords: Related text discovery · Text filtering · Imbalanced data
Bagging · Logistic regression

1 Introduction

Related text discovery or filtering is a process of matching an incoming text stream to a topic or profile of user's interests to detect or recommend the texts according to that topic or profile [1]. So, it is just a binary text classification which divides the input texts into two categories 'related' and 'unrelated'. In fact, this text filtering problem has been investigated from two different communities: machine learning (ML) and information retrieval (IR) [5].

In the IR community, most studies are based on Rocchio algorithm [6], which was developed under the framework of the vector space model. If all the texts are ranked for a query to the topic or profile, an ideal query should rank all the related texts above all unrelated texts. In Rocchio algorithm, an optimal query is defined to maximize the difference between the average scores of the related

Z. Shi et al. (Eds.): ICIS 2018, IFIP AICT 539, pp. 211–220, 2018.
https://doi.org/10.1007/978-3-030-01313-4_22

and unrelated texts. In this way, an optimal query vector is the difference vector of the centroid vectors for the related and unrelated texts: $Q_{opt} = \frac{1}{R} \sum_{i \in Rel} t_i - \frac{1}{N-R} \sum_{i \notin Rel} t_i$, where t_i denotes the weighted term vector for text i, $R = |Rel|$ is the number of related texts in Rel, and N is the total number of texts in the data set. All the negative components of the resulting query are assigned a zero weight. According to the actual performance, the recall of Rocchio algorithm is high, but its precision is very low, which means that the retrieved texts do not contain many related texts.

In the ML community, the most successful method is the boosting algorithm, especially the AdaBoost algorithm [9]. Schapire [7] proposed the boosting method and further proved that a weak learner can be turned into a strong learner in the sense of probably approximately correct learning framework. Actually,

AdaBoost is the most representative algorithm for text classification.

In fact, AdaBoost uses the whole training data to train each classifier serially, but after each round, it brings about more impact to difficult instances, with the goal of correctly classifying examples in the next iteration that are incorrectly classified during the current iteration. Hence, it gives more impact to examples that are harder to classify, the quantity of impact is measured by a weight, which is initially equal for all instances. After each iteration, the weights of misclassified instances are increased; on the contrary, the weights of correctly classified instances are decreased. Furthermore, another weight is assigned to each individual classifier depending on its overall accuracy which is then used in the test phase, i.e. more confidence is given to more accurate classifiers. Finally, when a new instance is submitted, each classifier gives a weighted vote, and the class label is selected by majority [8].

From the actual performance of the Adabost algorithm, most of retrieved texts are related, but the number of retrieved texts is very small, that is, it loses many texts that the user wants. In recent years, Liu et al. [2] constructed some different classifiers with contextual features to complete the ensemble learning for the text filtering; Lu et al. [4] utilized the RNN with attention mechanism and combine it; Kang et al. [3] built the text-based hidden Markov models and aggregated such models to obtain a final classifier. However, these methods have not paid attention to the ratio of the negative and positive examples.

Generally speaking, we want to know the conditional probability of the form:

$$Pro(y = 1|\mathbf{w}, \mathbf{x}) \tag{1}$$

from a set of training examples $D = \{(\mathbf{x}_1, y_1), (\mathbf{x}_2, y_2), \cdots, (\mathbf{x}_N, y_N)\}$.

For a given classifier, the parameters are generally estimated by the principle of maximum likelihood estimation (MLE):

$$\max L = \max \Pi_{i=1}^{N} (Pro(y = 1|\mathbf{w}, \mathbf{x}))^{y_i} (1 - Pro(y = 1|\mathbf{w}, \mathbf{x}))^{1-y_i} \tag{2}$$

If the number of the negative texts is far greater than that of the positive texts, we will get a very weak classifier by MLE. Therefore, the key step is to find a good ratio of the negative and positive texts. We will solve this problem by the consecutive filtering procedure with the supervised learning process.

2   Methodology

The concept of bootstrap aggregating (bagging) [10] is useful to construct the ensemble. It consists in training different classifiers with bootstrapped replicas of the original training data. That is, a new data is formed to train each classifier by randomly drawing (with replacement) instances from the original data (usually, maintaining the original data size). Hence, the diversity is obtained with the resampling procedure by the usage of different data subsets. Finally, when an unknown instance is presented to each individual classifier, a majority or weighted vote is used to infer the class membership. However, this random undersampling method ignores the supervised information in the training data. Moreover, this is a fusion method based on certain local classifiers. Here, we try to use the supervised learning method to consecutively filter out the negative texts with the higher degree of confidence so that the ratio of the negative and positive texts decreases stably. In this way, we bag the texts with a better ratio of the negative and positive ones consecutively and finally leads to a tree-like bagging or filtering and recognition system.

Unlike Bagging's classifier $H_T(x) = sign(\sum_{t=1}^{T} h_t(x))$, where $h_t(x)$ is the trained classifier in the tth stage.

We retain a series of classifiers: $H_T = \{h_1, h_2, \cdots, h_T\}$ with the corresponding thresholds $\Delta_T = \{\delta_1, \delta_2, \cdots, \delta_T\}$. In each previous or bagging stage, the classifier can remove the examples that has been classified into the unrelated texts with the threshold. In the final stage, the classifier makes the final decision of the related text. Algorithm 1 shows the pseudocode of the consecutive supervised bagging (CSB) algorithm.

Algorithm 1. Consecutive Supervised Bagging

Input: A set of positive class examples $\mathcal{P}$.
        A set of negative class examples $\mathcal{N}$.
        The finally ratio of negative class and positive class, α.
        Sampling speed factor, λ(generally choose 0.1).
1: $i = 1$
2: **repeat**
3:    $P_i = |\mathcal{P}|$, $N_i = |\mathcal{N}|$, $\alpha_i = \frac{P_i}{N_i}$
4:    Learn classifier h_i using $\mathcal{P}$ and $\mathcal{N}$
5:    Calculate the number of examples that will be removed from $\mathcal{N}$:

$$n_i = \text{int}(\lambda \cdot \alpha_i \cdot P_i)$$

6:    Through the classifier h_i, select the n_i^{th} small $Pro(y = 1|\mathbf{x}_i \in \mathcal{N})$ as threshold δ_i
7:    Remove from $\mathcal{N}$ all examples that satisfy $Pro(y = 1|\mathbf{x}_i \in \mathcal{N}) \leq \delta_i$
8:    $i \leftarrow i + 1$
9: **until** $\alpha_i \leq \alpha$
Output: An ensemble classifier $H_T = \{h_1, h_2, \cdots, h_T\}$with thresholds
        $\Delta_T = \{\delta_1, \delta_2, \cdots, \delta_T\}$($T$ is the number of iterations).

In Line 5, the number of removed examples n_i is changed, and decreases in the direct proportion to the ratio of the negative and positive data α_i in i^{th} iteration. In Line 6, we rank the prediction probabilities from small to large through the classifier h_i, and select the probability of n_i^{th} small negative example. The reason for Line 6 is that the positive examples are so few, and the number of positive examples which prediction probabilities are lower than δ_i is almost 0. When select δ_i as threshold, almost positive examples can be retained. In Line 7, the number of removed examples may be higher than n_i. Because, all negative examples that have the same probabilities equal to δ_i will be removed in i^{th} iteration. Through the CSB, we get $H_T = \{h_1, h_2, \cdots, h_T\}$ with thresholds $\Delta_T = \{\delta_1, \delta_2, \cdots, \delta_T\}$.

In Baaging, when an unknown instance is presented to each individual classifier, a majority or weighted vote is used to infer the class. But, we take another way to predict an unknown instance in CSB. In each iteration, instances which have been predicted to positive class through classifier h_{i-1} can be send to classifier h_i. So the finally results of instances is

$$y_i^{pred} = \begin{cases} 1, \text{if } i \in I^{(T)} \text{ and } y_i^{(T)} = 1 \\ 0, \qquad\qquad \text{otherwise} \end{cases} \tag{3}$$

where, T is the number of all iterations, $I^{(T)}$ is the instances index set in the Tth iteration and $y_i^{(T)}$ is prediction of $\mathbf{x}_i$ in the Tth iteration. This criterion means that the last prediction which is positive can be predicted positive. Figure 1 shows the process of prediction.

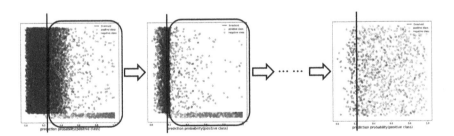

Fig. 1. The black vertical line represents the threshold in each classification. In the black box, all instances can be predicted by the next classifier.

Until now, we have not solved how to find a good ratio of the negative and positive examples. The Ratio Adjustment algorithm which is based on CSB can easily solve this problem. Algorithm 2 shows the pseudocode for the Ratio Adjustment.

Algorithm 2. Ratio Adjustment

Input: A set of positive class examples $\mathcal{P}$.
A set of negative class examples $\mathcal{N}$.
Sampling speed factor, λ(generally choose 0.1).
Adjust speed factor, μ(generally choose 0.1).
1: $i = 1$, $\alpha_i = 1$, marker$_i = 0$
2: **repeat**
3:    H_{T_i}, $\Delta_{T_i} = $ **Supervised Undersampling**$(\mathcal{P}, \mathcal{N}, \alpha_i)$
4:    **for all** $j = 1$ to T_i **do**
5:       $f_j = $ **Prediction**$(\mathcal{P} \cup \mathcal{N}, H_{T_i}, \Delta_{T_i}, j)$
6:    **end for**
7:    **if** $\{f_1, f_2, \cdots, f_j\}$ firstly increase, then decrease **then**
8:       $\alpha_i = \alpha_i + \mu$, marker$_i = 1$
9:    **else**
10:       $\alpha_i = \alpha_i - \mu$, marker$_i = -1$
11:    **end if**
12: **until** marker$_i$ · marker$_{i-1} = -1$
Output: The best ratio of positive class and negative class(in training data), α^*

In Line 5, f_j is f-measure of positive class. In Line 7–10, we make sure that the trend of $\{f_1, f_2, \cdots, f_{T_i}\}$ has only two ways. One is that f_j firstly increases, and then decreases. Another is that f_j increases. Because, the ratio determines the number of iterations. The ratio decreases, the number of iterations increases, more positive instances are judged to negative. Figure 2 shows this phenomenon.

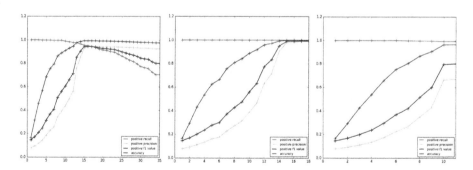

Fig. 2. The abscissa axis is the number of iterations T. Blue line is the positive class f-measure, red line is the positive class recall, green line is the positive class precision and black line is the accuracy. The left figure is $\alpha = 0.4$ and $T = 35$; The middle figure is $\alpha = 2.5$ and $T = 18$; The right figure is $\alpha = 5$ and $T = 11$. $\alpha = 2.5$ is appropriate, $\alpha = 5$ makes overly filtering and $\alpha = 0.5$ is just on the contrary.

3 Experimental Results

3.1 Data Sets

We use two standard text categorization test collections or data sets for our experiments. One is English texts–20NewsGroups, while another is Chinese texts–THUCNews.

1. **20NewsGroups** consists of roughly equal-sized samples of postings to 20 Usenet newsgroups. Some postings may belong to more than 1 newsgroup, and we treat the data as specifying 20 binary classification problems(One class as positive class, the others as negative class). We use the 'by date' training/test split (see http://people.csail.mit.edu/people/jrennie/20Newsgroups/) giving 11,314 training documents and 7,532 test documents.

2. **THUCNews** consists of 14 corpus. We treat the data as specifying 14 binary classification problems.

Table 1 shows the detail information of data sets.

Table 1. Basic information of 20NewsGroups and THUCNews

	20NewsGroups			THUCNews		
class	name	train	test	name	train	test
1	alt.atheism	480	319	sports	1000	500
2	comp.graphics	584	389	entertainment	1000	500
3	comp.os.ms-windows.misc	591	394	home	1000	500
4	comp.sys.ibm.pc.hardware	590	392	lottery	1000	500
5	comp.sys.mac.hardware	578	385	house property	1000	500
6	comp.windows.x	593	395	education	1000	500
7	misc.forsale	585	390	fashion	1000	500
8	rec.autos	594	396	politics	1000	500
9	rec.motorcycles	598	398	constellation	1000	500
10	rec.sport.baseball	597	397	games	1000	500
11	rec.sport.hockey	600	399	society	1000	500
12	sci.crypt	595	396	technology	1000	500
13	sci.electronics	591	393	stock	1000	500
14	sci.med	594	396	finance and economics	1000	500
15	sci.space	593	394			
16	soc.religion.christian	599	398			
17	talk.politics.guns	546	364			
18	talk.politics.mideast	564	376			
19	talk.politics.misc	465	310			
20	talk.religion.misc	377	251			

3.2 Text Representation

First, we break texts into tokens, that is, individual terms, such as words or phrases. Then, we tabulate the number of occurrences of each distinct term t in a text i and across texts. Finally, we compute a numeric weight for each term with respect to each text.

Now, we represent each text as a vector of such weights. We use a form of TF-IDF (term frequency times inverse document frequency)[11] term weighting.

$$TF_{ij} = \frac{n_{i,j}}{\sum_{i=1}^{|V|} n_{i,j}} \quad IDF_i = \log \frac{|D|}{|\{j : t_i \in d_j\}|}$$

$$TFIDF_{i,j} = TF_{i,j} \times IDF_i$$

where, $n_{i,j}$ is the number of occurrences of term i (term i is in vocabulary V.) in a text j, $|V|$ is the total number of vocabulary V (the length of vocabulary can be choose with feature selection.), $|D|$ is the number of training texts, $\{j : t_i \in d_j\}$ is the number of texts that contains term i.

3.3 Feature Selection

The chi-squared measure [12] chooses the features that are least independent from the class label and is widely used in text categorization. This measure is based on a 2×2 contingency table between a predictor term t and a predicted category label s. Let a be the number of training texts in s and containing t, let b be the number in s but not containing t, let c be the number not in s but containing t, and let d be the number neither in s nor containing t. Let $n = a + b + c + d$ be the total number of training texts. Then the chi-squared measure is

$$\chi^2 = \frac{n(ad - bc)^2}{(a + b)(c + d)(a + c)(b + d)}$$

Table 2. Confusion matrix

	Positive class	Negative class
Retrieval	True positive (TP)	False positive (FP)
Not retrieval	False negative (FN)	True negative (TN)

3.4 Performance Measures

From confusion matrix (Table 2), we have following measures:

$$recall = \frac{TP}{TP + FN} \quad precision = \frac{TP}{TP + FP}$$

$$f\text{-}measure = \frac{2 \times recall \times precision}{recall + precision}$$

The recall measures the ratio of correctly classified positive instances and all positive instances. The precision measures the ratio of correctly classified positive instances and all instances which are classified as positive. The f-measure metric combines precision and recall as a single measurement to indicate the effectiveness of a classifier [13].

For multiple classification, we usually use *macro-recall*, *macro-precision* and *macro-f*:

$$macro\text{-}recall = \frac{1}{n}\sum_{i=1}^{n} recall_i \quad macro\text{-}precision = \frac{1}{n}\sum_{i=1}^{n} precision_i$$

$$macro\text{-}f = \frac{2 \times macro\text{-}recall \times macro\text{-}precision}{macro\text{-}recall + macro\text{-}precision}$$

3.5   Parameter Setting

In our experiments, we choose logistic regression as basic classifier with log loss function and L1 penalty. In Algorithms 1 and 2, Sampling speed factor $\lambda = 0.1$ and Adjust speed factor $\mu = 0.1$. In vocabulary, we choose 5000 terms with high chi-squared measure.

3.6   Performances and Comparisons

From Table 3, we can see that our method is faster than Rocchio and AdaBoost in training and testing. Because, as the number of iterations increases, the number of retained data which will used to train classifier decreases.

The experimental results well reflect the properties of Rocchio and AdaBoost: Rocchio method's recall is high, but the precision is very low. AdaBoost method's precision is high, but the recall is low. In Table 4, the results show that our proposed model's macro-recall is a little lower than Rocchio, but the macro-precision is significantly lager than Rocchio in 20NewsGroups. Although, our model is better than Rocchio and AdaBoost in 20NewsGroups, there are still some classes that are not very good, like class 2,12,19,20. we guess some instances (postings) can belong to more than one class (1 newsgroup). In THUCNews, we found that our proposed model outperformed the others in the index of macro-recall, macro-precision and macro-f.

Table 3. The comparison of running time(second)

Data name	Training			Testing		
	Rocchio	Adaboost	Ours	Rocchio	Adaboost	Ours
20NewsGroups	975.92	1003.52	**891.30**	82.43	81.72	**80.43**
THUCNews	853.81	946.80	**846.71**	56.84	55.70	**49.74**

Table 4. The testing results of 20NewsGroups

	Recall			Precision			F-measure		
Class	Rocchio	Adaboost	Ours	Rocchio	Adaboost	Ours	Rocchio	Adaboost	Ours
20NewsGroups									
1	0.7492	0.4953	**0.8433**	0.4121	0.7822	**0.8054**	0.5317	0.6065	**0.8239**
2	**0.8740**	0.4139	0.7789	0.2794	0.6518	**0.7519**	0.4234	0.5063	**0.7652**
3	0.9213	0.4137	**0.9264**	0.1899	0.6443	**0.9288**	0.3148	0.5039	**0.9276**
4	**0.8801**	0.3776	0.8316	0.3200	0.6091	**0.7837**	0.4694	0.4661	**0.8069**
5	**0.9013**	0.5351	0.8883	0.3788	0.8110	**0.8486**	0.5334	0.6448	**0.8680**
6	**0.9316**	0.5544	0.9266	0.2471	0.7110	**0.9196**	0.3907	0.6230	**0.9231**
7	**0.9308**	0.5513	0.9282	0.4421	0.8238	**0.9258**	0.5995	0.6605	**0.9270**
8	**0.9141**	0.5000	0.9040	0.5552	0.7586	**0.8585**	0.6908	0.6027	**0.8807**
9	0.9347	0.7312	**0.9397**	0.7686	**0.9268**	0.9078	0.8435	0.8174	**0.9235**
10	**0.8992**	0.6574	0.8615	0.5107	0.7885	**0.8301**	0.6515	0.7170	**0.8455**
11	0.9674	0.7794	**0.9975**	0.6380	0.9482	**0.9925**	0.7689	0.8556	**0.9950**
12	**0.8283**	0.3232	0.5903	0.2965	0.5100	**0.5370**	0.4251	0.3956	**0.5624**
14	0.8359	0.5581	**0.8510**	0.4117	**0.8435**	0.8425	0.5517	0.6717	**0.8467**
15	0.9086	0.7107	**0.9365**	0.5967	0.9003	**0.9111**	0.7203	0.7943	**0.9237**
16	0.8970	0.7638	**0.9548**	0.5518	0.8889	**0.9429**	0.6833	0.8216	**0.9488**
17	0.8819	0.5687	**0.9341**	0.3929	0.6656	**0.8924**	0.5436	0.6133	**0.9128**
18	0.7846	0.6569	**0.9096**	0.8967	**0.9216**	0.9096	0.8369	0.7671	**0.9096**
19	**0.7516**	0.4194	0.6484	0.3807	0.6075	**0.6401**	0.5054	0.4962	**0.6442**
20	0.7092	0.2869	**0.7211**	0.2132	0.4865	**0.5710**	0.3278	0.3609	**0.6373**
macro-	**0.8626**	0.5505	0.8610	0.4584	0.7599	**0.8314**	0.5987	0.6384	**0.8460**
THUCNews									
1	0.9360	0.9680	**1.0000**	0.8357	0.9878	**0.9980**	0.8830	0.9778	**0.9990**
2	0.9580	0.8900	**0.9980**	0.8523	0.9488	**0.9980**	0.9021	0.9185	**0.9980**
3	**0.9260**	0.6680	0.9160	0.6044	0.8743	**0.9034**	0.7314	0.7574	**0.9160**
4	0.9060	0.9400	**0.9920**	0.9340	0.9792	**0.9861**	0.9198	0.9592	**0.9890**
5	0.9160	0.9460	**1.0000**	0.6050	0.9854	**1.0000**	0.7287	0.9653	**1.0000**
6	0.5980	0.6580	**0.8920**	0.8214	**0.9139**	0.8124	0.6921	0.7651	**0.8503**
7	0.9680	0.9080	**0.9820**	0.5538	0.9153	**0.9840**	0.7045	0.9116	**0.9830**
8	0.9420	0.6980	**0.9640**	0.6937	0.9160	**0.9659**	0.7990	0.7923	**0.9650**
9	0.9980	0.9960	**1.0000**	0.9559	0.9746	**1.0000**	0.9765	0.9852	**1.0000**
10	0.9040	0.8880	**0.9860**	0.8794	0.9548	**0.9554**	0.8915	0.9202	**0.9705**
11	0.9420	0.6740	**0.9700**	0.5621	0.8686	**0.9700**	0.7040	0.7590	**0.9700**
12	**0.9640**	0.8180	0.9600	0.6667	0.9191	**0.9449**	0.7882	0.8656	**0.9524**
13	0.9740	0.8620	**0.9980**	0.5650	0.8620	**0.9881**	0.7151	0.8620	**0.9930**
14	0.7980	0.6980	**0.9060**	0.7151	**0.8596**	0.8118	0.7543	0.7704	**0.8563**
macro-	0.9093	0.8294	**0.9689**	0.7318	0.9257	**0.9513**	0.8109	0.8749	**0.9600**

4   Conclusion

We have established a consecutive supervised bagging method to related or topic-based text discovery that is adaptive for the dataset without any human intervention. Related or topic-based text discovery is generally a strongly imbalanced classification problem which is very challenging in both machine learning and information retrieval. Our proposed consecutive supervised bagging method can effectively solve this problem from the consecutively decreasing the ratio of the negative and positive texts with the supervised learning method. The experimental results on 20NewsGroups data (English data) and THUCNews (Chinese data) demonstrate that our proposed method is much better than AdaBoost and Rocchio.

Acknowledgment. This work is supported by the Natural Science Foundation of China for Grant 61171138. We also acknowledge Zhengzhou Shuneng Science and Technology Limited Company for the contribution of the data set THUCNews.

References

1. Soboroff, I., Nicholas, C.: Combining content and collaboration in text filtering. In: IJCAI 1999 Workshop: Machine Learning for Information Filtering, pp. 86–91 (1999)
2. Liu, Y., Jiang, C., Zhao, H.: Using contextual features and multi-view ensemble learning in product defect identification from online discussion forums. Decis. Support. Syst. **105**, 1–12 (2018)
3. Kang, M., Ahn, J., Lee, K.: Opinion mining using ensemble text Hidden Markov Models for text classification. Expert. Syst. Appl. **94**, 218–227 (2018)
4. Lu, Z., Liu, W., Zhou, Y., et al.: An effective approach for Chinese news headline classification based on multi-representation mixed model with attention and ensemble learning. In: National CCF Conference on Natural Language Processing and Chinese Computing, pp. 339–350 (2017)
5. Schapire, R.E., Singer, Y., Singhal, A.: Boosting and Rocchio applied to text filtering. In: Proceedings of the 21st Annual International ACM SIGIR Conference on Research and Development in Information Retrieval, pp. 215–223 (1998)
6. Rocchio, J.J.: The SMART Retrieval System: Experiments in Automatic Document Processing. Relevance Feedback in Information Retrieval, pp. 313–323 (1971)
7. Schapire, R.E.: The strength of weak learnability. Mach. Learn. **5**, 197–227 (1990)
8. Galar, M., Fernandez, A., Barrenechea, E., et al.: A review on ensembles for the class imbalance problem: bagging-, boosting-, and hybrid-based approaches. IEEE Trans. Syst., Man, Cybern. Part C (Appl. Rev.) **42**(4), 463–484 (2012)
9. Freund, Y., Shapire, R.E.: Experiments with a new boosting algorithm. In: 13th ICML, pp. 148–156 (1996)
10. Breiman, L.: Bagging predictors. Mach. Learn. **24**, 123–140 (1996)
11. Salton, G., Buckley, C.: Term-weighting approaches in automatic text retrieval. Inf. Process. Manag. **24**(5), 513–523 (1988)
12. Zheng, Z., Wu, X., Srihari, R.: Feature selection for text categorization on imbalanced data. ACM SIGKDD Explor. Newslett. **6**(1), 80–89 (2004)
13. Rong, T., Gong, H., Ng, W.W.Y.: Stochastic sensitivity oversampling technique for imbalanced data. In: Wang, X., Pedrycz, W., Chan, P., He, Q. (eds.) ICMLC 2014. CCIS, vol. 481, pp. 161–171. Springer, Heidelberg (2014). https://doi.org/10.1007/978-3-662-45652-1_18

Natural Language Semantics and Its Computable Analysis

Zhao Liang[1($\boxtimes$)] (iD) and Chongli Zou[2]

[1] Tianjin University of Finance and Economics, Tianjin 300222, China
zlcass@163.com
[2] Chinese Academy of Social Sciences, Beijing 100005, China
chlizou@263.net

Abstract. With an illustrative example of $5 \times 3 = 5 \times 3$, the first section explained the major problem in the natural language interpretation process. The second section introduced Frege's famous idea of 'sense'. And this paper argues that Frege's definition is not constructive. The Third section contributes a subtle amendment to the syntax of IL (Intentional Logic). Traditionally, the three basic types e, t and s were not treated as a part of vocabulary of IL, while this paper argues that this treatment has allowed a much explicit way to coding programs in real algorithm. Given this treatment, "an expression α of type τ" is shortened to "$\alpha|\tau$". In the end of section three, this paper gave a natural language sentence to illustrate the intention and extension operators. And furthermore, there comes the analyses of information preservation ability of these operators.

Keywords: Sense · Reference · Meaning · Intention · Extension

1 Introduction

The concept of artificial intelligence has been around ever since the advent of computer. While it is not until GPU is exploited in the deep learning algorithm at the beginning of the 21st century, did study on artificial intelligence boomed like an overnight sensation. Thanks to the development of computer science and engineering, even private laptops are quite efficient to run a variety of artificial intelligence algorithms on them. On the other hand, there is a treasure trove of information stored on the internet publicly to feed the data-starved learning machines all over the globe. With the right tool and limitless materials, the prosperity of artificial intelligence cannot be failed.

It seems that artificial intelligence serves as the Holy Grail of almost everything, but it is actually not. Artificial intelligence is catching up with human beings in more ways than one, especially in pattern recognition of voices, faces, handwritings etc. The best AI does even better than the best human board game

Supported by The National Social Science Fund of China.

master. But even the smartest machine can barely understand human language properly. Human language, also known as natural language, is a promising candidate for the last citadel of Homo sapiens' half ludicrous, half pathetic sense of superiority over machine. Speaking a foreign language is seemingly not so hard as becoming a chess master, and speaking mother language is even much easier and more natural, which is the name, natural language, came after. While nine-year-old children could regale themselves with tall tales, the bedtime stories still do not make much sense to the best AI. Natural language processing with statistical methods and machine learning algorithms is a successful part of artificial intelligence, whereas there are inherent defects in this magic touch. The crux of the problem resides in the interpretation of expressions, which statistical methods alone do not help so much, because natural language is intentional rather than extensional.

Mathematics is the absolute epitome of extensional language. $5 \times 3 = 3 \times 5$ is an expression declares the equality between two terms. This expression hold true, because both 5×3 and 3×5 refers to the same number 15. However, for a five years old boy, he has three apples in five bags each, and he is sure all these apples is all these apples, that is $15 = 15$, but he may not sure how many larger bags will do to take all these apples with five apples each. He might be surprised that three larger bags will just do to take all these apples, because $5 \times 3 = 3 \times 5$ update his knowledge with new information. This boy knows that all these apples are all these apples, that is $5 \times 3 = 5 \times 3$, and both 5×3 and 3×5 refer to the same number 15, thus $5 \times 3 = 3 \times 5$ contribute nothing new information to $5 \times 3 = 5 \times 3$. The question is that where the new information came from that updated the boy's knowledge. It is the meaning of 5×3 that differs from the meaning of 3×5, which in the knowledge context that gave rise to the new information. The context where the equality hold true, whenever the replacement of same reference happens, is an extensional context. The context where equality failed by replacement of expressions with same reference is an intentional context. In most cases, natural language is applied in an intentional context. As long as intentional context cannot be took into account properly, artificial intelligence will stay artificial retarded.

2   The Meaning of Natural Language

This $5 \times 3 = 3 \times 5$ is essentially raised the question of what identity should be treated in the context of cognition. The seminal work of Frege [1] preferred the items on both sides of the equal sign treated as names or signs of objects rather than the object itself. In programming language, C for example, there is an obvious difference between the assignment operator "=" and the equality operator "==" [2]. But the assignment versus equality issue is quite another different story. Practically both assignment and equality refer to the items as the objects themselves. In the statement of "a = b", a is assigned the same information stored on the storage devices as b has already designated, but not the other way round. If a and b has already labeled the same information stored on the storage devices, "a = b" could be removed from the code without messing up the programming anyway. And "a == b" is only served as the condition checking of the fact that a and b has already labeled the same information on the storage devices. Both the assignment operator "=" and the equality operator "==" has nothing to do with the cognitive significance in the case of natural language which was so-called Frege's Puzzles as in the case of $5 \times 3 = 3 \times 5$. Actually the Frege's Puzzle raises everywhere in natural language:

– Mark Twain is Samuel Clemens.
– Rachel is Bill's sister.
– Beijing is the capital of China.

Historically, the most famous example is the following two sentences [3]:

S1: The Morning Star is the Morning Star.
S2: The Morning Star is the Evening Star.

The ancient Greeks noticed 'two' bright stars in the sky at different times, which turned out to be the same star we now known as planet Venus. Both The Morning Star and the Evening Star refers to Venus. If the meaning of a word is all about its reference, the two sentences reduce to one:

Venus is Venus.

While it is obvious that S2 giving more information than S1. Anyone without any astronomical knowledge could agree with S1. It should be noted that to agree with S2 one does not have to know both The Morning Star and the Evening Star refers to planet Venus, they just need the confidence of the fact that the two stars refers to the same thing. While reference is important to meaning, there is more than that. There comes the 'sense' that Frege used to describe the meaning of a word other than its reference. "the Morning Star" and "the Evening Star" both refer to the same thing, while they do have different senses. When it comes to the equation $5 \times 3 = 3 \times 5$, the two different senses are quite obvious: the sense of 5×3 is 'three times five', and the sense of 3×5 is 'five times three'.

Thanks to Frege's remarkable insights, it is much clear that there is something in the meaning which is quite different from its reference. However, Frenge's definition of sense is not constructive. His main idea is that there are two parts of meaning, and the sense is the rest part of meaning that the reference part can not help.

SENSE = MEANING – REFERENCE

'Meaning' is an umbrella term, which contains more than sense and reference. A sentence could be uttered more than once. Any particular utterance may have its own programmatic importance. It should be noted that all the programmatic role of 'Meaning' is beyond this paper. The 'meaning' discussed in this paper has nothing to do with emotion and imperative motives in any particular utterance of a sentence.

3 The Intensional Logic

Intensional logic is one of the most promising schemes to define sense constructively. The syntax of Intensional logic is practically Propositional calculus alongside the type theory part [4].

IL (Intensional Logic) is a Lambda Calculus with constants. The semantics of IL is practically an embedded part of IL which cannot be changed and no need to change. Every syntactic well formed expression is fixed with a semantic interpretation. The IL here will be furnished with a modal operator and two tense operators. The intension of an expression α will be interpreted in terms of possible worlds and possible times [5].

Definition 1. *The <u>vocabulary</u> of IL*

(1) An countable infinite set **VAR** $= v_1, v_2, v_3, v_4, \cdots$
(2) An nonempty set **CON**
(3) e, t, s
(4) $\neg, \vee$
(5) $\forall$
(6) $(, \langle, |,), \rangle$
(7) λx, for every $x \in$ **VAR**

It should be noticed that "λx" as a whole is a symbol in the vocabulary of IL. The x in "λx" can never be treated as a variable in **VAR**. Actually it is much less misleading to use "λ_x" instead of "λx". The nonempty set **CON** is the set of all constants. When possible worlds and possible times are concerned **CON** are bound to be uncountably infinite simply because the moments on the time line is as uncountably infinite much as the real numbers.

Definition 2. *The <u>type part</u> of an expression of IL*

TP refers to the set of all type parts. And then a member of set TP can be derived by applying the following rules finitely:

(1) $e \in TP$, and $t \in TP$
(2) If $\tau_1 \in TP$, and $\tau_2 \in TP$, then $\langle \tau_1, \tau_2 \rangle \in TP$
(3) If $\tau \in TP$, then $\langle s, \tau \rangle \in TP$

Definition 3. *The <u>expression</u> of IL*

WE refers to the set of all well-formed expressions. Then a member of set WE can be obtained by applying the following rules finitely:

(1) If $a \in TP$, then $v|a$ and $c|a$ are expressions, where $v \in \mathscr{V}$ and $c \in \mathscr{C}$.
(2) If $\alpha|\langle \tau_1, \tau_2 \rangle$ and $\beta|\tau_1$ are expressions, then $(\alpha \ \beta)|\tau_2$ is an expression.
(3) If $\alpha|\tau_1$ is an expression and $x|\tau_2$ is an variable, $\lambda x \alpha|\langle \tau_2, \tau_1 \rangle$ is an expression.
(4) If $\phi|t$ and $\psi|t$ are expressions, then so are $(\neg \phi)|t$ and $(\phi \vee \psi)|t$.
(5) If $\phi|t$, and $x|\tau$ are expressions, where x is a variable, then $\forall x \phi|t$ is an expression.

(6) If $\alpha|\tau_1$ and $\beta|\tau_2$ are expressions, then $(\alpha = \beta)|t$ is an expression.

(7) If $\phi|t$ is an expression, then $\Box\phi|t$ is an expression.

(8) If $\phi|t$ is an expression, then $\mathbf{F}\phi|t$ is an expression.

(9) If $\phi|t$ is an expression, then $\mathbf{P}\phi|t$ is an expression.

(10) If $\alpha|\tau$ is an expression, then $^\wedge\alpha|\langle s, \tau\rangle$ is an expression.

(11) If $\alpha|\langle s, \tau\rangle$ is an expression, then $^\vee\alpha|\tau$ is an expression

Then "$\alpha|\tau \in WE$" means "$\alpha|\tau$ is an expression". τ is the **type part** of the expression $\alpha|\tau$, and α the **expression part** of the expression $\alpha|\tau$. The last two rules introduce the intentional operator and extensional operator respectively. $^\wedge\alpha|\langle s, \tau\rangle$ could serve as the intention of α, and $\alpha|\langle s, \tau\rangle$ the extension of α.

WE_τ refers to the set of all the expressions with type part τ. VAR_τ refers to the set of all the expressions with expression part as a variable and type part as τ. CON_τ refers to the set of all the expressions with expression part as a constant and type part as τ.

The semantics of IL could simply be defined step by step along with the reductive definition of syntax. The first step is to built the domain D_τ of every type τ. Next step is to specify all the constants with valuation function f. Finally for any assignment function g there is an interpretation function $\| \ \|^g$ which interpreted all the well-formed expressions of IL.

Epistemology is the fancy name for the theory of knowledge, which gave rise to a whole bunch of intentional contexts. And epistemic logic is the logic of knowledge, which could be treated as a part of Epistemology. While initially the study of intentional logic begins with modal logic rather than epistemic logic. And Kripke semantics still plays a major role in the intentional logic. That's because the intention and extension of an expression can be treated as two operators to form new expressions $^\wedge\alpha$ and $^\vee\alpha$ respectively, where the explanation of an intentional expression is modeled in Kripke's possible-world logic. The same expression could be interpreted into different means in different conditions. This ingenious idea is largely credited to Montagueşáŕs work, and Morrill [6] "gave a technical refinement in this idea".

The building blocks of semantics of IL are four non-empty sets:

s individuals	A
truth values	$\{0, 1\}$
possible worlds	W
possible times	T

With these original building blocks it is possible to build any domain D_τ with type τ by recursive definition on types.

type	domain
t	$D_t = \{0, 1\}$
e	$D_e = A$
$\langle \tau_1, \tau_2 \rangle$	$D_{\langle \tau_1, \tau_2 \rangle} = D_{\tau_2}^{D_{\tau_1}}$
$\langle s, \tau \rangle$	$D_{\langle s, \tau \rangle} = D_\tau^{W \times T}$

Let the model is set with A, $\{0,1\}$, W, T, and the *universal valuation function* f. Given any *universal assignment function* g, and then it is all set to interpret every well-formed expressions of IL. Interpretation function $\| \ \|^{w,t,g}$ is defined recursively over expressions. The routine procedure given as follows [6]:

(1) If $c \in CON$, then $\|c\|^{w,t,g} = f(c)(\langle w,t \rangle)$
It should be noticed that the interpretation of a constant has nothing to do with the assignment function g.

(2) If $x \in VAR$, then $\|x\|^{w,t,g} = g(x)$
It should be noticed that the interpretation of a variable has nothing to do with the spacetime $\langle w,t \rangle$.

(3) If $\alpha | \langle \tau_1, \tau_2 \rangle$ and $\beta | \tau_1$, then $\|(\alpha \ \beta)\|^{w,t,g} = \|\alpha\|^{w,t,g}(\|\beta\|^{w,t,g})$

(4) If $\alpha | \tau_1$ and $x | \tau_2$, $\|\lambda x \alpha\|^{w,t,g} \in D_{\tau_1}^{D_{\tau_2}}$, and $m \in D_{\tau_2}$, then $\|\lambda x \alpha\|^{w,t,g}(m) \in D_{\tau_1}$, such that

$$\|\lambda x \alpha\|^{w,t,g}(m) = \|\alpha\|^{w,t,g^{[x \mapsto m]}}$$

where

$$g^{[x \mapsto m]}(y) = \begin{cases} g(y) & y \neq x \\ m & y = x \end{cases}$$

(5) If $\phi | t$, and then

$$\|\neg \phi\|^{w,t,g} = \begin{cases} 1 & \|\phi\|^{w,t,g} = 0 \\ 0 & \|\phi\|^{w,t,g} = 1 \end{cases}$$

(6) If $\phi | t$ and $\psi | t$, then

$$\|(\phi \vee \psi)\|^{w,t,g} = \begin{cases} 1 & else \\ 0 & \|\phi\|^{w,t,g} = 0 \ and \ \|\psi\|^{w,t,g} = 0 \end{cases}$$

(7) If $\alpha | \tau$ and $\beta | \tau$, then $(\alpha = \beta)|t$ such that

$$\|(\alpha = \beta)\|^{w,t,g} = \begin{cases} 1 & \|\alpha\|^{w,t,g} = \|\beta\|^{w,t,g} \\ 0 & else \end{cases}$$

(8) If $\phi | t$, and $x | \tau$, where x is a variable, then

$$\|\forall x \phi\|^{w,t,g} = \begin{cases} 1 & else \\ 0 & \exists m \in D_\tau \ such \ that \ \|\phi\|^{w,t,g^{[x \mapsto m]}} = 0 \end{cases}$$

(9) If $\phi | t$, then

$$\|\Box \phi\|^{w,t,g} = \begin{cases} 1 & for \ all \ \langle w',t' \rangle \in W \times T \ such \ that \ \|\phi\|^{w',t',g} = 1 \\ 0 & else \end{cases}$$

(10) If $\phi | t$, then

$$\|\mathbf{F}\phi\|^{w,t,g} = \begin{cases} 1 & \exists t' > t \ such \ that \ \|\mathbf{F}\phi\|^{w,t',g} = 1 \\ 0 & else \end{cases}$$

(11) If $\phi | t$, then

$$\|\mathbf{P}\phi\|^{w,t,g} = \begin{cases} 1 \ \exists t' < t \ \text{such that} \ \|\mathbf{P}\phi\|^{w,t',g} = 1 \\ 0 \ \text{else} \end{cases}$$

(12) If $\alpha | \tau$, then $\|^{\wedge}\alpha\|^{w,t,g} \in D_{\tau}^{W \times T}$ such that

$$\|^{\wedge}\alpha\|^{w,t,g}(\langle w', t' \rangle) = \|\alpha\|^{w',t',g}$$

(13) If $\alpha | \langle s, \tau \rangle$, then $\|^{\vee}\alpha\|^{w,t,g} \in D_{\tau}$ such that

$$\|^{\vee}\alpha\|^{w,t,g} = \|\alpha\|^{w,t,g}(\langle w, t \rangle)$$

The last two items interpret the intentional operator and extensional operator respectively. $^{\wedge}\alpha$ refers to the "intention" of α, and $^{\vee}\alpha$ the "extension" of α. It shows that the interpretation of $^{\wedge}\alpha$ has nothing to do with the spacetime $\langle w, t \rangle$. In other words, the interpretation of $^{\wedge}\alpha$ is always the same regardless of possible worlds and times. And the interpretation of $^{\wedge}\alpha$ is a function defined on $W \times T$, which practically multiplied all the spacetimes in each and every spacetime. Initially, suppose there were n spacetimes in the model, whenever "intention" is concerned it is actually n^2 spacetimes in picture. The interpretation of $^{\vee}\alpha$ has simply the opposite effect, such that only each of the spacetime $\langle w, t \rangle$ in spacetime $\langle w, t \rangle$ is left.

It is convenient to go back to the simplified model to illustrate the major point. Let $W = \{w_1, w_2\}$ and $T = \{t_1, t_2\}$. And then there are only four spacetimes in this model:

$$\frac{(w_1, t_2) | (w_1, t_1)}{(w_2, t_1) | (w_2, t_2)}$$

Actually, any specific wold or time is not concerned. The four different spacetimes could simply be treated as four possible worlds. And then the model is $\langle A, \{0, 1\}, W, f \rangle$, where $W = \{w_1, w_2, w_3, w_4\}$:

$$\frac{w_2 | w_1}{w_3 | w_4}$$

Let $\|\alpha\|^{w_1,t_1,g} = x_1$; $\|\alpha\|^{w_1,t_2,g} = x_2$; $\|\alpha\|^{w_2,t_1,g} = x_3$; $\|\alpha\|^{w_2,t_2,g} = x_4$, and then the interpretation of α could be illustrated as

$$\frac{x_2 | x_1}{x_3 | x_4}$$

Then the interpretation of $^{\wedge}\alpha$:

$$\frac{\frac{x_2 | x_1}{x_3 | x_4} \ \| \ \frac{x_2 | x_1}{x_3 | x_4}}{\frac{x_2 | x_1}{x_3 | x_4} \ \| \ \frac{x_2 | x_1}{x_3 | x_4}}$$

The process from $^\wedge\alpha$ to $^{\vee\wedge}\alpha$:

$$\frac{\dfrac{(x_2)\,|\,x_1}{x_3\,|\,x_4}\,\Big\|\,\dfrac{x_2\,|\,(x_1)}{x_3\,|\,x_4}}{\dfrac{x_2\,|\,x_1}{(x_3)\,|\,x_4}\,\Big\|\,\dfrac{x_2\,|\,x_1}{x_3\,|\,(x_4)}} \;\rightarrow\; \frac{x_2\,|\,x_1}{x_3\,|\,x_4}$$

As a matter of fact it has illustrated $\|^{\vee\wedge}\alpha\|^{w,t,g} = \|\alpha\|^{w,t,g}$ in a small-scale case.

It is much convenient and intuitive to illustrate this counterexample in diagrams. This counterexample is an simplified model with only four spacetimes which could be labeled with A, B, C and D:

$$A = \langle w_1, t_1\rangle, B = \langle w_1, t_2\rangle, C = \langle w_2, t_1\rangle, D = \langle w_2, t_2\rangle$$

And then the diagram could looked like this

$$\frac{B\,|\,A}{C\,|\,D}$$

Suppose A, B, C and D are four different countries with different currencies. The following sentence could simulate the expression in the counterexample.

currency has purchasing power

In different countries the "currency" refers to different currencies. For instance, in country A the "currency" refers to "currency of country A". It is true that currency of country A has purchasing power in country A, while it is false in country B, C and D. A fat stack of cash of country A does you no good in country B, C and D. This interpretation of "currency" is only the case of country A:

$$\frac{0\,|\,1}{0\,|\,0}$$

If the interpretation of "currency" in country B, C and D is included, the bigger picture looks like this:

$$\frac{\dfrac{1\,|\,0}{0\,|\,0}\,\Big\|\,\dfrac{0\,|\,1}{0\,|\,0}}{\dfrac{0\,|\,0}{1\,|\,0}\,\Big\|\,\dfrac{0\,|\,0}{0\,|\,1}}$$

As sentence "currency has purchasing power" simulates the expression α, the following diagram could illustrate the application of extensional operator and intentional operator successively:

$$\|\alpha\|\frac{\dfrac{1\,|\,0}{0\,|\,0}\,\Big\|\,\dfrac{0\,|\,1}{0\,|\,0}}{\dfrac{0\,|\,0}{1\,|\,0}\,\Big\|\,\dfrac{0\,|\,0}{0\,|\,1}} \rightarrow \|^\vee\alpha\|\frac{1\,|\,1}{1\,|\,1} \rightarrow \|^{\wedge\vee}\alpha\|\frac{\dfrac{1\,|\,1}{1\,|\,1}\,\Big\|\,\dfrac{1\,|\,1}{1\,|\,1}}{\dfrac{1\,|\,1}{1\,|\,1}\,\Big\|\,\dfrac{1\,|\,1}{1\,|\,1}}$$

It is clear that the interpretation of $^{\wedge\vee}\alpha$ is quite different from the interpretation of α. Extensional operator was applied first, and then the interpretation of $^\vee\alpha$ shows that it is true in all the four countries. And the sentence "currency has purchasing power" does seem like a plain truth. However, the truth is not the whole truth. It is essentially lost some information in the extensional interpretation, and the following intentional interpretation could not restore it. The lost information is just the intention of the sentence, i.e. the currency of a particular country has purchasing power in this particular country. The intentional interpretation practically multiplied the whole piece of information faithfully in every spacetimes, and thus immuned to extensional interpretation once. Thus the interpretation of $^{\vee\wedge}\alpha$ was expected to be just the same as the interpretation of α.

4 Conclusion

The introduction part explains the crux of the problem in natural language processing. The example of $5 \times 3 = 5 \times 3$ takes most of the credit to illustrate this problem. With this illustrative example this paper has clarified the distinction between the assignment operator "=" and the equality operator "==" has nothing to with the cognitive significance of sense. The Third section deals with syntax and semantic of IL. It has showed that the three basic types e, t and s could be included in the vocabulary of IL. The result is that "an expression α of type τ" [7] is shortened to expression "$\alpha|\tau$", where α is the *expression part* and τ the *type part*. And then the following recursive definitions became explicit and economic. This amendment has allowed intentional logic much explicit to coding programs in real algorithm. In the end of section three, this paper gave an natural language sentence to illustrate the intention and extension operators. And it shows that intentional interpretation preserved the whole piece of information in the meaning, whereas extensional interpretation are bound to lose some. In this sense Intentional Logic is not only an expansion of Extensional Logic, but also enjoys much power of interpretation.

References

1. Frege, G.: Sense and reference. Philos. Rev. **57**(3), 209–230 (1948)
2. Damas, L.: Type assignment in programming languages. The University of Edinburgh (1984)
3. Kratzer, A., Heim, I.: Semantics in Generative Grammar, vol. 1185. Blackwell, Oxford (1998)
4. Zhao, L., Zou, C.L.: Intentional logic as a possible scheme to sematics computation. J. Chongqing Univ. Technol. (Soc. Sci.) **32**(6), 9–10 (2018)
5. Morrill, G.V.: Type Logical Grammar. Kluwer Academic Publishers, Dordrecht (1994)
6. Morrill, G.V.: Categorial Grammar. Oxford Univ. Press, New York (2011)
7. Gamut, L.T.F.: Logic, Language, and Meanimg. The University of Chicago Press, Chiacago and London (1991)

Can Machines Think in Radio Language?

Yujian Li[(✉)]

College of Computer Science, Faculty of Information Technology,
Beijing University of Technology, 100 Pingleyuan, Beijing 100124, China
liyujian@bjut.edu.cn

Abstract. People can think in auditory, visual and tactile forms of language, so can machines principally. But is it possible for them to think in radio language? According to a first principle presented for general intelligence, i.e. the principle of language's relativity, the answer may give an exceptional solution for robot astronauts to talk with each other in space exploration. This solution implies a great possibility to realize high-level machine intelligence other than brain-like intelligence.

Keywords: The principle of language's relativity · First principle
Radio language · Space exploration

1 Introduction

In computer science, one of the biggest unsolved problems is to develop intelligent machines. Since a seminal paper by Turing on the topic of artificial intelligence [1], the central question "Can machines thinks?" began to excite interest in building systems that learn and think like people. That is a fascinating dream! Recently, the interest has been renewed again because of impressive progress with deep learning [2], in spite of great difficulties such as the Character Challenge and the Frostbite Challenge [3], to perform a variety of tasks as rapidly and flexibly as people do.

What does it mean for a system to learn and think like a person? Lake et al. argued that this system should build causal models of the world, ground learning in intuitive theories of physics and psychology, and harness compositionality and learning-to-learn [3]. They claimed that these key ideas of core ingredients would play an active and important role in producing human-like learning and thought. Undoubtedly, their claim is attractive and promising for the ultimate dream of implementing machines with human-level general intelligence. However, the claim says little about a person's ability to communicate and think in natural language, which is clearly vital for human intelligence [4]. The question is, how to develop a capacity of language for machines? In this paper, it will be discussed and answered from a first principle for general intelligence, i.e. the principle of language's relativity. According to the answer, an exceptional solution may be obtained for robot astronauts to talk each other in space exploration.

Z. Shi et al. (Eds.): ICIS 2018, IFIP AICT 539, pp. 230–234, 2018.
https://doi.org/10.1007/978-3-030-01313-4_24

2   The Principle of Language's Relativity

Language is a basic tool in human society, playing an essential role in communication and thought. People are accustomed to thinking in sound language (sound thinking). In different countries, people generally speak different languages. There are about 5000–7000 languages spoken all over the world, 90% of them used by less than 100000 people. As estimated by UNESCO (The United Nations Educational, Scientific and Cultural Organization), the most widely spoken languages are: Mandarin Chinese, English, Spanish, Hindi, Arabic, Bengali, Russian, Portuguese, Japanese, German and French. In practice, a language usually takes forms of speech and text, but can also be encoded into whistle, sign or braille. This may lead to an interesting question, can machines think in language with other forms, e.g. radio?

To all appearance, a Chinese can think in Chinese, an American can think in English, a Spanish can think in Spanish, and so on. From the viewpoint of daily life, all these forms of language, even including other forms such as whistle, sign and braille, should be equivalent for people to think about the world. In my opinion, this quite common point can be generalized as a first principle to establish a theory of mind (and intelligence more broadly), termed the principle of language's relativity, or the principle of symbolic relativity [5]. The principle is described as follows.

All admissible forms of language are equivalent for an intelligent system to think about the world.

In the principle of language's relativity, an admissible form means that the system can use it for thinking, i.e. the formulation of thoughts about the world. Therefore, the principle can be stated in other words,

All admissible forms of language are equivalent with respect to the formulation of thoughts about the world.

Note that this principle is named with inspiration from the principle of relativity in physics [6], namely,

All admissible frames of reference are equivalent with respect to the formulation of the fundamental laws of physics.

That is, physic laws are the same in all reference frames - inertial or noninertial. By analogy, it can be stated that, thoughts about the world are the same in all language forms - speech, text, whistle, sign or braille. Therefore, in this sense a language form can also be regarded as a reference frame to formulate thoughts. However, note that different forms of language may not be as easily implemented as each other from the viewpoint of engineering.

3   Significance of the Principle for Intelligence

If taken as a first postulate of intelligence theory, the principle of language's relativity implies that language is independent of modality. This explains why any human language can be encoded into a lot of different media such as using auditory, visual, and tactile stimuli. Moreover, it gives profound and original insights to guide engineering future generations of intelligent machines. For example, principally robots can think in radio language. Such robots would be tremendously useful in space exploration, where radio language is much more convenient for them to talk with each other than sound language for lack of air. Since no person has an inborn ability to receive and send radio waves, the radio form of language is not admissible for human. Thus, radio language is a novel and creative idea for robots to think, although radio itself is certainly very ordinary for information transmission and remote control. Clearly, thinking in radio language (radio thinking), is a different way to implement intelligence than people can. One may argue that, even without language, artificial intelligence (e.g. by residual network [7], deep Q-network [8] and AlphoGo [9]) could equal or even beat human intelligence in deep learning performance of some tasks such as object recognition, video games and board games. In practical realization, it is still reasonable to require that an intelligent machine be able to communicate through language. It goes without saying, language is an essential ability for general intelligence.

Yet nobody is quite sure of what intelligence is. Perhaps in most general purpose, intelligence measures an agent's ability to achieve goals in a wide range of environments [10]. Nonetheless, this informal definition, together with the mathematic description [10], plays a limited role in design of intelligent machines, albeit bringing together some key features from many expert definitions of human intelligence. To understand the nature of intelligence, not only a far-reaching definition is further expected, but also a comprehensive theory.

What does this theory look like? At the very least, it should contain just a few first principles at the system level. These few principles must be fundamental and independent in all phenomena of intelligence, and cannot be deduced from any other principles in physics, chemistry and biology. Although the principles may not lead to anything like Maxwells equations or $E = mc^2$, they should capture the essential characteristics of intelligence comprehensively in perspectives between science and philosophy. More importantly, they should be able to make a guide to engineering intelligent machines, especially with some different intelligence from human. Human-like implies imitation before grasping the nature of intelligence clearly, while human-different implies creation after understanding genuinely.

One of such principles is the principle of language's relativity, others to be certain. Obviously, the principle is independent of physics, chemistry and biology. Furthermore, it can account for modality-independence of language, and give rise to a revolutionary idea of radio thinking. As a high-level intelligence envisaged for future robot astronauts, the importance of radio thinking should be emphasized again. It may overturn a public view of how robonauts talk with each other in space exploration. Imagine two robonauts are talking about the earth on the moon (see Fig. 1). Traditionally, people think that they would talk in sound language, just as could be seen in some science fiction films. Nevertheless, in reality they cannot do so at all without air. Alternatively, they can talk with each other in a Morse-code radio language that lacks any neural mechanisms.

This is an excellent thought experiment to show that, although human language is an ability developed on the basis of neural mechanisms in the brain, intelligent machines may have a capacity of radio language not based on these mechanisms. Thereby, high-level intelligence may not be brain-like. The brain-like intelligence tries to achieve intelligence as demonstrated by brains [11], preferably of highly evolved creatures. However, the nature of intelligence should be understood in a brain-different way, just like the secret of how to fly. Indeed, without flapping its wings, an airplane can fly in a bird-different way based on the theory of aerodynamics, not on the bird's brain and control.

Fig. 1. Two robonauts are talking about the earth in a Morse-code radio language on the moon. One says "THE EARTH IS OUR HOME", the other "IT IS TRULY BEAUTIFUL". This is a potential application of Morse code to artificial intelligence in the future, beyond classical transmission of text information between people.

For space exploration, autonomous robonauts would be extremely helpful on the moon or the other planets. Since the environments change, there are a lot of barriers to implement these robonauts. But they should require certain kinds of human-different intelligence for high-level autonomy in actions. For example, radio language helps them to think, communicate and collaborate. Kirobo is the world's first talking robot sent into space [12]. But it was tasked to be a companion, not an explorer. It could talk in sound language only inside the spacecraft. Radio language may be an exceptional solution for two or more robonauts to talk with each other outside.

4   Conclusion

It is a terrific endeavour to engineer machines with general intelligence. This endeavour should be based on a theory of intelligence, which requires a few first principles, for example, the principle of language's relativity. Based on this principle, an exceptional solution has been given for robot astronauts to talk with each other in space exploration. This solution implies a great possibility to realize high-level machine intelligence other than brain-like intelligence.

References

1. Turing, A.M.: Computing machinery and intelligence. Mind **147**, 433–460 (1950)
2. LeCun, Y., Bengio, Y., Hinton, G.E.: Deep learning. Nature **521**, 436–444 (2015)
3. Lake, B.M., Ullman, T.D., Tenenbaum, H.B., et al.: Behavioral and Brain Sciences, vol. 253, pp. 1–72 (2017)
4. Mikolov, T., Joulin, A., Baroni, M.: A roadmap towards machine intelligence. In: Gelbukh, A. (ed.) CICLing 2016. LNCS, vol. 9623, pp. 29–61. Springer, Cham (2018). https://doi.org/10.1007/978-3-319-75477-2_2
5. Li, Y.: Reveal the secrets of consciousness - also on theory of cognitive relativity. In: Li, X., et al. (eds.) 100 Interdisciplinary Science Puzzles of the 21st Century. Science Press, Beijing (2005)
6. https://en.wikipedia.org/wiki/Principle_of_relativity
7. He, K., Zhang, X., Ren, S., et al.: Deep residual learning for image recognition. In 2016 IEEE Conference on Computer Vision and Pattern Recognition, pp. 770–778. IEEE Press, New York (2016)
8. Mnih, V., Kavukcuoglu, K., Silver, D., et al.: Human-level control through deep reinforcement learning. Nature **518**, 529–533 (2015)
9. Silver, D., Huang, A., Maddison, C.J., et al.: Mastering the game of Go with deep neural networks and tree search. Nature **529**, 484–489 (2016)
10. Legg, S., Hutter, M.: Universal intelligence: a definition of machine intelligence. Minds Mach. **17**, 391–444 (2007)
11. Sendhoff, B., Körner, E., Sporns, O.: Creating brain-like intelligence. In: Sendhoff, B., Körner, E., Sporns, O., Ritter, H., Doya, K. (eds.) Creating Brain-Like Intelligence. LNCS (LNAI), vol. 5436, pp. 1–14. Springer, Heidelberg (2009). https://doi.org/10.1007/978-3-642-00616-6_1
12. http://www.telegraph.co.uk/news/science/space/10221399/Talking-robot-astronaut-blasts-into-space.html

Language Understanding of the Three Groups of Connections: Management Innovation Dynamic Mechanism and Intelligent Driving Environment

Guangsheng Wang[1], Hanglin Pan[2], and Xiaohui Zou[3($\boxtimes$)] (iD)

[1] Chinese Academy of Social Sciences, No. 5, Jianguomen Inner Street, Beijing 100005, China
wanggs@cass.org.cn
[2] Hangzhou Ziya Intelligent Technology Co., Ltd., Hangzhou, China
[3] Sino-American Searle Research Center, Beijing, China
949309225@qq.com

Abstract. The purpose of this paper is to show how to improve people's understanding of the dynamics of management innovation and intelligent drive environment. The method is as follows: First, through the intelligent text analysis, refine the essence of management innovation dynamic mechanism and intelligent driving environment. Then, it expounds the progress of research on the dynamic mechanism of management innovation. Finally, it discusses the construction of intelligent driving environment, and takes "physical, transaction, insight" and its application scenarios as examples. The result is: not only the research on the dynamic mechanism of management innovation and the concise expression of the construction of intelligent driving environment, that is, the result of intelligent text analysis. The significance lies in: from the perspective of the three kinds of connections: "physical, transactional, and epiphany", it gives a more accurate language understanding paradigm for both the management innovation dynamic mechanism and the intelligent driving environment.

Keywords: Management innovation dynamic mechanism · Root theory analysis
Internet of Things · Business contact forms · Enlightened connections
Intelligent driving environment

1 Introduction

1.1 Background

Since Schumpeter's innovative theory, entrepreneurship has been highlighted. This is a good echo with the market role emphasized by Adam Smith's "invisible hand". Combined with the practice of 40 years of China's reform and opening-up, we found that there is a management innovation problem for companies as legal drafters [1]. It involves not only the dynamic mechanism but also the driving environment. Therefore, the three teams in this study have come together from each other's perspectives. Below, we explain the focus of common concerns one by one.

© IFIP International Federation for Information Processing 2018
Published by Springer Nature Switzerland AG 2018. All Rights Reserved
Z. Shi et al. (Eds.): ICIS 2018, IFIP AICT 539, pp. 235–242, 2018.
https://doi.org/10.1007/978-3-030-01313-4_25

1.2 Purpose

The purpose of this paper is to show how to improve people's understanding of the management innovation dynamic mechanism and intelligent drive environment. The method is: First, through the intelligent text analysis, refine the essence of management innovation dynamic mechanism and intelligent driving environment. Then, it is on the management innovation dynamic mechanism research. Finally, it discusses the construction of intelligent driving environment, and takes "physical, transaction, insight" and its application scenarios as examples.

2 Research on Management Innovation Dynamic Mechanism

This study mainly answers two questions. First, what is the formation mechanism of the management innovation dynamic mechanism? What are the power and resistance, respectively? On this basis, try to construct a conceptual model of management innovation mechanics. Our theoretical group believes that we must first clarify the definition of the dynamic mechanism of management innovation.

2.1 Theoretical Discussion

As far as this study is concerned, the so-called management innovation dynamic mechanism mainly refers to the sum of the interconnection, interaction of various factors that promote the emergence of management innovation and the development of management innovation in the process of management innovation system [2, 3]. Based on the definition of management innovation dynamic mechanism, its characteristics can be summarized as follows: First, systemic, various dynamic elements of management innovation are interconnected, mutually restrictive, and interact. These elements constitute an organic system of management innovation [4]. The formation and development of the dynamic mechanism depends on various dynamic factors, but also on the systematic role between these elements. The second is holistic. Chinese classical philosophy advocates the idea of "the unity of nature and human". He believes that the world is a unified whole. Only by understanding the whole can grasp the overall connotation and extension. In the typical complex system of management system, the dynamic mechanism is a metaphysical abstract concept, which is similar to the so-called "Tao" of Laozi, which is deeply hidden behind the appearance and affects the operation of things. The third is rationale. Any artificial system has a certain purpose. The purpose of management innovation dynamic mechanism is to form a strong driving force and promote enterprises to carry out management innovation activities. To achieve this goal, various power elements must be coordinated and have relatively consistent purpose. The fourth is dynamic. It mainly means that the dynamic elements of management innovation are dynamic development rather than static. In different stages of management innovation development, the interaction and driving force between power elements are constantly changing [5].

The management innovation dynamic mechanism has the following effects for an enterprise. The first is the driving role, which mainly refers to various internal and external factors that promote enterprise management innovation. Through interaction, it promotes the occurrence and development of enterprise management innovation, which is also the core of the management innovation dynamic mechanism. The second is the role of selection. It mainly refers to the management innovation process. In each period of time, the management innovation driving force and resistance are all working, thus forming a dynamic adjustment of termination, suspension and continuing management innovation. The dynamic mechanism of management innovation will be in the comparison of various choices, enterprises form a choice that is conducive to management innovation. The third is to strengthen the role. Through the management innovation dynamic mechanism, the effect of enterprises carrying out management innovation should be better than the result of not carrying out management innovation. Therefore, the management innovation has a positive strengthening function and promotes management innovation activities to continue to develop.

2.2 Theoretical Construction

This study uses a single case rooted research method to analyze the management innovation dynamic mechanism. In this analytical framework, the case collections are related to the utility and conclusions of the research, and the typicality of the case and the degree of access to the data need to be considered [6]. Based on the above factors, this study selected Taobao as a case.

This research collects more than 30 articles on the company's official website, news reports, academic literature, and corporate leadership interviews, and combines experiential research as the main material of this case. After data verification, the acquired information is highly consistent. In order to analyze the management innovation dynamic mechanism, this study based on the Taobao case to carry out rooted research, followed by open coding, associative coding, selective coding, story line analysis [7].

Based on the existing literature, this study considers that the management innovation dynamic mechanism can be summarized as enterprise Model (Fig. 1).

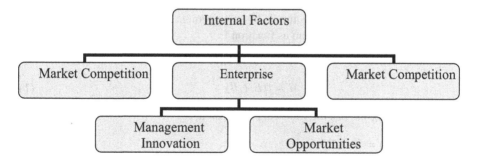

Fig. 1. Management innovation dynamics model diagram

Enterprises need to overcome the resistance of management innovation in order to promote the smooth implementation. The measures to overcome the resistance of management innovation are as follows: First, to strengthen incentives, the key to promoting innovation in management is to stimulate the initiative of employees through various means such as economic means; Formulate rules and regulations that are conducive to management innovation, standardize and sustain management innovation through system construction; third, increase investment, provide sufficient financial guarantee for management innovation, and provide good conditions for employees to implement management innovation [8].

In the context of the technological revolution and industrial transformation, the internal mechanism of management innovation dynamics has the following characteristics: First, it is the most complex period of the external environment of the enterprise, and the fastest period of change.

The most direct and important motivation is managing innovation activities. Secondly, as the initiator, organizer and implementer of management innovation, the enterprise itself plays a central role in the whole process of management innovation. The enterprise itself pursues the interests and solves the actual needs of problems encountered in its development [9]. It is management innovation. The important force that can be implemented and deepened, the enterprise's own driving force for management innovation is an important part of enterprise management innovation.

On the basis of the above-mentioned exploration of the management innovation dynamic mechanism, combined with the system mechanics thought, we can try to construct the system motivation theory model. As mentioned above, the enterprise management innovation power system mainly includes external environmental factors and internal factors of enterprises. These factors interact and influence each other to jointly promote the management innovation of enterprises, and the innovation of enterprise management also has corresponding effects and indirect influence on management innovation [10].

In order to express the relationship between these factors by mathematical formula, you can set:

The external environmental factor is E (environmental factor)
The internal factor of the enterprise is I (internal factor)
The overall benefit of the enterprise is B (benefit cost)
Enterprise management innovation is M (management innovation)
Management innovation system as function f.

Then there is the following expression or equation:

$$M = f(E, I, B) \tag{1}$$

Its system diagram has the following expression:

$$M(t) = (X(t), Y(t))$$
$$X(t) = \{M(t), E(t), I(t), B(t)\},$$
$$Y(t) = \{M(t)E(t), M(t)I(t), M(t)B(t), E(t)I(t), E(t)B(t), I(t)B(t)\}$$

Analysis of the variables of the above expressions shows that the force of environmental factors is positively related to management innovation, so it can be expressed as follows: $M(t)\ E(t) \to +$, similarly: $M(t)\ I(t) \to +$, $M(t)\ B(t) \to +$, $E(t)\ I(t) \to +$, $E(t)\ B(t)$

→+, I(t) B(t) →+. Therefore, a mapping expression can be established: F(t):X(t) → {+}. Therefore, the model of enterprise management innovation dynamic mechanism can be expressed as:

$$P(t) = (X(t), Y(t), F(t)) \tag{2}$$

From the perspective of the direct impact on management innovation, the combination of external environmental factors such as market demand, competition, laws and regulations, etc., generates the power of enterprise management innovation. Through management innovation and self-development, enterprises also have a reactionary influence on the market and policies, and even guide market demand, intensify market competition, and seek the inclination of laws and regulations for enterprises (see Fig. 2: Management innovation dynamic mechanism circuit diagram).

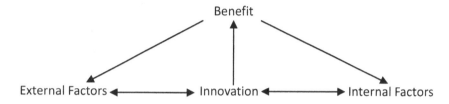

Fig. 2. Management innovation dynamic mechanism circuit diagram

3 The Creation of Intelligent Driving Environment

Our operations team believes that it is important to focus on the innovative drive environment. They believe that information to big data, knowledge to AI, and go on with intelligence to a new paradigm iteration of object number fusion.

3.1 Actual Build: Environment

Our Sino-German cooperation team believes that as a result, such three sets of connections are brewing.

It can be seen from Table 1 that the three groups of A, B, and C are based on the "phase: the size of the data, the type of data, the mode of the data, the processing method, the thinking mode, the program" to produce corresponding development and linkage.

Table 1. A, B, C as the three groups of connections

Phase	A	B	C
Size	MB/GB	TB/PB/ZB	TB/PB/ZB
Type	Structured data	Multi-structured data	Unstable IOT data
Mode	One size fits all	No size fits all	Not always fit
Method	Scientific thinking	Artistic thinking	Design thinking
Program	Open source	Knowledge graph	Artificial intelligence

A: Traditional database era: Small data, MB/GB; Less, structured data; Can be pre-determined, Data is generated after the first data mode, Data mode is relatively fixed; One Size Fits All, commercial database; Scientific thinking, data convergence, copy and well done; open source big data tools combination application;

B: The age of informationization knowledge: big data, TB/PB/ZB; multi-, including structure, and the proportion of semi-structured and unstructured data is increasing; can not be predetermined, static data is the main mode After the data appears, it can be determined that the data model evolves as the data grows; No Size Fits All, there are relatively common open source solutions; artistic thinking, business agility is the focus, creative but doable; based on knowledge graph development;

C: Intelligence after future data fusion: Very large data, TB/PB/ZB, increased after cross-indexing; more, in addition to stable structured, semi-structured, unstructured data, and more extensive and unstable IOT data; streaming data, The proportion of real-time data is getting bigger and bigger, the data model is mixed, the physical digital fusion mode is fast iterative; Not Always Fit, it is necessary to develop new solutions to adapt to changes; design thinking, focusing on digital asset integration, innovation, sometimes Research; integration of overseas outstanding core artificial intelligence team.

3.2   Actual Optimization: Intelligent

Combined Fig. 3 with Table 1, the intelligent graphic analysis of the ABC three-group connection can be advanced to a new level. In particular, combined with scientific thinking, artistic thinking and design thinking or innovative thinking, information collection and data aggregation; data cleaning, process grooming; rule mining, knowledge mapping; decision service, intelligence reuse; search engine, intelligent system; network, value transfer; data, processes, services, and statutes are handled uniformly.

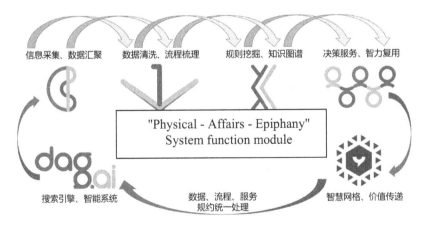

Fig. 3. "Physical – Affairs – Epiphany" System function module (Hanglin Pan, etc., Knowledge computing architecture of bio industry, *Bio Processing Business*, 2018.1)

Design is the forerunner and preparation of human purposeful innovation and creation of practical activities. It is itself creative, creative and innovative. Creative and innovative design has always been the forerunner and starting point of all human innovation and practice activities. Good design not only creates economic value, but also creates social, cultural and ecological values, which is one of the core factors for the competitiveness of enterprises and countries [11]. Design should be proven to be the "third wisdom" and ability of human beings not to be destroyed in the future, in addition to science and art [12]. Innovation is the basic and core feature of design. Without innovation, it is not a design. Innovative design is a kind of design method and technology. It specializes in the conceptual system of product innovation and studies various strategies, methods, techniques and tools to promote product innovation [13]. Social enterprises are the main path to achieve social innovation. Different from traditional enterprises, social enterprises are putting social responsibility first in the pursuit of profits [14].

4 Conclusion

4.1 Dynamic Mechanism and Driving Environment

At present, there are not many researches on the enterprise management innovation dynamic mechanism, and it mainly focuses on the determination of the dynamic factors of management innovation, but the dynamic mechanism has not been systematically discussed. Based on the above research status, this study starts with the definition and connotation of the management innovation dynamic mechanism. Through the root analysis of Taobao management innovation, the system theory is used to conduct a more in-depth and systematic discussion on the management innovation dynamic mechanism, and try to build management innovation. The kinetic model has achieved research results with certain value. First, on the basis of the existing research literature, clarify the definition and connotation of the dynamic mechanism of management innovation, and expound the definition of management innovation dynamics in the perspective of this research. Second, from the Taobao management innovation practice, in the absence of any pre-requisites, through the grounded theory analysis, the case is open-coded, correlated coding, selective coding, story line analysis, and the management innovation dynamic factors are analyzed. Thirdly, on the basis of grounded analysis, research on the external environmental power and internal dynamics of the enterprise, and try to construct a dynamic model of management innovation.

4.2 Significance

Combining the "Physical-Affairs-Epiphany" System function module to construct a driving environment is unique. Breaking through the new technologies of German Industry 4.0 and the new generation of artificial intelligence in the United States and Japanese robots, as another kind of driving environment, the driving environment function brought about by China's future management innovation cannot be underestimated. Nevertheless, whether it is within the enterprise or the domestic research on the management innovation dynamic mechanism, it has special significance.

Next, we will do further research on the relationship between "management innovation, dynamic mechanism and driving environment" and "economic innovation, smart integration (involving knowledge-based block-chains), and financial communication (involving virtual currency-based block-chains)". In particular, it is a systematic study of the relationship between Schumpeter's "economic innovation" and Adam Smith's "invisible hand" with "market innovation." It is also necessary to use the "human brain and computer (two kinds of hardware - biological and physical)" and "the intellectual research in psychology and the artificial intelligence research in computer science" (two kinds of software - in the scope of logic, mathematics and linguistics can be combined into one) [15] "deep questions, put on the agenda" [16].

References

1. Lü, Y.: What are the reasons for the innovation motivation of state-owned enterprise managers? Oper. Manag. **2**, 10–11 (2010)
2. Zhang, G., Feng, D., Wei, J.: Discussion on dynamic theory of enterprise technology innovation. Technol. Innov. Manag. **31**(1), 23–26 (2010)
3. Hu, Z.: Research on Dynamic Mechanism and Model of Enterprise Green Technology Innovation. Central South University (2006)
4. Lewin, K.: A Dynamic Theory of Personality: Selected Papers. Read Books Ltd, Redditch (2013)
5. Jean, M.: Group Dynamics. The Commercial Press, Beijing (1997)
6. Zhang, J.: The enlightenment of grounded theory to curriculum research. J. Comp. Educ. **10**, 81–85 (2010)
7. Strauss, A.L.: Qualitative Analysis for Social Scientists. Cambridge University Press, Cambridge (1987)
8. Lü, Y.: Factor flow in enterprise technology innovation system. Sci. Technol. Manag. Res. **32**(4), 9–11 (2012)
9. Chen, Y.: On musical therapy and Freud's psychoanalysis. Artist. Ocean **2**, 48 (2010)
10. Wang, L., Li, W., Wang, Z.: Research on the motivation and management of organizational inertia. Forecast **23**(6), 1–4 (2004)
11. Lu, Y.: Rethinking on design evolution. Mechanical Engineering Trends, Issue 2 (total issue 171), pp. 3–5. National People's Congress Standing Committee (2014)
12. Liu, G.: Design is the "third wisdom" that humanity will not be destroyed in the future. Mechanical Engineering Trends, No. 2 (total 171), pp. 6–23. Tsinghua University Academy of Fine Arts (2014)
13. Feng, P.: Design, innovation and development environment. Mechanical Engineering Trends, No. 2 (total issue 171), pp. 24–28. Zhejiang University (2014)
14. Liu, Y.: Innovative design and new value: system design thinking for transferring economy. Mechanical Engineering Trends, No. 2 (total issue 171), pp. 9–31 (2014)
15. Zou, X., Zou, S., Ke, L.: Fundamental law of information: proved by both numbers and characters in conjugate matrices. In: Proceedings, vol. 1, p. 60 (2017)
16. Zou, X., Zou, S.: Bilingual information processing method and principle. J. Comput. Appl. Softw. **11**, 69–76 (2015). Article no 102

Perceptual Intelligence

CSSD: An End-to-End Deep Neural Network Approach to Pedestrian Detection

Feifan Wei[✉], Jianbin Xie, Wei Yan, and Peiqin Li

School of Electronic Science, National University of Defense Technology,
Changsha 410073, Hunan, People's Republic of China
wff0316@foxmail.com

Abstract. Single Shot Multibox Detector (SSD) provides a powerful framework for detecting objects using a single deep neural network. The detection framework is one of the top object detection algorithms in both accuracy and speed which processes a large set of object locations sampled across an image. However, this framework does not behave well for the task of pedestrian detection since the images in popular pedestrian datasets have multiple objects occlusion problem and contain lots of small objects. In this paper, we incorporate deconvolution and downsampling unit into the SSD framework allowing detection network to recycle feature maps learned from images. The enhanced performance was obtained by changing the structure of classifier network, e.g., by replacing VGGNet with DenseNet. The contribution of this paper is a one-stage approach to compose a single deep neural network for pedestrian detection task in real-time. This approach addresses the typical difficulty of detecting different scale pedestrian at only one layer by providing a novel channel fusion. To solve small objects problem, base network has been replaced with more powerful one. This approach outperforms competing one-single methods on standard Caltech pedestrian dataset benchmark. It is also faster than all the other methods.

Keywords: Deep learning · Pedestrian detection · One-stage detector

1 Introduction

Pedestrian detection is one of main areas of researches in computer vision, due to its importance for a number of human-centric applications, such as video surveillance, autonomous driving, person identification and robotics [1]. Real-time accurate detection of pedestrians is a key for these systems.

In recent years, convolutional neural networks (CNNs) have been applied to pedestrian detection algorithms in various ways, to improve the accuracy and speed of pedestrian detection [2–5]. As illustrated in [6], the detection algorithms based on deep learning can be divided into two-stage detectors and one-stage detectors, and one-stage detectors has the potential to achieve faster and better results. As shown in Fig. 1, one-stage detectors can be divided into two parts: backbone network and detection network.

Among various one-stage detection methods, Single Shot Multibox Detector (SSD) [7] is one of the few algorithms that can guarantee robustness in real-time

Z. Shi et al. (Eds.): ICIS 2018, IFIP AICT 539, pp. 245–254, 2018.
https://doi.org/10.1007/978-3-030-01313-4_26

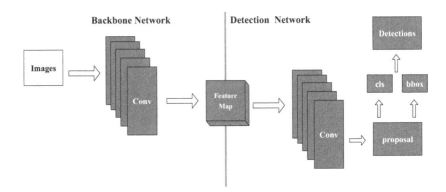

Fig. 1. Overview of the proposed CSSD framework.

detection, because it adopts multiple convolution layers for detection. Although the conventional SSD performs well in object detection, it still has a few problems when applying it to pedestrian detection since the images in popular pedestrian datasets have multiple objects occlusion problem and contain numerous small objects.

Firstly, each layer in the feature pyramid of SSD is used independently as an input to the classifier network. Thus, the method can detect pedestrians of different scales in one picture, meanwhile the same pedestrian can be detected in multiple scales. However, SSD looks at only one layer for each scale, so it does not consider the relationships between the different scales. For example, in Fig. 2, SSD finds two boxes for one person. Pedestrians in front of images tend to have higher confidence than later ones. After applying the NMS algorithm, the detection boxes of pedestrians behind are often suppressed. This is why the SSD algorithm does not perform well for occlusion problems.

(a)   (b)

Fig. 2. Detection results of SSD, boxes with person score of 0.5 or higher is drawn: (a) SSD with two boxes for the right person. (b) after applying NMS, only the right person can be detected.

Secondly, SSD is not robust to small-scale pedestrian detection. This problem is ubiquitous in current detection algorithms. In an image, small-scale pedestrians have smaller receptive fields and it is more difficult to extract their features. To solve this problem, there have been many attempts such as increasing the number of channels in one layer or replacing the basic network with more powerful one.

In this paper, these problems are tackled as follows: At first, the backbone network in the SSD is replaced with the improved DenseNet [8]. As shown in Table 1, the network structure has been deepened, thereby improving feature extraction capabilities, with fewer parameters and faster convergence. Then, a circular feature pyramid is set up by deconvolution and downsampling units, after which the algorithm Cycle Single Shot Detector (CSSD) is named. The circular feature pyramid can make full use of the information between each layer to accurately predict the pedestrian detection boxes. Finally the algorithm performs better on occlusion than the others.

Table 1. CSSD backbone network

Layers		Output size	Structure
Stem	Convolution	150×150	3×3 conv, stride 2
	Convolution	150×150	3×3 conv, stride2
	Convolution	150×150	3×3 conv, stride2
	Pooling	75×75	2×2 avg pool, stride2
Dense block (1)		75×75	$\begin{bmatrix} 1 \times 1 \text{ conv} \\ 3 \times 3 \text{ conv} \end{bmatrix} \times 6$
Transition layer (1)		75×75	1×1 conv
		38×38	2×2 avg pool, stride 2
Dense block (2)		38×38	$\begin{bmatrix} 1 \times 1 \text{ conv} \\ 3 \times 3 \text{ conv} \end{bmatrix} \times 8$
Transition layer (2)		38×38	1×1 conv
		19×19	2×2 avg pool, stride 2
Dense block (3)		19×19	$\begin{bmatrix} 1 \times 1 \text{ conv} \\ 3 \times 3 \text{ conv} \end{bmatrix} \times 8 \begin{bmatrix} 1 \times 1 \text{ conv} \\ 3 \times 3 \text{ conv} \end{bmatrix} \times 8$
Transition layer (3)		19×19	1×1 conv
Dense block (4)		19×19	$\begin{bmatrix} 1 \times 1 \text{ conv} \\ 3 \times 3 \text{ conv} \end{bmatrix} \times 8$
Transition layer (4)		19×19	1×1 conv

As shown in Fig. 1, the proposed detection algorithm can be divided into two parts: backbone network and detection network. DenseNet and CSSD are used as the backbone network and detection network, respectively.

The contributions made by this paper can be summarized as follows:

- The DenseNet network have been modified to improve the feature extraction capabilities of the backbone network and to enhance the ability of the algorithm to detect small-scale pedestrians.

- A circular feature pyramid has been designed to make full use of the information in each layer of the network to detect pedestrian boxes more accurately, thereby improving the algorithm's robustness to occlusion.
- The state-of-the-art performance has been achieved in the one-stage detectors on Caltech pedestrian dataset.

2   CSSD

In this section, Cycle Single Shot Detector (CSSD) is introduced. As shown in Fig. 1, the CSSD is divided into backbone network for extracting features and detection network for the generation and classification of candidate boxes.

We have improved SSDs in three ways:

- A brand-new backbone network is designed following the DenseNet framework.
- The deconvolution and downsampling modules are used to construct a circular feature pyramid in order to make full use of the information in each layer of the network.
- The number of channels in each layer is reduced with fewer parameters and higher efficiency.

2.1   Backbone Network

Our base network is a variant of DenseNet. The network structure of the entire CSSD is shown in Table 1. The backbone network consists of one Stem block, four Dense blocks and four transition layers. Inspired by the Inception-v4 [9] network, we used three 3 * 3 convolutional layers and a 2 * 2 pooling layer are adopted to form the Stem block. The Stem block replaces the 7 * 7 convolutional layer in DenseNet, which increases the network depth and improves the network feature extraction capability while ensuring the same receptive field. By comparing with other classification networks, the performance of the network in the experimental part will be demonstrated.

2.2   Circular Feature Pyramid

In order to make full use of the information in each layer and integrate more high-level semantic information in detection, as shown in Fig. 3, a circular feature pyramid using down-sampling and deconvolution modules is set up. The CSSD model is improved from SSD with modified DenseNet. The number of channels in each layer is reduced during downsampling. The two reasons why a paradigm that predicts every layer is not used in the FPN [10] algorithm are as follows: On the one hand, pedestrian detection is a fundamental task in a system, which needs to provide enough information for the downstream task. Therefore, speed is a key to the algorithm. Making predictions on each layer means the time for inference will increase several times. This is not acceptable for a fast pedestrian detection algorithm. On the other hand, a deconvolution module can reduce the number of parameters, improve efficiency, and make full use of

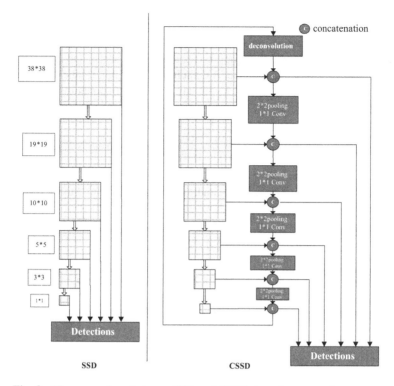

Fig. 3. The comparison between SSD and CSSD detection network structure.

information between layers. The more accurate detection boxes can be obtained, the less impact of occlusion will be on detection.

2.3   Connection Modules

As shown in Fig. 3, there are two types of connection modules in the CSSD network. The first is that a module has been designed to combine the downsampling layer with a 1 * 1 convolution. The downsampling connection units has the following two roles:

- The upper feature size is processed by the downsampling unit to be consistent with the size of the underlying feature map;
- The channel fusion of the upper feature map keeps the number of output channels within a reasonable range.

The second type is, the deconvolution unit has been adopted to fuse the feature map of the last layer of the detection module with the feature map of the first layer. In this way the information flow of the network can be recycled. Each layer can obtain information from other layers. Similar to DSSD, the deconvolution unit consists of 3 convolution layers, 3 Batch Normalization layers, 1 deconvolution layer, 2 Relu activation functions, and a connection unit. The main difference is that we are connected to the first and last layer of feature pyramid.

2.4   Training Objective

The CSSD training objective is derived from the SSD [7] objective but just handles one object category. A set of prior bounding boxes at different scales and aspect ratios are generated at each position in the image. A default bounding box is labeled as positive if it has a Jaccard overlap greater than 0.5 with any ground truth bounding box, otherwise negative.

$$labels = \begin{cases} 0, & otherwise \\ 1, & if \ \frac{A_d \cap A_g}{A_d \cup A_g} > 0.5 \end{cases} \tag{1}$$

where A_d and A_g represent the default bounding box and the ground truth, respectively. The training objective is given as Eq. (2):

$$L = \frac{1}{N} \left(\alpha L_{loc} + L_{conf} \right) \tag{2}$$

where N is the number of matched default boxes, and α is a constant weight term to keep a balance between the two losses. L_{conf} is the softmax loss over person category confidence.

$$L_{conf} = - \sum_{i \in Pos}^{N} x_{ij} \log(c_i) - \sum_{i \in Neg} \log(c_i)$$
$$where \quad c_i = \frac{\exp(c_i)}{\sum_p \exp(c_i)} \tag{3}$$

where Pos and Neg represent the positive and negative default boxes, respectively. $x_{ij} = \{1, 0\}$ is an indicator for matching the i-th default box with the j-th ground truth box of category person. L_{loc} is the Smooth L1 loss [11], not modified, for more details about the loss please refer to [7].

3   Experiments

3.1   Dataset and Evaluation

The Caltech pedestrian dataset is currently the most commonly adopted and challenging dataset in pedestrian detection. The pedestrian detection algorithm proposed in this paper was evaluated on the Caltech dataset. A total of approximately 250,000 frames 350,000 rectangular frames and 2300 pedestrians were marked. The original frame size is 480 × 640. The FPPI standard is proposed by the Caltech dataset to evaluate the algorithm.

3.2   Training, Testing Settings and Results

The pedestrian detector proposed in this paper is implemented in the caffe framework. Most training strategies follow the SSD algorithm, including data augment and loss function settings. Due to the difference between the scale ratio of the pedestrian and the common object in the object recognition, the scale setting of the prior box has been adjusted in the algorithm. In addition, because the algorithm is trained from scratch, the setting of hyperparameters such as learning rate will also be different. This algorithm is trained on **NVIDIA 1080Ti GPU**.

The Caltech datasets includes a total of 11 video segments (s0–s10), of which the first 6 video segments are training sets and the last 5 video segments are used as testing sets. The original size of these images is 480 * 640. In the setting of the candidate box and aspect ratio, the method of F-DNN has been basically followed. Experiments show that more candidates can be obtained through the method. Therefore the matches to the ground truth are not lost. By using the SGD training method, a batchsize of 2, the learning rate of 10^{-5}, and all weights are randomly initialized and trained from scratch.

CSSD has achieved an impressive 9.5% missing rate on the Caltech dataset using reasonable setting. Compared with SSD-ours, this algorithm achieved a 32% improvement. SSD-ours is a learning algorithm based on SSD framework for pedestrians. As shown in Fig. 4, the ROC plot of missing rate against FPPI is shown for the current top performing methods reported on Caltech. To the best of our knowledge, this detector is the first true one-stage pedestrian detector, and has achieved similar performance to the state-of-the-art algorithm. Not surprisingly, compared with SSD-ours, this algorithm achieved a 56% improvement using the Occ.partial setting and 77% improvement using far setting. In other words, our algorithm performs extremely well in dealing with pedestrian occlusion and small-scale pedestrian.

4   Results Analysis

4.1   Effectiveness Analysis

Through ablation experiments, the role of backbone network and detection network has been explored in the CSSD network. Figure 4 visualizes the results of SSD and our method. At the beginning, we used the original SSD algorithm to train and test on the Caltech dataset, achieving a 14.3% missing rate in the reasonable setting. After replacing the backbone network with DenseNet, the performance of the algorithm has improved from 14.3% to 11.6%. After the backbone network employed VGG, and the detection structure has been replaced with the proposed CSSD, the performance of the algorithm has improved from 14.3% to 12.4%. Having adopted DenseNet and the detection network CSSD, the miss rate finally reached 9.56%, as shown in Table 2.

4.2   Runtime Analysis

As shown in Table 3, compared with the recent state-of-the-art algorithms, our algorithm framework uses only a single convolutional framework. Thus the processing

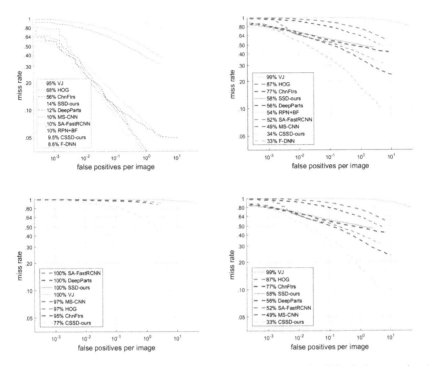

Fig. 4. Comparison of CSSD with the state-of-the-art methods on the Caltech dataset using the reasonable, Occ.partial setting, far and medium setting.

Table 2. Effectiveness of the backbone network and detection network

Method	Reasonable
VGG + SSD	14.3%
DenseNet + SSD	11.6%
VGG + CSSD	12.4%
DenseNet + CSSD	9.50%

Table 3. A comparison of speed among the state-of-the-art models

Method	Caltech (%)	Runtime (GPU)
DeepParts	11.89	1 s
ComACTr-Deep	11.75	1 s
MS-CNN	9.95	0.4 s
SA-FastRCNN	9.68	0.59 s
RPN + BF	9.58	0.60 s
F-DNN	8.65	0.30 s
SSD	13.06	0.06 s
CSSD (ours)	**9.50**	**0.08** s

speed of an image is only 0.08 s, which completely meets the needs of real-time processing. In the currently known algorithms, no other pedestrian detection algorithm can achieve the processing speed achieved by the CSSD algorithm with the same accuracy. This algorithm is designed for pedestrian detection in real-world scenarios. Since it is possible to train from scratch, the CSSD algorithm has a very strong migration capability and can be used in a very wide range of applications. This is one of the advantages of the CSSD algorithm.

5 Conclusions

An efficient and robust one-stage pedestrian detector has been proposed based on a single DNN trained from scratch. A brand new network CSSD for pedestrian detection is designed. To the best of our knowledge, this algorithm is state-of-the-art one-stage pedestrian detector on Caltech datasets. Furthermore, CSSD has great potential on special domain scenario like military district early warning, night guard, etc.

For future work, the pedestrian detection system based on semantic segmentation has achieved the best detection results. Due to the versatility of the CSSD network, semantic segmentation modules can be easily integrated into the network and will achieve better results. This part of the work will be the focus of our future research.

References

1. Geiger, A., Lenz, P., Urtasun, R.: Are we ready for autonomous driving? The KITTI vision benchmark suite. In: Conference on Computer Vision and Pattern Recognition (CVPR), vols. 1, 2, 5, 6 (2012)
2. Hosang, J.H., Omran, M., Benenson, R., Schiele, B.: Taking a deeper look at pedestrians. CoRR, abs/1501.05790 (2015)
3. Du, X., El-Khamy, M., Lee, J., Davis, L.S.: Fused DNN: a deep neural network fusion approach to fast and robust pedestrian detection. arXiv preprint arXiv:1610.03466, vols. 2, 3, 6 (2016)
4. Tian, Y., Luo, P., Wang, X., Tang, X.: Deep learning strong parts for pedestrian detection. In: Proceedings of the IEEE International Conference on Computer Vision, vols. 2, 6, pp. 1904–1912 (2015)
5. Zhang, L., Lin, L., Liang, X., He, K.: Is faster R-CNN doing well for pedestrian detection? In: Leibe, B., Matas, J., Sebe, N., Welling, M. (eds.) ECCV 2016. LNCS, vol. 9906, pp. 443–457. Springer, Cham (2016). https://doi.org/10.1007/978-3-319-46475-6_28
6. Lin, T.Y., Goyal, P., Girshick, R., et al.: Focal loss for dense object detection, pp. 2999–3007 (2017)
7. Liu, W., et al.: SSD: Single Shot Multibox Detector. In: Leibe, B., Matas, J., Sebe, N., Welling, M. (eds.) ECCV 2016. LNCS, vol. 9905, pp. 21–37. Springer, Cham (2016). https://doi.org/10.1007/978-3-319-46448-0_2
8. Huang, G., Liu, Z., Weinberger, K.Q., van der Maaten, L.: Densely connected convolutional networks. In: CVPR, vols. 1, 2, 3, 4 (2017)

9. Szegedy, C., Ioffe, S., Vanhoucke, V., et al.: Inception-v4, inception-ResNet and the impact of residual connections on learning (2016)
10. Lin, T.-Y., et al.: Feature pyramid networks for object detection. In: CVPR (2017)
11. Girshick, R.: Fast R-CNN. In: International Conference on Computer Vision (ICCV) (2015)

Predicting Text Readability
with Personal Pronouns

Boyang Sun and Ming Yue[✉]

School of International Studies, Zhejiang University,
Hangzhou, People's Republic of China
yueming@zju.edu.cn

Abstract. While the classic Readability Formula exploits word and sentence length, we aim to test whether Personal Pronouns (PPs) can be used to predict text readability with similar accuracy or not. Out of this motivation, we first calculated readability score of randomly selected texts of nine genres from the British National Corpus (BNC). Then we used Multiple Linear Regression (MLR) to determine the degree to which readability could be explained by any of the 38 individual or combinational subsets of various PPs in their orthographical forms (including *I*, *me*, *we*, *us*, *you*, *he*, *him*, *she*, *her* (the Objective Case), *it*, *they* and *them*). Results show that (1) subsets of plural PPs can be more predicative than those of singular ones; (2) subsets of Objective forms can make better predictions than those of Subjective ones; (3) both the subsets of first- and third-person PPs show stronger predictive power than those of second-person PPs; (4) adding the article *the* to the subsets could only improve the prediction slightly. Reevaluation with resampled texts from BNC verify the practicality of using PPs as an alternative approach to predict text readability.

Keywords: Readability · Personal Pronouns · Linear Regression

1 Introduction

The history of predicting textual readability quantitatively dates back to the 1940s when several linguists including Rudolf [1], George [2], Dale and Chall [3] introduced readability formulas into the field of research, thus unleashing a wave of researches and applications. Until 2017, Web of Science has published more than 11,000 researches on readability and its applications have moved from the field of education to fields of administration, commerce, computers, military, scientific research, etc. [4–6].

Traditional readability studies usually start with vocabulary and sentence complexity. For instance, the most widely recognized Flesch Reading Ease Formula uses word length (in terms of syllable) and sentence length (in terms of word count) as variables to calculate readability; the Dale-Chall Readability Formula exploits numbers of words that are not in the Dale-Chall 3000 Vocabulary and sentence length as criteria for predicting readability; the Gunning Fog Formula [7] and the SMOG Formula [8] employ number of polysyllabic words and sentence length as measures of readability. As computer technologies improve, many other factors are taken into account, such as type-token ratio, numbers of affixes, prepositional phrases and clauses, cohesive ties,

© IFIP International Federation for Information Processing 2018
Published by Springer Nature Switzerland AG 2018. All Rights Reserved
Z. Shi et al. (Eds.): ICIS 2018, IFIP AICT 539, pp. 255–264, 2018.
https://doi.org/10.1007/978-3-030-01313-4_27

other linguistics features [9], and even L2 learner's reading experience, etc. [10]. While these studies are valuable and significant, they usually involve multiple indirect indices that are subjectively defined or difficult to calculate in large-scale analysis. For example, it is hard to tell whether a word such as *factory* with two or more phonetic variants should be counted as 2 syllables (/ˈfæktrɪ/) or 3 syllables (/ˈfæktəri/). Besides, most of the classic formulae target for texts in English (and some other syllabic language), their applicability for non-syllabic languages such as Chinese remain untested.

In this research, we hope to test whether Personal Pronouns (hereinafter referred to as PPs) alone can have any predictive power for readability or not. There are several reasons for us to try them: (1) Given that PPs are always monosyllabic words used to replace full personal names or noun phrases, their usage in a text would affect its total word number, average sentence length as well as average word length; (2) PPs are often anaphorically used and can thus serve as cohesive ties to reduce redundancy and improve comprehension; (3) PPs were only tested collectively in [11] and [12] as part of linguistic features or cohesive ties, and consequently reached different conclusions on the role PPs play in readability prediction.

Since most languages have pronouns, we therefore propose that PPs could be promising candidate indicators of readability across languages and deserve further investigation. In this study, we will use a corpus-based approach to test the utility of individual PP forms in English texts of different genres. Specific research questions are as follows:

(1) Which person (first-, second-, or third- person, hereinafter referred to as 1P, 2P and 3P respectively) of PPs can predict text readability most accurately?
(2) Which number (Singular and Plural) of PPs can predict text readability more accurately?
(3) Which case (Subjective and Objective, with Possessive temporarily excluded) of PPs can predict text readability more accurately?

Sections 2 and 3 will introduce our research methods and data processing, Sect. 4 will report the data results from 5 aspects, Sect. 5 will reevaluate the results and Sect. 6 will summarize our major findings and limitations.

2 Materials and Methodology

This research uses corpus-based method and examines the predictability of various subsets of the PP forms (as shown in Table 1) on text readability in terms of Person, Number and Case.

It should be noted that the Possessive Case is not taken into consideration in this research. Nor will this paper look into the gender issue. So (*he* + *she*) and (*him* + *her*) will be considered as individual Subjective and Objective singular forms of 3P + HUMAN respectively; *it* be considered as the individual singular form of 3P-HUMAN with unclear Case; and *you* as the only 2P form with unclear Number and Case.

Consequently, there are 38 reasonable subsets of PP forms: 10 subsets with only individual PP forms, and 28 others with various Person/Number/Case combinations.

Table 1. Personal pronoun forms studied in this project

	1P		2P	3P		
	Singular	Plural	Singular/plural	Singular		Plural
				+ HUMAN	− HUMAN	
Subjective	*I*	*we*	*you*	*he + she*	*it*	*they*
Objective	*me*	*us*		*him + her*		*them*

2.1   Corpus Data

British National Corpus (BNC) was chosen as our research object for the following reasons:

(1) All text materials in BNC were collected from native speakers as representative samples of Standard British English. So errors in pronoun use by non-native speakers have been excluded to a large extent; variations in geographical and social dialects should have been reasonably controlled or avoided as well.
(2) BNC contains approximately 100 million words, 90% of which are written materials collected from nine domains (also referred to as "genres" hereinafter) namely: (a) Arts; (b) Belief; (c) Commerce; (d) Imaginative; (e) Leisure; (f) World affairs; (g) Natural science; (h) Social science; and (i) Applied science. Due to the different effects of genres on usages of PPs [13], proper sampling of this balanced general corpus allows for control over the genre variable that may affect readability.

Text materials used in this study (Corpus I) consist of 1,091,347 words in total, which are randomly selected from each of the nine domains. Corpus II consists of 972,490 words in total.

2.2   Readability Formula

In the present study, we choose the Flesch Reading Ease Score, which is recognized as the most widely used and the most tested and reliable formula [6], as approximants of real text readability to native readers. The specific formula is as follows:

$$Reading\ Ease\ Score = 206.835 - (1.015 * ASL) - (84.6 * ASW)$$

Where ASL = average sentence length (total word number divided by total sentence number), ASW = average word length (total syllable number divided by the total word number). The correlation coefficient between the Flesch readability formula and the Mc-Crabbs Reading Test was 0.7 [1].

3 Data Processing

Data processing are divided into 4 steps:

(1) Use Perl program to count word and sentence length;
(2) Calculate the Flesch Reading Ease scores of sample texts of nine genres respectively;
(3) Use AntConc to count numbers of PP forms. Tokens of *US* as the abbreviation of the United States and tokens of the Possessive *her* are excluded during the retrieval. After that, the densities of the individual pronouns ($D_{(I)}$, $D_{(we)}$, etc.) based on the total word number of each text domain are calculated respectively;
(4) Use SPSS for multivariate regression analysis. Take the density of each subset of PPs as an independent variable, and the Flesch Reading Ease score as the dependent variable. Use Sig., correlation coefficient (R^2), as well as the adjusted correlation coefficient (adjusted R^2) values to determine which subset(s) of PPs may have better predictability. The criteria and process for determining moderate and strong fitting subsets are shown in Fig. 1.

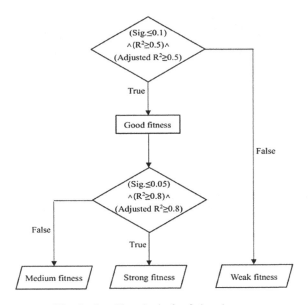

Fig. 1. Specific criteria for fitting degrees

4 Results and Discussion

4.1 Readability Results of Random Texts of Nine Genres

Table 2 shows that texts from Belief, Arts and Imagination domains are easiest to understand with highest readability scores among all texts from the nine domains; texts of Commerce, Natural Science, Applied Science and World Affairs are most difficult to read with lowest scores.

Table 2. Readability results for nine domains in BNC

Domain	Flesch Reading Ease Score	Difficulty level
Belief	87.829	Easy
Arts	87.053	Easy
Imaginative	80.811	Easy
Leisure	67.623	Standard
Social science	51.449	Moderately difficult
Commerce	49.712	Difficult
Natural science	47.571	Difficult
Applied science	44.922	Difficult
World affairs	44.829	Difficult

4.2   Fitness Results

Individual Pronoun Forms and Readability There are 10 subsets with individual PP forms as listed in Table 3. According to Fig. 1, it can be concluded that the subsets (*him + her*) and *them* present significant linear relations with readability and can explain almost 80% of variance ($R^2 \approx 0.8$), indicating strong predictive power. Additionally, *us* also shows a significant linear relation (Sig. = 0.024) with accounting for about 50% of variance ($R^2 = 0.543$), showing that in contrast with individual Subjective PPs, individual Objective ones show fairly strong predictability on readability.

Person and Readability. The 38 individual and combinational subsets of PPs can be divided into seven groups according to Person (1P: 9 subsets; 2P: 1 subset; 3P: 12 subsets; 1P + 2P: 1 subset; 1P + 3P: 11 subsets; 2P + 3P: 2 subsets; 1P + 2P + 3P: 2 subsets).
Results in Fig. 2 show that the 3P group has the best fitting degrees, with 5 subests (over 40%) of strong fitting and 2 (nearly 10%) of medium fitting subsets. The mixed (1P + 3P) group performs similarly well, with 3 subsets (nearly 30%) of strong fitting and another 2 (nearly 10%) of good fitting subsets, way better than 1P and 2P subsets do. Therefore, it can be concluded that 3P subsets perform better than 1P and 2P subsets do in both individual and mixed subsets, which means that adding 1P and 2P subsets into the 3P subsets will lowered their predictability.

Number and Readability. The 38 individual and combinational subsets of PPs can be divided into three groups according to Number (singular PPs: 12 subsets, plural PPs: 9 subsets, singular + plural PPs: 17 subsets).
Figure 3 shows that 50% of the singular-Number group offer good predication (with strong and/or medium fitness); and nearly 45% (11.1% + 33.3%) of the plural-Number group show good prediction. The mixed-number group performs not as well.

Case and Readability. The 38 individual and combinational subsets of PPs can be divided into three groups according to Case (Subjective PPs: 9 subsets; Objective PPs: 9 subsets; Subjective + Objective PPs: 20 subsets).

Table 3. Results for predictability of individual personal pronouns on readability

Pronoun forms	Case	Regression formulas	Sig.	R^2	Adjusted R^2
I	S	R = 1503.801 * D$_{(I)}$ + 53.669	0.091	0.354	0.261
we	S	R = 2328.992 * D$_{(we)}$ + 53.662	0.343	0.129	0.004
you	S + O	R = 2722.168 * D$_{(you)}$ + 53.735	0.102	0.336	0.241
he + she	S	R = 1208.909 * D$_{(he)}$ + 728.189*D$_{(she)}$ +52.435	0.414	0.254	0.006
they	S	R = 11937.951 * D$_{(they)}$ + 30.955	0.101	0.338	0.243
me	O	R = 6757.564 * D$_{(me)}$ + 54.975	0.076	0.383	0.295
US	O	R = 14042.402 * D$_{(us)}$ + 50.635	**0.024**	**0.543**	**0.478**
him + her	O	R = 25621.555 * D$_{(him)}$ − 10026.320 * D$_{(her)}$ + 39.606	**0.013**	**0.768**	**0.690**
them	O	R = 34550.512 * D$_{(them)}$ + 13.615	**0.000**	**0.863**	**0.843**
it	S + O	R = 4275.486 * D$_{(it)}$ + 25.507	0.064	0.408	0.324

Note: S stands for the Subjective Case; O for the Objective case. R for readability score; D() for word density in the text. The significances of bold refer to the subsets that perform very well (with strong fitness) in both Corpus I and Corpus II.

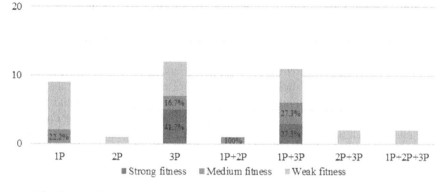

Fig. 2. Results for predictability of different Persons on readability in Corpus I

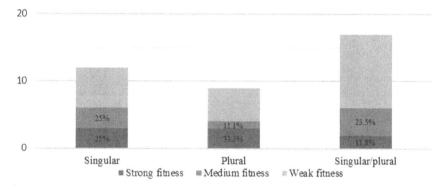

Fig. 3. Results for predictability of different Numbers on readability in Corpus I

Figure 4 shows that Objective PP group has much stronger predictability than the Subjective group and the mixed-Case group, in both good and strong fitting area.

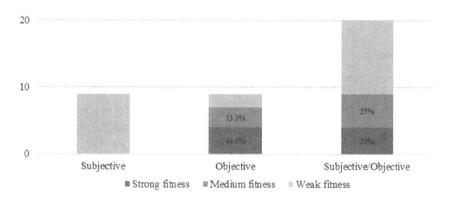

Fig. 4. Results for predictability of different Cases on readability in Corpus I

The* and Readability.** Since the definite article ***the in English has similar deictic/specifying function as pronouns do, we will test and see if this particular word and its combination with some of the PP subsets would have any predictive power on readability.

First, we use $D_{(the)}$ to predict text readability and gain a medium performance (Sig. = 0.019, R^2 = 0.570, Adjusted R^2 = 0.509). Results in Fig. 5 show that subsets with ***the*** included perform slightly better than those without ***the*** in good and in strong fitting ranges. To test whether there is a significant difference while adding ***the*** in PPs, we use chi-square tests and draw the conclusion that the improvement is not significant (Chi-square value = 0.213, df = 2, p = 0.899 > 0.05).

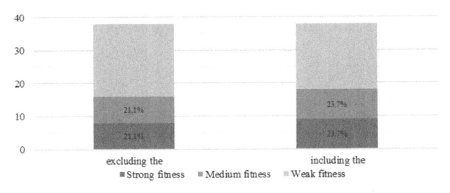

Fig. 5. Results for predictability of including and excluding ***the*** on readability

5   Reevaluation for Strong Fitting Subsets

All the subsets with a strong fitting degree are shown in Table 4. To explore whether subsets with strong predicting power can perform consistently, we repeated the procedures in Sect. 3 with re-sampled texts from BNC (Corpus II) and recalculated the pronoun and readability data in the new corpus. Test results from both Corpus I and II are shown in Table 4.

Table 4. Personal pronoun subsets with strong fitness in Corpus I and II

Personal pronoun subsets	Corpus I			Corpus II		
	Sig.	R^2	Adjusted R^2	Sig.	R^2	Adjusted R^2
them	0.000	0.863	0.843	0.070	0.670	0.623
us + them	0.002	0.871	0.827	0.036	0.670	0.560
him + her + them	0.001	0.959	0.934	0.110	0.872	0.796
me + us + him + her + them	0.024	0.962	0.899	0.128	0.879	0.677
he + him + she + her + it	**0.022**	**0.964**	**0.905**	**0.005**	**0.986**	**0.963**

(continued)

Table 4. (*continued*)

Personal pronoun subsets	Corpus I			Corpus II		
	Sig.	R^2	Adjusted R^2	Sig.	R^2	Adjusted R^2
I + me + he + him + she + her + it	**0.021**	**0.999**	**0.999**	**0.036**	**0.999**	**0.998**
he + him + she + her	0.004	0.964	0.928	0.036	0.887	0.774
they + them	0.002	0.869	0.825	0.035	0.672	0.563

The significances of bold refer to the subsets that perform very well (with strong fitness) in both Corpus I and Corpus II.

Table 4 shows that there are still two subsets with strong fitting degree in Corpus II, namely *"he + him + she + her + it"* and *"I + me + he + him + she + her + it"*. Although the other subsets have some changes in the fitting degree, they are almost in the moderate fitting range, indicating fair predictability.

6 Conclusion

A corpus-based approach is used in research to explore the readability predictability of 77 subsets with various personal pronoun forms and the definite article *the*. The results show that: (1) *them* has the best predictive power among individual pronoun forms; (2) 3P and 1P make better predictions than 2P; (3) plural PPs outperforms singular ones only in strong fitting range; (4) Objective PPs can predict more accurately than Subjective ones; (5) definite article *the* may only improve subsets' predictability slightly; (6) Retesting results are consistent for those PP subsets with good predictability. Therefore, we believe that using specific subsets of PPs to predict text readability appears practical.

However, large-scale tests are needed before any solid conclusion can be drawn concerning the applicability of PPs for readability prediction. Detailed investigation into the predictability of Possessive PPs, and *it* in Subjective and Objective Cases may be needed as well. Besides, it needs to be verified on whether texts in other geographical varieties such as American English are similar to their British matches.

References

1. Flesch, R.: A new readability yardstick. J. Appl. Psychol. **32**, 221–233 (1948)
2. Klare, G.R.: Measures of the readability of written communication: an evaluation. J. Educ. Psychol. **43**(7), 385–399 (1952)
3. Dale, E., Chall, J.S.: A formula for predicting readability: instructions. Educ. Res. Bull. **2** (27), 37–54 (1948)
4. Meppelink, C.S., van Weert, J.C.M., Brosius, A., Smit, E.G.: Dutch health websites and their ability to inform people with low health literacy. Patient Educ. Couns. **11**(100), 2012–2019 (2017)
5. Botas, S., Veiga, R., Velosa, A.: Bond strength in mortar/ceramic tile interface-testing procedure and adequacy evaluation. Mater. Struct. **50**(5), 211 (2017)

6. Dubay, W.H.: Smart Language: Readers, Readability and the Grading of Text. Costa Mesa, California (2007)
7. Gunning, R.: The Technique of Clear Writing. McGraw-Hill, New York (1952)
8. Laughlin, G.H.M.: SMOG grading-a new readability formula. J. Read. **12**(8), 639–646 (1969)
9. Pitler, E., Nenkova, A.: Revisiting readability: a unified framework for predicting text quality. In: EMNLP (2008)
10. Kotani, K., Yoshimi, T.: Measuring readability for learners of English as a foreign language by linguistic and learner features. In: Hasida, K., Purwarianti, A. (eds.) Computational Linguistics. CCIS, vol. 593, pp. 211–222. Springer, Singapore (2016). https://doi.org/10.1007/978-981-10-0515-2_15
11. Brinton, J.E., Danielson, W.A.: A factor analysis of language elements affecting readability. Journal. Q. **35**(4), 420–426 (1958)
12. Todirascu, A., Ois, T.F., Bernhard, D.: Are cohesive features relevant for text readability evaluation? pp. 987–997 (2017)
13. Biber, D.: Spoken and written textual dimensions in English: resolving the contradictory findings. Language **2**(62), 384–414 (1986)

The Influence of Facial Width-to-Height Ratio on Micro-expression Recognition

Siwei Zhang, Jinyuan Xie, and Qi Wu[✉]

Cognition and Human Behavior Key Laboratory of Hunan Province,
Department of Psychology, Hunan Normal University, Changsha, China
sandwich624@yeah.net

Abstract. The aim of the present study was to uncover the potential impact of facial width-to-height ratio (fWHR) on micro-expression and macro-expression recognition. The JACBART paradigm was used for the presentation of facial expressions. Participants were asked to recognize six kinds of basic expressions (sadness, surprise, anger, disgust, fear, happiness) on high fWHR faces or on low fWHR faces under 67 ms, 333 ms and 600 ms duration conditions respectively. The results indicated that, the fWHR did not affect the recognition of macro-expressions which were presented for 600 ms in the present study, but the fWHR could influence the recognition accuracy of micro-expressions of surprise and happiness. Specifically, participants could identify the facial expression of surprise more effectively on high fWHR faces than on low fWHR faces under the condition of 67 ms. And participants also could recognize the facial expression of happiness more accurately on high fWHR faces in the conditions of 67 ms and 333 ms. These results revealed the facial expressions of happiness and surprise on high fWHR faces may have an early processing advantage in micro-expression recognition. And the result also demonstrated that individuals spontaneously use fWHR as a clue to recognize micro-expressions.

Keywords: Emotion and affection · Mechanism of cognition
Essence of perception · Facial Width-to-Height Ratio · Micro-expression
Micro-expression recognition

1 Introduction

We often say a person should never judge a book by its cover. But in real life, we usually make a judgment about his/her age or personal character, according to individuals' appearance, especially the facial features, when we see unfamiliar faces on the moment. Humans are very good at using other people's facial structures to interpret their nonverbal signals, like facial expressions, helping themselves to achieve a better social interaction.

Indeed, numerous studies in the past have shown that the facial features of the expressers, such as eyes, mouths, and other facial structures (Blais et al. 2012; Deska et al. 2017; Sacco and Hugenberg 2009), can influence facial expression recognition.

© IFIP International Federation for Information Processing 2018
Published by Springer Nature Switzerland AG 2018. All Rights Reserved
Z. Shi et al. (Eds.): ICIS 2018, IFIP AICT 539, pp. 265–272, 2018.
https://doi.org/10.1007/978-3-030-01313-4_28

The present study focused on whether individuals can identify fleeting micro-expressions effectively through other people's facial structures.

Facial width-to-height ratio (fWHR) is a stable facial structure. A number of previous studies investigated the predictive effects of fWHR on individual traits and behaviors (Geniole and Mccormick 2015; Haselhuhn et al. 2015). For instance, fWHR can influence individual's facial expression recognition (Deska et al. 2017). But no studies have examined the impact of fWHR on micro-expression recognition. Therefore, we used the six basic facial expressions (happiness, sadness, anger, fear, disgust, surprise) on high fWHR faces or on low fWHR faces made by Facegen Modeller (www.facegen.com), to investigate the effect of fWHR on facial expression recognition under different duration conditions (67 ms, 333 ms, 600 ms).

1.1 Micro-expressions of Emotion

Individual's emotion is impacted by the unconscious impulses which stems from the brain's mood centers (Frank and Svetieva 2015), so that the control of body by emotion is stronger than that by logical thinking in some cases. The expression of emotions can be indicated by changes in facial expression, body posture, and other manners (Ekman et al. 1983).

Micro-expressions and macro-expressions are both the manifestation of one's internal emotion. Macro-expression is a direct expression of an 'emotion', which is the most common expression in our daily life and it usually lasts for 1/2 s to 4 s at least (Yan et al. 2013). It is also easy to be controlled and falsified. But the micro-expression is the barometer of authentic human emotional activities and can help people to spot the lies (Frank and Svetieva 2015; Vrij and Ganis 2014). Micro-expressions are actually very short and uncontrollable facial expressions that show up when human beings are trying to suppress or hide real emotions (Ekman 2009; Yan et al. 2013). More specifically, micro-expressions usually last no longer than 1/2 s (Matsumoto and Hwang 2011; Yan et al. 2013), and can often be neglected during the daily conversations (Frank and Svetieva 2015). In general, they occur under stressful situations where people want to control the feeling of fear and guilty, such as lying (Porter and ten Brinke 2008). In such pressure circumstances, the ability to better recognize micro-expression may be indispensable. In order to read micro-expressions correctly, a Micro-expression Training Tool (METT) has been developed (Ekman 2002; Hurley et al. 2014).

Currently, micro-expression recognition has been widely applied in the field of medicine, political psychology, national security and justice (ten Brinke et al. 2012; Shen et al. 2012; Weinberger 2010). Even in the field of education, some researchers found that the changes of micro-expression can be used as effective indicators to measure the intellectual level and the knowledge conceptual paradox of students. It can also be used to monitor the teaching efficiency (Chiu et al. 2014). However, due to its characteristics, micro-expression recognition is challenging for humans and the recognition accuracy is about 45 ~ 59% (Matsumoto and Hwang 2011). Therefore, it is advisable for researchers to develop automatic micro-expression recognition tools (e.g., Guo et al. 2017; Wu et al. 2011) and investigating the potential factors that affect the recognition of micro-expressions in humans would help us to better achieve that purpose.

1.2 Facial Width-to-Height Ratio (FWHR)

As an anthropometric measure of facial shape, fWHR is calculated by measuring the ratio between bizygomatic width and the distance from the upper lip to the mid-brow (Hehman et al. 2015; Weston et al. 2007). The higher the fWHR, the wider the face. The smaller the fWHR, the longer the face (Haselhuhn and Wong 2012). Abundant studies found that high fWHR is associated with aggression and dominance (Geniole and Mccormick 2015). Meta-analytic evidence also strongly indicate that there is a positive relationship between high fWHR and aggression traits (Haselhuhn et al. 2015).

Furthermore, there were findings suggested that the anger—the state most closely related to aggression—increased fWHR (Marsh et al. 2014). And some researchers also found that high fWHR may share phenotypic overlap with angry expressions (Deska et al. 2017). The recent evidence also showed that fWHR significantly influences the recognition of macro-expressions of anger, fear and happiness (Deska et al. 2017). For example, researchers demonstrated that when neutral expressions were presented, high fWHR and low fWHR neutral expression were respectively perceived as signaling anger and fear. When individuals' faces display emotional expressions, such as happiness, fear and anger, fWHR can bias ascriptions of emotion by facilitating the recognition of fear and happiness on low fWHR faces and increasing the recognition of anger on faces with high fWHR (Deska et al. 2017).

2 The Current Study: Effects of Facial Width-to-Height Ratio on the Recognition Micro-expressions

Macro-expression is an important source of social interactions that appear over a single or various regions of the face. It is easily recognized by human. Like macro-expression, micro-expression has three rendering stages: onset, peak and offset (Yan et al. 2013). But unlike macro facial expressions, micro-expression displays more rapidly and transiently (less than 1/2 s) which makes it hard to be identified. And few can fake a micro-expression (Frank and Svetieva 2015).

Macro-expression recognition has already been studied a lot by researchers. We found that macro facial expression recognition is easily to be influenced by our facial structure, such as larger eyes facilitated the recognition of fearful facial expressions (Sacco and Hugenberg 2009). The latest study also found that the fWHR can influence the recognition of macro facial expression of anger, fear and happiness (Deska et al. 2017). Previous study has also illustrated that the recognition of micro-expression is affected by facial features, such as the mouth movements of a face inhibited conscious detection of all types of micro-expressions in that face (Iwasaki and Noguchi 2016). But it is unclear whether the fWHR can also influence the recognition of micro-expressions. Therefore, the present study investigated the relationship between fWHR and the recognition of micro-expressions and explored the differential influence of fWHR on macro-expression and micro-expression recognition.

3   Method

3.1   Participants and Design

We used the Gpower software (V3.1; Faul et al. 2009) to estimate our sample size. This analysis suggested we collect at least 168 participants to obtain 0.8 power. Consequently, one hundred sixty-eight undergraduate students as participants were recruited from Hunan Normal University. All the participants were Chinese, $M_{age} = 21.75$, $SD = 1.75$, 47.6% female. Participants had normal or corrected-to-normal vision and never had experience with such an experiment. After the experiment, the participants were given partial course credit or monetary reward.

A 2 (fWHR: high fWHR, low fWHR) × 3 (duration: 67 ms, 333 ms, 600 ms) × 6 (facial expression: sadness, surprise, anger, disgust, fear, happiness) mixed-model experimental design was used, with fWHR being the between-subjects factor while facial expression and duration being the within-subjects factors.

3.2   Stimuli

Emotion images used in the practice trials were from Nimstim database (Tottenham et al. 2009), which could help participants to get familiar with the experimental procedure. In the formal test, we used the FaceGen Modeller software to generate 24 models (12 females, 12 males; age 21–30) and transformed them into images of seven facial expressions (neutral, sadness, surprise, anger, disgust, fear, happiness) with high fWHR faces (fWHR > 1.9) and low fWHR faces (fWHR < 1.7) (Chae et al. 2014; Haselhuhn and Wong 2012). The height of faces from one person were consistent. And the intensity of basic expressions was set to maximum (1). The number of images was 336 in total, the size of each image was 400 × 400 pixels.

4   Apparatus and Procedure

Each participant was randomly assigned to the high fWHR face group ($n = 84$) or the low fWHR face group ($n = 84$). Before the formal test, participants were familiarized with the procedure and finished six practice trials. The stimulus was presented centrally with a 60-Hz refresh rate on a laptop. The E-Prime software controlled stimulus presentation and data collection. In the practice trials, six images of basic expressions would be presented. In the formal test, the duration of expression was divided into three conditions (67 ms, 333 ms and 600 ms) which consisted of 144 trials, with 48 trials per condition. The 24 models were randomly assigned to three duration conditions, with 8 models per condition. The order of the combination of the models and the duration conditions was counterbalanced across participants by balanced Latin square. In each condition, the six micro-expressions of each model were randomly displayed.

This study adopted the JACBART paradigm (Ekman 2002; Svetieva and Frank 2016) to present the facial expressions. Firstly, a fixation cross (500 ms) was presented in the center of the screen, then the facial expression image was sandwiched in between two 1000 ms presentations of the same model's neutral expression. Facial expression images randomly presented for 67 ms, 333 ms or 600 ms. When the facial expression disappeared, participants were asked to choose one of six emotion expression labels (i.e., sadness, surprise, anger, disgust, fear, happiness) and "none of the above" from a list to describe the expression on the screen (pressing one of seven buttons 1–7 on a response keyboard). The next trial was presented after the participants made the judgment (unlimited reaction time).

5   Results

We conducted a 2 (fWHR: high fWHR, low fWHR) × 3 (duration: 67 ms, 333 ms, 600 ms) × 6 (facial expression: sadness, surprise, anger, disgust, fear, happiness) analysis of variance (ANOVA) on recognition accuracy. The analysis yielded a significant main effect on facial expression [F (5, 830) = 356.93, $p < 0.01$, $\eta_p^2 = 0.68$], and the duration [F (2, 332) = 125.39, $p < 0.01$, $\eta_p^2 = 0.43$]. Generally, the recognition accuracy of happy, surprise and sad expressions was higher than other expressions in three duration conditions, and the recognition accuracy of all facial expressions in the condition of 600 ms duration was much higher than the other two duration conditions. Descriptive statistical results were shown in Table 1. The main effect of fWHR was not significant [F (1, 166) = 2.33, $p > 0.05$, $\eta_p^2 = 0.01$]. The interaction between facial expression and duration was significant [F (10, 1660) = 15.15, $p < 0.01$, $\eta_p^2 = 0.08$]. However, there were no significant interactions of fWHR × facial expression and fWHR × duration [F (5, 830) = 0.86, $p > 0.05$, $\eta_p^2 = 0.01$; F (2, 332) = 1.26, $p > 0.05$, $\eta_p^2 = 0.01$]. The three-way interaction among the three factors was significant [F (10, 1660) = 2.15, $p < 0.05$, $\eta_p^2 = 0.01$].

Table 1.   Descriptive statistical results from this study

	67 ms High fWHR		67 ms Low fWHR		333 ms High fWHR		333 ms Low fWHR		600 ms High fWHR		600 ms Low fWHR	
	M	SD	M	SD	M	SD	M	SD	M	SD	M	SD
Sadness	0.69	0.33	0.71	0.35	0.75	0.34	0.74	0.34	0.79	0.31	0.75	0.33
Surprise	0.72	0.26	0.64	0.25	0.78	0.21	0.73	0.24	0.76	0.23	0.77	0.23
Anger	0.12	0.17	0.14	0.19	0.16	0.21	0.13	0.20	0.12	0.19	0.12	0.19
Disgust	0.07	0.13	0.08	0.14	0.09	0.20	0.09	0.19	0.08	0.17	0.11	0.19
Fear	0.40	0.27	0.40	0.29	0.59	0.33	0.54	0.34	0.60	0.30	0.57	0.33
Happiness	0.78	0.22	0.67	0.27	0.90	0.18	0.83	0.20	0.90	0.17	0.85	0.22

Further simple effects analysis revealed that, participants were more sensitive to the facial expressions of happiness and surprise on high fWHR faces than on low fWHR faces under the condition of 67 ms [F (1, 166) = 8.31, $p < 0.01$, $\eta_p^2 = 0.05$; F (1, 166) = 4.14, $p < 0.05$, $\eta_p^2 = 0.02$]. But under the condition of 67 ms, the fWHR did not influence the recognition of other micro-expressions (i.e., sadness, anger, disgust, fear) [$Fs < 1$, $ps > 0.43$]. In addition, under the condition of 333 ms, the recognition accuracy of the expression of happiness was higher on high fWHR faces than on low fWHR faces [F $(1, 166) = 5.11$, $p < 0.05$, $\eta_p^2 = 0.03$]. But the fWHR also did not influence the recognition of other micro-expressions (i.e., sadness, surprise, anger, disgust, fear) [$Fs < 1.8$, $ps > 0.19$]. The fWHR also did not influence the recognition of macro-expression in the condition of 600 ms (i.e., sadness, surprise, anger, disgust, fear, happiness) [$Fs < 3.4$, $ps > 0.07$].

6 Discussion

This study compared the recognition of facial expression with 6 basic expressions on high fWHR faces or on low fWHR faces under different duration conditions (67 ms, 333 ms, 600 ms), and investigated the differential influence of facial width-to-height ratio on macro and micro expressions. To be specific, the duration of expression did influence the recognition of facial expression, which was consistent with previous studies (Shen et al. 2012). And the results also showed that, the fWHR did not affect the recognition of macro expression which was presented for 600 ms in the present study. In contrast, the fWHR influenced the micro-expressions of happiness and surprise. More specifically, participants identified the facial expression of surprise more effectively on high fWHR faces than on low fWHR faces under the condition of 67 ms. And participants also recognized the facial expression of happiness more accurately on high fWHR faces in the condition of 67 ms and 333 ms duration. It suggests that the expression of surprise and happiness on high fWHR faces may have an early advantage in micro-expression recognition.

A recent eye-tracking study indicated that individuals pay more attention to the eyes area in large faces, while the distribution of gaze towards small faces is still in the nose center (Wang 2018). The expression of surprise is accentuated by the eyes, such as raised eyebrows and wide eyes (Ekman 1978). In a very short period, individuals are likely to notice the eyes area first. So the micro-expression of surprise on high fWHR face may be more likely to be unconsciously processed by individuals. The reason for fWHR can influence the recognition of happiness on high fWHR faces might be related to the processing characteristics of the expression of happiness. The expressed regions of happiness in face is mainly concentrated on the mouth, which is the prominent region to convey information (Blais et al. 2012). Studies have indicated that the lower face plays more important role than the upper face in micro-expression recognition (Iwasaki and Noguchi 2016). So high fWHR means the lower face is large, the participants were more likely to focus on the mouths, and thus were more accurately to recognize the expression of happiness. The present study, did not find a similar results as previous research that the fWHR affects the recognition of macro-expression (Deska et al. 2017). However, in the previous studies (Deska et al. 2017), researchers had only provided two labels (anger versus fear or happiness) for participants to choose. This procedure would definitely overactive the concepts of anger thereby participants could easily establish a link between fWHR and anger. But in the current experiment, participants were asked to choose correct response from more than six expression labels. Such a design would reduce the activation of emotional semantic concepts that may be related to fWHR. The interaction between language and fWHR should be investigated in future works.

In summary, the results demonstrated that there are differences between the recognition of micro-expressions and macro-expressions, and that fWHR can influence the processing of micro-expression of happiness and surprise on high fWHR faces. The current study only found that fWHR can affect the micro-expression recognition, but the underlying mechanisms behind this phenomenon—such as how the fWHR changes the processing of faces—have not been clarified yet. In addition, our participants mainly were from China. More cross-cultural studies should be considered in future.

Acknowledgements. The authors wish to express sincere appreciation to Shengjie Zhao (Hunan Normal University, Changsha, China) for his assistance with data collection. This work was supported by the National Natural Science Foundation of China (Grant No. 31300870) and the Hunan Normal University (Grant No. 13XQN01 2015yx08).

References

Blais, C., Roy, C., Fiset, D., Arguin, M., Gosselin, F.: The eyes are not the window to basic emotions. Neuropsychologia **50**(12), 2830–2838 (2012)

Chae, D.H., Nuru-Jeter, A.M., Adler, N.E., Brody, G.H., Lin, J., Blackburn, E.H., et al.: Discrimination, racial bias, and telomere length in African–American men. Am. J. Prev. Med. **46**(2), 103–111 (2014)

Chiu, M.H., Chou, C.C., Wu, W.L., Liaw, H.: The role of facial microexpression state (FMES) change in the process of conceptual conflict. Br. J. Edu. Technol. **45**(3), 471–486 (2014)

Deska, J.C., Lloyd, E.P., Hugenberg, K.: The face of fear and anger: facial width-to-height ratio biases recognition of angry and fearful expressions. Emotion **18**(3), 453–464 (2017)

Ekman, P.: Facial Action Coding System. Network Information Research, Salt Lake City, UT (1978)

Ekman, P.: Micro Expression Training Tool. University of California, San Francisco (2002)

Ekman, P.: Lie catching and microexpressions. In: The Philosophy of Deception, pp. 118–133 (2009)

Ekman, P., Levenson, R.W., Friesen, W.V.: Autonomic nervous system activity distinguishes among emotions. Science **221**(4616), 1208–1210 (1983)

Faul, F., Erdfelder, E., Buchner, A., Lang, A.G.: Statistical power analyses using G* Power 3.1: tests for correlation and regression analyses. Behav. Res. Methods **41**(4), 1149–1160 (2009)

Frank, M.G., Svetieva, E.: Microexpressions and deception. In: Understanding Facial Expressions in Communication, pp. 227–242 (2015)

Geniole, S.N., Mccormick, C.M.: Facing our ancestors: judgements of aggression are consistent and related to the facial width-to-height ratio in men irrespective of beards. Evol. Hum. Behav. **36**(4), 279–285 (2015)

Haselhuhn, M.P., Ormiston, M.E., Wong, E.M.: Men's facial width-to-height ratio predicts aggression: a meta-analysis. PLoS ONE **10**(4), e0122637 (2015)

Haselhuhn, M.P., Wong, E.M.: Bad to the bone: facial structure predicts unethical behaviour. Proc. Biol. Sci. **279**(1728), 571–576 (2012)

Hehman, E., Leitner, J.B., Deegan, M.P., Gaertner, S.L.: Picking teams: when dominant facial structure is preferred. J. Exp. Soc. Psychol. **59**, 51–59 (2015)

Hurley, C.M., Anker, A.E., Frank, M.G., Matsumoto, D., Hwang, H.C.: Background factors predicting accuracy and improvement in micro expression recognition. Motiv. Emot. **38**(5), 700–714 (2014)

Iwasaki, M., Noguchi, Y.: Hiding true emotions: micro-expressions in eyes retrospectively concealed by mouth movements. Sci. Rep. **6**, 22049 (2016)

Guo, J., Zhou, S., Wu, J., Wan, J., Zhu, X., Lei, Z., et al.: Multi-modality network with visual and geometrical information for micro emotion recognition. In: IEEE International Conference on Automatic Face and Gesture Recognition IEEE, pp. 814–819 (2017)

Marsh, A.A., Cardinale, E.M., Chentsova-Dutton, Y.E., Grossman, M.R., Krumpos, K.A.: Power plays expressive mimicry of valid agonistic cues. Soc. Psychol. Pers. Sci. **5**(6), 684–690 (2014)

Matsumoto, D., Hwang, H.S.: Evidence for training the ability to read microexpressions of emotion. Motivation and Emotion **35**(2), 181–191 (2011)

Porter, S., ten Brinke, L.T.: Reading between the lies: identifying concealed and falsified emotions in universal facial expressions. Psychol. Sci. **19**(5), 508–514 (2008)

Sacco, D.F., Hugenberg, K.: The look of fear and anger: facial maturity modulates recognition of fearful and angry expressions. Emotion **9**(1), 39–49 (2009)

Shen, X., Wu, Q., Fu, X.: Effects of the duration of expressions on the recognition of microexpressions. J. Zhejiang Univ. Sci. B Biomed. Biotechnol. **13**(3), 221–230 (2012)

Svetieva, E., Frank, M.G.: Empathy, emotion dysregulation, and enhanced microexpression recognition ability. Motiv. Emot. **40**(2), 309–320 (2016)

ten Brinke, L., MacDonald, S., Porter, S., O'Connor, B.: Crocodile tears: facial, verbal and body language behaviours associated with genuine and fabricated remorse. Law Hum Behav. **36**(1), 51–59 (2012)

Tottenham, N., Tanaka, J.W., Leon, A.C., Mccarry, T., Nurse, M., Hare, T.A., et al.: The nimstim set of facial expressions: judgments from untrained research participants. Psychiatry Res. **168**(3), 242–249 (2009)

Vrij, A., Ganis, G.: Theories in deception and lie detection. Credibility Assessment, pp. 301–374 (2014)

Wang, S.: Face size biases emotion judgment through eye movement. Sci. Rep. **8**(1), 317 (2018)

Weinberger, S.: Airport security: intent to deceive? Nature **465**(7297), 412–415 (2010)

Weston, E.M., Friday, A.E., Liò, P.: Biometric evidence that sexual selection has shaped the hominin face. PLoS ONE **2**(8), e710 (2007)

Wu, Q., Shen, X., Fu, X.: The machine knows what you are hiding: an automatic micro-expression recognition system. In: D'Mello, S., Graesser, A., Schuller, B., Martin, J.-C. (eds.) ACII 2011. LNCS, vol. 6975, pp. 152–162. Springer, Heidelberg (2011). https://doi.org/10.1007/978-3-642-24571-8_16

Yan, W.J., Wu, Q., Liang, J., Chen, Y.H., Fu, X.: How fast are the leaked facial expressions: the duration of micro-expressions. J. Nonverbal Behav. **37**(4), 217–230 (2013)

Shortest Paths in HSI Space for Color Texture Classification

Mingxin Jin, Yongsheng Dong$^{(\boxtimes)}$, Lintao Zheng, Lingfei Liang, Tianyu Wang, and Hongyan Zhang

School of Information Engineering, Henan University of Science and Technology, Luoyang, China
jinmingxin0501@163.com, dongyongsheng98@163.com

Abstract. Color texture representation is an important step in the task of texture classification. Shortest paths was used to extract color texture features from RGB and HSV color spaces. In this paper, we propose to use shortest paths in the HSI space to build a texture representation for classification. In particular, two undirected graphs are used to model the H channel and the S and I channels respectively in order to represent a color texture image. Moreover, the shortest paths is constructed by using four pairs of pixels according to different scales and directions of the texture image. Experimental results on colored Brodatz and USPTex databases reveal that our proposed method is effective, and the highest classification accuracy rate is 96.93% in the Brodatz database.

Keywords: Texture analysis · Shortest paths · Graph · HSI

1 Introduction

Texture analysis is an active research topic in computer vision and pattern recognition. Its application is very extensive, including texture classification, segmentation, synthesis and retrieval. A texture is generally defined as a complex visual pattern composed of entities or sub-patterns with specific size brightness and slope etc. As an important step in texture classification, texture feature extraction is to extract a discriminative feature from these complex visual pattern. Many scholars have made contributions to feature extraction that is usually divided into four major categories: statistical, signal processing, structural and model based methods. Statistical based methods such as Haralick et al. used the co-occurrence matrices to model the texture pattern [13]. Signal processing methods is also called transform-based methods. The widely used transforms include Gabor transform [14], wavelet transforms [15,20,23,24], contourlet transforms [21] and shearlet transforms [22]. The typical structural approaches are the local binary pattern-based methods including complete local binary pattern (CLBP) [16] and scale selective local binary pattern (SSLBP) [17] and so on. The model based methods that depend on stochastic models to interpret image texture.

© IFIP International Federation for Information Processing 2018
Published by Springer Nature Switzerland AG 2018. All Rights Reserved
Z. Shi et al. (Eds.): ICIS 2018, IFIP AICT 539, pp. 273–281, 2018.
https://doi.org/10.1007/978-3-030-01313-4_29

In addition, some scholars put forward other methods, such as Martinez et al. used deterministic tourist walk to analyze and classify texture [5], Backes et al. proposed fractal descriptors [6], complex network theory proposed by Costa et al. [7] and simplified gravitational systems [8,9].

However, most of these analysis methods are on gray-scale image. In the real world, color texture is the main form of existence. The recognition of color texture is more consistent with human vision. Therefore, color texture classification is still a challenge in the field of texture classification. In the past literature, we know that many scholars have done a lot of research on color texture analysis. For example, Drimbarean et al. extended the three gray-scale related texture analysis methods to color image [3], Harvey et al. compared different color texture classification methods on the theoretical and experimental results [4]. Recently, Li et al. used Gaussian copula models of Gabors wavelets to analysis color texture [18], Napoletano et al. compared hand-crafted and learned descriptors for color texture classification [19].

In this paper, we extend the analysis methods proposed in [1] and [2] respectively in RGB and HSV color space to HSI color space. Particularly, two undirected graphs are used to model the H channel and the S and I channel respectively in order to model a color texture image. First-order statistic of the shortest path is constructed by using four pairs of vertices according to different directions is calculated as features of the texture (called Shortest Paths in Graphs method − SPG method).

The remainder of the paper is organized as follows. Section 2 shows how a color texture can be modeled as graph in HSI color space. Section 3 presents the shortest path as the texture descriptors. In Sect. 4, we describe experiments in which our approach is compared against some traditional methods and the results achieve by our proposed methods. Finally, Sect. 5 gives a brief conclusion.

2   Model Texture to Graph

2.1   Graph and Shortest Path

In graph theory, a graph $G = (V, E)$ is composed of a vertex set V and an edge set E. Due to that an digital image is expressed by its pixels, it is reasonable that any pixel in a given image texture can be considered as a vertex and relationship between any pair of pixels can be represented by an edge. As a classical problem in graph theory, the shortest path problem aims at finding the minimum path form the initial vertex to target vertex in a given weighted directed graph or undirected graph.

2.2   Modeling Texture to Graph

Modeling an Undirected Graph from Texture. Graphs can be divided into directed graphs and undirected graphs. The difference between them is whether they are directional. In undirected graphs, the edge of the link vertex v_i and v_j

lacks orientation, $(v_i, v_j) = (v_j, v_i)$. On the other hand, there is a clear starting vertex and end vertex, $(v_i, v_j) \neq (v_j, v_i)$. We aim to build a discriminant feature that characterizes an input texture image based on its undirected graph representation, which the first step of this processing is constructing the undirected graph that represents the neighborhood relation of the given image. A graph $G = (V, E)$ is composed of the set of vertices V and edges E. We first set up the vertices set V. In this model, each pixel $I(x, y), x = 1, \ldots, M$ and $y = 1, \ldots, N$ is viewed as a vertex belonging to the set V. The location of this vertex in the graph is the same as in the original image. Secondly, we propose to represent the neighborhood relationship by using the edges set E, connecting to each pair of vertices where the Chebyshev distance between them is shorter than or equal to a threshold value t ($t=1$ in our paper):

$$E = \left\{ e = (v, v^0) \in V * V \mid \max \left(|x - x_0|, |y - y_0| \right) \leq 1 \right\} \tag{1}$$

where x and y are the Cartesian coordinates of the pixel $I(x, y)$ associated to the vertex v. Finally, weight $w(e)$ assigned to each $e \in E$, which is defined as:

$$w(e) = |I(x, y) - I(x_0, y_0)| + \frac{1}{2} [I(x, y) + I(x_0, y_0)] \tag{2}$$

where $I(x, y) = g, g = 0, \ldots, L$ represents the intensity value of the corresponding pixel. Figure 1 shows an undirected graph on a gray-scale image.

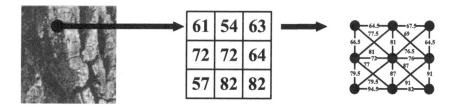

Fig. 1. Undirected graph on gray-scale image

Graph Based Representation of an HSI Color Texture. HSI color space is similar to RGB color space, each color is composed of three color components. But the difference is that the three color components of HSI are Hue (H), Saturation (S), and Intensity (I), where H represents the attribute of the pure color, S measures the extent of pure being diluted by the white light, and I is a subject description and a key parameter for human perception of color. The difference between the HSI color space and the HSV color space includes the difference in the representation of the first model, and second in the HSV, the component Value (V) represents the degree of bright color, and finally the calculation of the component V and the component I form the RGB color space is different.

In the process of building a graph. First, we regard pixels in each color channel as a vertex, and connect each pixel in a specific channel. This process is similar to the color texture analysis method introduced in article [1], [2]. Let $I_H(x,y)$, $I_S(x,y)$ and $I_I(x,y)$ be each color component in an HSI color texture. According to the method described in the previous section, each color components is constructed into an undirected graph. In this way, three separate undirected graphs in HSI color space represent a color texture image.

Furthermore, we would explore the neighborhood relationship between the color components belonging to neighbor pixels. Because the HSI color space reflects the way people perceive the color of the visual system. Human vision is most intuitive to hue perception, we suggest that the H channel be regarded as an independent undirected graph to represent the attribute of the pure color, and the other two channels S and I can be connected to construct a new graph. By adding restrictions to Eq. (1), we create the edge that connects the S channel and I channel. The condition is $\{e = (v, v_0)|v \in S \wedge v_0 \in I\}$. As shown in Fig. 2, texture images are represented by two undirected graphs.

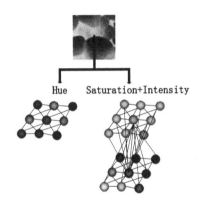

Fig. 2. Modeling a HSI color texture   **Fig. 3.** Weight equation analysis

2.3   Edge Weight Analysis

In order to calculate the shortest path, Eq. (2) is used to assign weights to each side of the undirected graph. The calculation of weights consists of two parts, each of which plays an important role in extracting feature vectors.

The first part of Eq. (2) is the difference between the intensities of two color components. We use the absolute value of the difference to avoid negative values of weights. As Fig. 3 shows above, exploring the shortest path from vertex V3 that maintains the intensity of the 0.1 color component, the shortest path will choose the edge between vertex $V3$ and $V1$. Obviously, the absolute value of the difference between the vertex $V3$ and $V1$ is smaller than the value of between vertex $V3$ and $V2$. This means that those two vertices are similar.

Moreover, the second part of Eq. (2) shows the average intensity between two vertices. In Fig. 3, the absolute value of the difference between the vertices $V3$ and $V1$ is equal to the absolute value of the difference between the vertices

$V2$ and $V1$. We have to consider the average intensity between two vertices as a second part of the weight equation. This part emphasizes exploring the lower levels of intensity in the texture image. By considering the two parts, we can obtain a balance absolute value of the difference and average intensity minimization during the computes of the shortest path in the image.

It is worth noting that in the undirected graph jointly represented by the S channel and I channel, the starting point and terminal point of the shortest path must be located in the same channel.

3   Express Color Texture Features with the Shortest Path

As mentioned above, we propose to construct two undirected graphs to represent the color texture images. One is an undirected graph composed of H-channel, and the other is an undirected graph composed of S-channel and I-channel. The shortest paths on two undirected graph are calculated as a feature vector. In order to hold as much as possible of information about the color texture, we recommend to calculate the shortest path between the four sets of vertices in different directions. The paths in these four directions are the diagonal direction (path $p_{45°}$ and $p_{-45°}$), horizontal (path $p_{0°}$) direction and vertical path (path $p_{90°}$), as shown in Fig. 4.

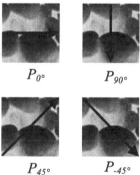

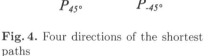

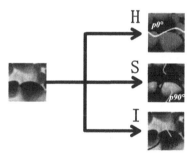

Fig. 4. Four directions of the shortest paths

Fig. 5. Examples of shortest path

The starting and ending vertices of the shortest path are respectively on the three color components. We can obtain a total of twelve paths. On the other hand, as both saturation and intensity correspond to one single undirected graph, the search for the shortest path according to a specified orientation will indeed explore both color channels as shown in Fig. 5.

In order to describe the local texture image information as much as possible, we suggest covering the texture with a grid of size $r*r$, where r is a divisor of the original texture size. For each grid block, we calculate the four shortest paths

$(p_{0^\circ}, p_{45^\circ}, p_{-45^\circ}, p_{90^\circ})$ on the three color components (for example, we can cover an image $128 * 128$ pixels with $32 * 32$ grid, each grid block size is $4 * 4$ pixels). We use the Dijkstra's algorithm to calculate the shortest path. Then, for each path direction d°, $d^\circ = \{0^\circ, 45^\circ, -45^\circ, 90^\circ\}$, we compute the first-order statistics in each direction as average μ_{d° and standard deviation σ_{d°. By following this strategy, the feature vector can be defined as:

$$\alpha_r^c = \left[\mu_{0^\circ}, \sigma_{0^\circ}, \mu_{45^\circ}, \sigma_{45^\circ}, \mu_{-45^\circ}, \sigma_{-45^\circ}, \mu_{90^\circ}, \sigma_{90^\circ}\right]$$

where r is the grid size of covering texture in a specific color channel $C = \{H, S, I\}$. Thus, one single vector that holds all the texture image texture is composed of these vectors α_r^H, α_r^S and α_r^I, which are the feature vectors characterizing each of the color channels:

$$\theta_r = \left[\alpha_r^H, \alpha_r^S, \alpha_r^I\right]$$

According to the different size of the grid scale covering the texture, a multi-scale analysis can be accomplished by concatenating previous feature vectors corresponding to different grid size:

$$\theta_{r_1, r_2, ..., r_n} = \left[\theta_{r_1}, \theta_{r_2}, ..., \theta_{r_n}\right]$$

4   Experiment and Results

In this section, various experiments are carried out to demonstrate the efficiency of our proposed texture classification method. The parameter r is used to define the grid size $r * r$ for covering texture image. In order to avoid dealing with overlapping grid, we choose the common divisors of the original texture image size ($128 * 128$ or $160 * 160$) as the set of grid size r values: $r=4, 8, 16, 32$. 1 values were omitted because it does not provide the desire information about the standard deviation, which makes the composition of the θ_r not possible. Moreover, when the grid size is $128 * 128$ or $160 * 160$, each grid block size is $1 * 1$ pixel, it is too small to provide meaningful information about the texture.

We use the one nearest neighbor (1NN) classifier to evaluate the accuracy of our proposed approach for discrimination. At the same time, we use leave-one-out cross-validation that is using each sample in the database for validation while the remaining samples are used as the training set. And we also use holdout strategy by considering 2/3 of the samples for training and 1/3 for testing, with 10 repetitions.

In this experiment, we evaluate our proposed texture classification on the colored Brodatz database and USPTex. Colored Brodatz not only preserves the advantages of the original Brodatz database rich texture content, but also has a variety of color content. For the experiment using this database, we consider a total of 1792 samples of $160 * 160$ pixels size grouped in to 112 texture classes (16 texture samples per class). USPTex database consists of common textures, we consider a total 2292 samples of $128 * 128$ pixels size grouped into 191 texture classes (12 texture samples per class).

The proposed method is evaluated by two types of analysis, namely single-scale and multi-scale. Firstly, we conduct single-scale analysis. As seen in the Table 1, our proposed method has achieved good results in the Brodatz, the highest accuracy can reach 96.93% and 97.38%. As the size of the grid increase, so does the accuracy. This is due to the fact that the increase in grid size corresponds to a decrease in the size of the grid blocks to ensure better local texture analysis performance.

The second time we conduct a multi-scale analysis, the experimental results were slightly lower than single-scale. Table 2 shows the multi-scale results.

Table 1. The results of a single-scale on two texture database

Accuracy (%)								
	1NN+holdout				1NN+leave-one-out			
Grid size	32 * 32	16 * 16	8 * 8	4 * 4	32 * 32	16 * 16	8 * 8	4 * 4
Brodatz	96.93	90.77	86.2	73.38	97.38	91.8	88.06	75.17
USPTex	57.8	31.13	28.18	24.06	59.81	33.03	29.93	24.04

Table 2. The results of a multi-scale on two texture database

Accuracy (%)			
	1NN+holdout		
Multi-scale	$\{32, 16\}$	$\{32, 16, 8\}$	$\{32, 16, 8, 4\}$
Brodatz	96.46	95.41	54.75
USPTex	54.75	50.9	45.55
	1NN+leave-one-out		
Brodatz	96.99	96.14	94.81
USPTex	56.46	52.53	47.21

Finally, we compare our proposed method with other representative texture classification methods including single-band SPG in RGB [1], HRF [10], MultiLayer CCR [11] and MSD [12] by the considering the feature vector θ_4 with 1NN+holdout on USPTex database. In particular, we compared with the single-band methods in the RGB color space introduced in the literature [1], all of which are the same size grid (32 * 32, each grid block size is 4 * 4 pixels). Additionally, we also tested our approach against a simple method of average values of each R, G and B. The results shows in Table 3.

Table 3. Compared with other classic method on USPTex

Accuracy (%)	
Method	1NN+holdout
Propose method	57.80
Single-band SPG in RGB [1]	54.39
HRF [10]	49.86
MultiLayer CCR [11]	82.08
MSD [12]	51.29
Average RGB	36.19

5   Conclusion

In this paper, we propose to use shortest paths in the HSI space to build a texture representation for classification. Two undirected graphs is used to model texture images, which are undirected graphs constructed by H channel and graphs constructed jointly by S channel and I channel. A texture representation vector can be obtained by computing the shortest path in four different directions in a specific channel. Experimental results show that our proposed color texture classification method is effective.

Acknowledgements. This work was supported in part by Program for Science & Technology Innovation Talents in Universities of Henan Province under Grant 19HASTIT026, and in part by the Training Program for the Young-Backbone Teachers in Universities of Henan Province under Grant 2017GGJS065.

References

1. Sá Junior, J.J.M., Cortez, P.C., Backes, A.R.: Color texture classification using shortest paths in graphs. IEEE Trans. Image Process. **23**(9), 3751–3761 (2014)
2. Moutaouakkil, M.E., Maliani, A.D.E., Hassouni, M.E.: A graph based approach for color texture classification in HSV color space. In: International Conference on Wireless Networks and Mobile Communications, pp. 1–5 (2017)
3. Drimbarean, A., Whelan, P.F.: Experiments in colour texture analysis. Pattern Recognit. Lett. **22**(10), 1161–1167 (2001)
4. Harvey, R.W.: Theoretical and experimental comparison of different approaches for color texture classification. J. Electron. Imaging **20**(4), 49–59 (2011)
5. Martinez, A.S., Bruno, O.M.: Texture analysis and classification using deterministic tourist walk. Pattern Recognit. **43**(3), 685–694 (2010)
6. Backes, A.R., Casanova, D., Bruno, O.M.: Color texture analysis based on fractal descriptors. Pattern Recognit. **45**(5), 1984–1992 (2012)
7. Costa, L. da F., Rodrigues, F.A., Travieso, G., Boas, P.R.V.: Characterization of complex networks: a survey of measurements. Adv. Phys. **56**(1), 167–242 (2007)
8. Sá Junior, J.J.M., Backes, A.R., Cortez, P.C.: A simplified gravitational model for texture analysis. J. Math. Imaging Vis. **47**(1–2), 70–78 (2013)

9. Sá Junior, J.J.M., Backes, A.R.: A simplified gravitational model to analyze texture roughness. Pattern Recognit. **45**(2), 732–741 (2012)
10. Paschos, G., Petrou, M.: Histogram ratio features for color texture classification. Pattern Recognit. Lett. **24**(1), 309–314 (2003)
11. Bianconi, F., Caride, D.: Rotation-invariant colour texture classification through multilayer CCR. Pattern Recognit. Lett. **30**(8), 765–773 (2009)
12. Liu, G.H., Li, Z.Y., Zhang, L.: Image retrieval based on micro-structure descriptor. Pattern Recognit. **44**(9), 2123–2133 (2011)
13. Haralick, R.M.: Statistical and structural approaches to texture. Proc. IEEE **67**(5), 786–804 (2005)
14. Casanova, D., et al.: Plant leaf identification using Gabor wavelets. Int. J. Imaging Syst. Technol. **19**(3), 236–243 (2010)
15. Unser, M.: Texture classification and segmentation using wavelet frames. IIEEE Trans. Image Process. **4**(11), 1549–1560 (1995)
16. Guo, Z.H., Zhang, L., Zhang, D.: A completed modeling of local binary pattern operator for texture classification. IEEE Trans. Image Process. Publ. IEEE Signal Process. Soc. **19**(6), 1657–1663 (2010)
17. Guo, Z.H., Wang, X.Z., Zhou, J., You, J.: Robust texture image representation by scale selective local binary patterns (SSLBP). IEEE Trans. Image Process. Publ. IEEE Signal Process. Soc. **25**(2), 687–699 (2015)
18. Li, C.R., Huang, Y., Zhu, L.: Color texture image retrieval based on Gaussian copula models of Gabor wavelets. Pattern Recognit. **64**, 118–129 (2017)
19. Napoletano, P.: Hand-crafted vs learned descriptors for color texture classification. In: Bianco, S., Schettini, R., Trémeau, A., Tominaga, S. (eds.) CCIW 2017. LNCS, vol. 10213, pp. 259–271. Springer, Cham (2017). https://doi.org/10.1007/978-3-319-56010-6_22
20. Dong, Y., Ma, J.: Wavelet-based image texture classification using local energy histograms. IEEE Signal Process. Lett. **18**(4), 247–250 (2011)
21. Dong, Y., Ma, J.: Bayesian texture classification based on contourlet transform and BYY harmony learning of Poisson mixtures. IEEE Trans. Image Process. **21**(3), 909–918 (2012)
22. Dong, Y., Tao, D., Li, X., Ma, J., Pu, J.: Texture classification and retrieval using shearlets and linear regression. IEEE Trans. Cybern. **45**(3), 358–369 (2015)
23. Dong, Y., Tao, D., Li, X.: Nonnegative multiresolution representation based texture image classification. ACM Trans. Intell. Syst. Technol. **7**(1), 4:1–4:21 (2015)
24. Dong, Y., Feng, J., Liang, L., Zheng, L., Wu, Q.: Multiscale sampling based texture image classification. IEEE Signal Process. Lett. **24**(5), 614–618 (2017)

The 3D Point Clouds Registration
for Human Foot

Yi Xie[1,2], Xiuqin Shang[2,3], Yuqing Li[2,4], Xiwei Liu[5,6], Fenghua Zhu[2],
Gang Xiong[2,3(✉)], Susanna Pirttikangas[7], and Jiehan Zhou[7]

[1] School of Electrical and Electronic Engineering, The University
of Manchester, Manchester M13 9PL, UK
xieyi94@hotmail.com
[2] The State Key Laboratory of Management and Control for Complex Systems,
Institute of Automation, Chinese Academy of Sciences, Beijing 100190, China
{xiuqin.shang,fenghua.zhu,gang.xiong}@ia.ac.cn,
13051108860@163.com
[3] Cloud Computing Center, Chinese Academy of Sciences, Dongguan,
Guangdong, China
[4] College of Information Science and Technology, Beijing University
of Chemical Technology, Beijing, China
[5] Qingdao Academy of Intelligent Industries, Qingdao, China
xiwei.liu@ia.ac.cn
[6] Beijing Engineering Research Center of Intelligent Systems and Technology,
Institute of Automation, Chinese Academy of Sciences, Beijing 100190, China
[7] The Center for Ubiquitous Computing, University of Oulu, Oulu, Finland
{susanna.pirttikangas,jiehan.zhou}@ee.oulu.fi

Abstract. For personalized design it is important to be able to collect, measure
and evaluate individual properties of human beings. This paper proposes registration for the point clouds during foot 3D scanning. In the experiment, we get
the point clouds of the human foot from the Artec 3D scanner and complete the
registration of the point clouds from different visual angles. Dealing with the
customized footwear, we choose a novel algorithm, which combines the NARF
key point detector and the FPFH descriptor, to improve the efficiency of the
initial iteration and reduce the computation burden of matching process.

Keywords: Registration for point clouds · NARF key point detector
FPFH descriptor

1 Introduction

Social manufacturing can transform traditional enterprises to intelligent enterprises that
can actively sense and respond to customers' individualized needs in massive scale [1].
For example, enterprises will be able to acquire the information or data of customers'
needs and properties, like the 3D data of their feet, and then design an individual
product through 3D printing. Our team focuses on the customized footwear and try to
complete scanning, registration, and reconstruction of foot point clouds swiftly and

© IFIP International Federation for Information Processing 2018
Published by Springer Nature Switzerland AG 2018. All Rights Reserved
Z. Shi et al. (Eds.): ICIS 2018, IFIP AICT 539, pp. 282–292, 2018.
https://doi.org/10.1007/978-3-030-01313-4_30

accurately to get the related parameters of feet. The process includes 3D scanning technology and getting the 3D data from the customers.

However, when acquiring point clouds data, usually it is impractical to get all the geometrical information of the object in one scan. This is because the 3D scanning is related to many factors like the contradiction between the size of the object and the field of the view. The self-occlusion problem from the scanning angle can also cause the problem. The information from the one-sided visual angle is just a part of the whole point clouds data so that we need to integrate all of the information collected. In addition, the mismatching problem caused by rotation and translation also needs to be solved [3]. To achieve the visualized operation of the whole point clouds data, we integrate the data from every visual angle and convert into the same coordinate system according to the visual angle relationship. This is the point clouds registration problem, which this paper focuses on.

Recently, the most common way to match clouds is Iterative Closest Point (ICP) method, proposed by Besl and Mckay [4]. The ICP algorithm establishes the correspondences through iterative search for the closest point between two-point clouds. However, the searching process is very time-consuming and may be trapped at local minimum. Therefore, this paper presents a method, only using the geometric feature data and through a coarse registration process, improving the efficiency of iterative search of ICP.

The remaining of this paper is organized as follows. The registration method is introduced in Sect. 2. The experiments and results are discussed in Sect. 3. Section 4 concludes the paper.

2 Method Review

This paper mainly aims to study 3D point cloud registration for foot 3D scan data. The methods we use and compare are *feature description* and *registration*. Our feature-based approach checks the rigid transformation through matching the same and important feature between two images. First, we extract the features from two pictures including points, lines and edges. Then we establish correspondences among the features. At last, we estimate the coordinate transformation parameters including rotation matrix T and translation vector R. The nature of point cloud registration is to choose a reasonable coordinate transformation, merge each angle point cloud to this same coordinate system, and perform the ICP algorithm [5] to form a complete point cloud. Figure 1 illustrates the corresponding steps.

This paper concentrates on rigid transformations. The rigid transformations have a relatively smaller computation burden and a more uniform model than non-rigid transformations. The key steps in this experiment reported in this paper include the following:

1. Acquire a point cloud from different angles under the rigid transformation with a handheld Artec Eva 3D scanner.
2. Use the Normal Aligned Radial Feature (NARF) algorithm to extract key points from the point cloud.

3. Put each key point on the origin and create a partial reference system. Estimate the Fast Point Feature Histograms (FPFH) descriptor.
4. Perform the SAmple Consensus Initial Alignment (SAC-IA) method to estimate the correspondences and reject the incorrect point pairs.
5. Calculate matrix T and translation vector R.
6. Use the ICP algorithm to get the registration result.

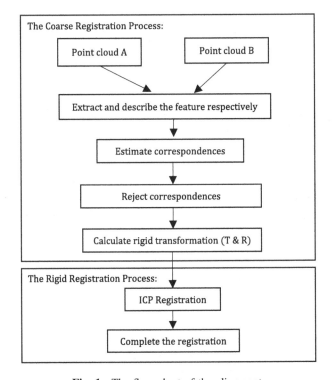

Fig. 1. The flow chart of the alignment.

2.1 The NARF Key Point

It is essential that the stability and distinctiveness in a point cloud are secured. The number of key points can be far less than the number of points in the original point cloud. In combination with the local features, the key point descriptor is usually used for the compact representation of the original point cloud without loss of representation and description, but will accelerate subsequent processing. In this paper, we choose the NARF algorithm to find the key points.

The NARF algorithm is a novel method of point-of-interest extraction that combines the feature descriptors of 3D range data. It aims at interest point detection and feature descriptor, which is applied in the object recognition in the range of 3D data calculation. This kind of method makes explicit use of borders information of the object and make a try to extract feature key points from the place where the surface is stable (to ensure a robust estimation of the normal) and the place where there are sufficient changes in the immediate vicinity [6]. The specific operations of extracting NARF key points from a range image are as follows:

1. Search substantial increases in the 3D distances between neighboring image points to find out edges in the range image, meaning non-continuous traversals from foreground to background.
2. Look at the local neighborhood of every image point and determine a score how much the surface changes at this position and a dominant direction for this change, incorporating the information about borders.
3. Look at the dominant directions in the surrounding of each image point and calculate an interest value that represents (i) how much these directions differ from each other and (ii) how much the surface in the point itself changes (meaning how stable it is).
4. Perform smoothing on the interest values.
5. Perform non-maximum suppression to find the final key points.

We have the following requirements for our interest point extraction procedure [7]: (i) The method must take information about borders and the surface structure into account; (ii) it must select positions that can be reliably detected even if the object is observed from another perspective; and (iii) the points must be on positions that provide stable areas for normal estimation or the descriptor calculation in general.

2.2 The FPFH Descriptor

For most of most real-time applications, calculation of closed point cloud through PFH algorithm is one of the biggest performance bottlenecks. This experiment uses the simplified calculation approach of PFH, which we name it FPFH [8].

2.2.1 The PFH Descriptor

As a result of so many feature points in the most scenes and these points have many same or similar feature values so point feature representation is an effective way to reduce the characteristics of all information. The PFH method gets it parameterization to consult the space difference between the point and its neighbors and form a multi-dimensional histogram describing the geometry property of the point's neighborhood [9]. The high dimensional hyperspace where the histogram is located provides a measurable information space for the feature expression. For 6D Object pose corresponding point cloud surface, it has invariance and it also has robustness under different sampling densities and neighbors' noise level.

The PFH expression method is based on the relationship between the point and its neighborhood and estimated normal. In brief, this approach takes the interaction among all estimated normal into consideration and try to grab the best surface changing situation with the described samples' geometry property. Therefore, making up the feature hyperspace depends on the quality of the point surface normal estimation. In Fig. 2, we see the PFH affected area of a query point (P_q). (P_q) was marked in red and put in the central position of globe (sphere in 3D), which the radius is r.

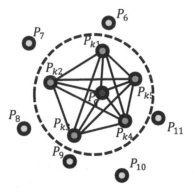

Fig. 2. The influence region diagram for a Point Feature Histogram.

The specific calculation method is as follows:

1. Go through all adjacent point in the k neighborhood of the sample point P_q.
2. For calculating every pair points P_i and $P_j (i \neq j)$ and the error among their corresponding normal vectors N_i and N_j. We define a local coordinate system UVW for one of the points, as in Fig. 3:

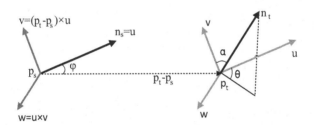

Fig. 3. Define a settled local coordinate system.

Using the above **uvw** frame, the difference between the two normals $\mathbf{n}_s$ and $\mathbf{n}_t$ can be expressed as a set of angular features as follows:

$$\alpha = \mathbf{v} \cdot n_t \# \tag{1}$$

$$\varphi = \mathbf{u} \cdot \frac{(P_t - P_s)}{\|P_t - P_s\|} \# \tag{2}$$

$$\theta = \arctan(\mathbf{w} \cdot n_t, \mathbf{u} \cdot n_t) \# \tag{3}$$

3. Calculate the quadruplet $\langle \alpha, \theta, \varphi, d \rangle$ of each pair of points in the k neighborhood, which reduces the related parameters of the pair and their normal from 12 (**uvw** coordinate values and normal information) to 4. The relation of two points in the

neighborhood can be expressed by α, θ and φ. When these three figures of every point have been calculated, the data would be put in the histograms based on a statistical method.

The paper uses PCL (Point Cloud Library)-features module of PCL to perform the PFH algorithm. The default PFH implementations uses 5 sections to classify. (E.g., each of the four feature values uses five sections to make statistics.) Then we get a feature vector with $125(5^3)$ floating-point numbers elements in a point type, making every pair can get into the interval of α, θ and φ. And we make statistics to every pair of points and finally get point feature histogram describing the local geometric features of the feature points.

2.2.2   The Principle of the FPFH Descriptor

As a simplified and swift algorithm of PFH, FPFH algorithm remains good robustness and identifying characteristic and meanwhile, through the way of simplifying and decreasing the computational complexity, FPFH improves the matching speed to make sure this approach is equipped with real-time [8].

We describe the histogram operating procedure as follows.

1. For each sample point, calculate the three feature values between this point and every point in its k neighborhood and then output a simplified point feature histogram (SPFH).
2. Check the k neighborhood of each point of the k neighborhood respectively and form their own SPFH.
3. Calculate the final SPFH using the followed formula,

$$\text{FPFH}(P_q) = \text{SPFH}(P_q) + \frac{1}{k}\sum_{i=1}^{k}\frac{1}{\omega_i} \cdot SPFH(P_i) \qquad (4)$$

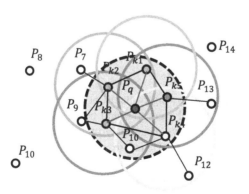

Fig. 4. The influence region diagram for a Fast Point Feature Histogram.

Fig. 5. Initial positions of point clouds.

where ω_i is weight coefficient, representing the reflecting range centered by the k neighborhood of P_q, meaning the distance between sample point P_q and the closed point P_j. As the Fig. 4 shown.

The influence region diagram for a Fast Point Feature Histogram.

The default FPFH implementations uses 11 subintervals. (e.g. each of the four feature values is divided by its parameter section into 11 parts.) The histogram is calculated respectively and be merged into a 33-element feature vector. We unify all SPFH of every pair of point of the sample point and finally we merge each SPFH into the FPFH of this sample point.

2.2.3 The Difference of PFH and FPFH

The paper [9] has introduced the distinctions of calculation method between PFH and FPFH:

1. The FPFH does not do statistics for the calculation parameters of all lined pair points so it may lose some important pairs of points, which might make contribution to the geometrical characteristics of the neighborhoods of the query points.
2. Because of the re-weight computing, FPFH recaptures important pairs of points again with the help of SPFH value.
3. FPFH has more possibilities to work in the real-time application in light of the decrease of computational complexity.

2.3 Improved FPFH Descriptors

As is easily known from the following contexts, at the same time that FPFH algorithm simplifies the calculation of PFH algorithm and boost the operating instantaneity and efficiency, the lost points may have meanings to the geometrical characteristics of the neighborhoods of the query points. Therefore, we added the NARF key points within the extraction of features points to make FPFH algorithm speed up the efficiency of initial iteration and reduce the matching calculation because of the decrease of descriptors dimensions. Therefore, we are able to use the features of few points to find similar parts in an object's or a scene's different visual angles. Figure 6 is the process of the improved FPFH method.

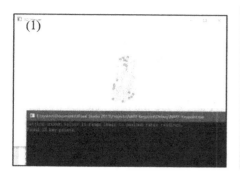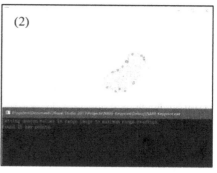

Fig. 6. The NARF key points results. (1) The key points of the point cloud from the single A. (2) The key points of the point cloud from the single B.

2.4 Registration Based on Descriptors

To achieve the rectification of the point clouds, we must find out the similar features between point cloud and determine the overlap sections among the data. We choose mutual correspond estimation algorithm to estimate the match of the corresponding points and take the intersections as the final results, which is able to boost the accuracy of estimation.

2.4.1 Sample Consensus Initial Alignment

Noise and other disturbing elements might have the matching process make mistakes, which would make transformation matrix appear error. To increase the accuracy of the transformation matrix, we work with Sample Consensus Initial Alignment (SAC-IA) algorithm to get rid of the wrong matching relation and we use ICP algorithm to complete the match in the end.

The step of SAmple Consensus Initial Alignment (SAC-IA) can be divided into two parts [11]: the greedy initial alignment and the consistency sampling method. The greedy initial alignment uses the feature of rotation invariance inside the cloud points so it is robust. However, the computational complexity of this alignment is high and it may only get a locally optimal solution. We choose the consistency sampling method, which can keep the same the geometrical relationship of the correspondences rather than get the data of the all correspondences' combinations.

SAmple Consensus Initial Alignment (SAC-IA):

1. Choose s samples points from point cloud A and make sure that their distance is greater than the given minimum value d_{min}.
2. Find the points, which satisfy the similarity conditions to s samples points, from point cloud B and deposit them into a list. Select some correspondences randomly.
3. According to the correspondences of point cloud A and B, calculate the rotation matrix.

2.4.2 Iterative Closest Point (ICP)

The rotation matrix result is able to make the two points clouds data match generally but the accuracy cannot meet the requirement of practical engineering application. Therefore, on the basis of coarse registration, we must go ahead with rigid registration. The ICP algorithm [5] is a common rigid alignment method. It requires closer distance of the two points clouds data. During each iterative process, firstly, determine the sets of correspondences P and Q, P ⊂ A, Q ⊂ B (A and B are the point clouds from two different visual directions) and the amount of the correspondences is n. Then use the least square method to calculate iteratively the best transformation of coordinates (rotation matrix T and translation vector R) make the error the least until it satisfies the error requirement.

The process of ICP can be divided into 4 main parts: (1) Search for correspondences. (2) Reject bad correspondences. (3) Estimate a transformation using the good correspondences. (4) Iterate.

3   Results and Discussion

In this research, we scanned a human foot by Artec3D Eva handed scanner and then aligned point clouds data of two different visual angles it. The development environment chooses Visual Studio 2013 + PCL1.7.2 and we coded with C++ to achieve the procedure of this researched method. The registration results and the analysis are as follow:

Figure 5 shows the initial positions of point clouds of both visual angles.

Then we use NARF algorithm to find out the key points in the single A and B respectively and we can see there are 18 key points in the first visual angle and 15 in the second one.

We tested the improved FPFH method and the FPFH algorithm respectively to complete the coarse registration. Then we used ICP algorithm to finish the rigid registration process. Figure 7 and Table 1 tells that although we added a step to find out the key points, the computing time of the improved FPFH method is much less than the time of the simple FPFH algorithm, while the accuracy of the former is even a little bit better than the latter. This is because calculating the descriptors of the NARF key points of the point clouds reduces the calculation and the key points keep the important information of the point clouds.

The rotation matrix T are
$$\begin{bmatrix} 0.993899 & -0.074168 & 0.081659 & -0.125493 \\ 0.074414 & 0.997230 & 0.000024 & -0.005049 \\ -0.08143 & 0.006053 & 0.996663 & 0.0140594 \\ 0 & 0 & 0 & 1 \end{bmatrix}$$

and
$$\begin{bmatrix} 0.781167 & -0.182765 & 0.596974 & -0.767557 \\ 0.164061 & 0.982681 & 0.086170 & -0.110012 \\ -0.602383 & 0.030627 & 0.797621 & 0.258679 \\ 0 & 0 & 0 & 1 \end{bmatrix}$$
respectively.

The number of the point clouds data of this robot model is approximately 3000. When the number of the point clouds rises to hundreds of thousands or several

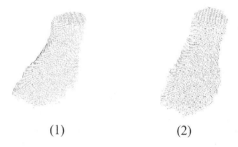

(1)   (2)

Fig. 7. The alignment result. (1) is the rigid registration result based on the improved FPFH method. (2) the rigid registration result based on the FPFH algorithm.

Table 1. The result of the rigid registration

Robot model	The improved FPFH method	The FPFH method
The alignment accuracy	$2.74126e^{-4}$	$2.83642e^{-4}$
The computing time of searching the NARF key points	21	N/A
The computing time of the coarse registration	47	141
The computing time of the rigid registration	30	37
The total time	98	178

millions, the method in this paper has a reliable performance and compact procedure. It also has practical value for the practical engineering requirement.

4   Conclusions

This paper presented a novel point clouds registration method used for creating process to smoothen a customized footwear desing. This method combines four algorithms: NARF, FPFH, SAC-IC and ICP, while the combination of the first two method reduce the computation burden of matching process and short the computing time. The experiment result illustrated the availability and practicability of the proposed method in this paper. Each algorithm has its advantages and disadvantages. The merge of specific algorithms is supposed to be a solution facing some special problems.

Acknowledgment. This work was supported in part by the National Key Research and Development Program of China (No. 2018YFB1004800), the National Natural Science Foundation of China under Grants 61773381, 61773382, 61533019 and 91520301; Chinese Guangdong's S&T Project (2016B090910001, 2017B090912001); Dongguan's Innovation Talents Project (Gang Xiong, Jian Lu); 2017 Special Cooperative Project of Hubei Province and Chinese Academy of Sciences.

References

1. Feiyue, W.: From social computing to social manufacturing: the coming industrial revolution and new frontier in cyber-physical-social space. Bull. Chin. Acad. Sci. **6**, 658–669 (2012)
2. Zhu, Y.J., Zhou, L.S.: Registration of scattered cloud data. J. Comput. Aided Des. Comput. Graph. **18**(4), 475–487 (2006). (in Chinese)
3. Rusu, R.B., Marton, Z.C., Blodow, N., Beetz, M.: Learning information point classes for the acquisition of object model maps. In: Proceedings of the 10th International Conference on Control, Automation, Robotics and Vision (ICARCV), Hanoi, Vietnam, 17–20 December 2008
4. Besl, P.J., Mckay, N.D.: Method for registration of 3-D shapes. In: Robotics: DL Tentative, pp. 239–256. International Society for Optics and Photonics (1992)
5. Zhang, Z.: Iterative point matching for registration of free-form curves and surfaces. Int. J. Comput. Vis. **13**(2), 119–152 (1994)
6. Radu, B.S., Rusu, R.B., Konolige, K., et al.: NARF: 3D range image features for object recognition (2010)
7. Steder, B., Rusu, R.B., Konolige, K., et al.: Point feature extraction on 3D range scans taking into account object boundaries, vol. 30, no. 1, pp. 2601–2608 (2011)
8. Rusu, R.B., Blodow, N., Beetz, M.: Fast point feature histograms (FPFH) for 3D registration. In: IEEE International Conference on Robotics and Automation, pp. 3212–3217. IEEE Press (2009)
9. Rusu, R.B., Blodow, N., Marton, Z.C., Beetz, M.: Aligning point cloud views using persistent feature histograms. In: Proceedings of the 21st IEEE/RSJ International Conference on Intelligent Robots and Systems (IROS), Nice, France, 22–26 September 2008
10. Lu, J., Peng, Z.T., Hong, D.L., et al.: The registration algorithm of point cloud based on optional extraction FPFH feature. J. New Ind. **7**, 75–81 (2014). (in Chinese)
11. Price, M., Green, J., Dickens, J.: Point-cloud registration using 3D shape contexts. In: Robotics and Mechatronics Conference of South Africa (ROBOMECH), pp. 1–5 (2012)

The Cognitive Philosophical Problems in Visual Attention and Its Influence on Artificial Intelligence Modeling

Jing-jing Zhao[✉]

College of Liberal Arts and Science,
National University of Defense Technology, Changsha, Hunan, China
jjzhao1983@126.com

Abstract. Human perception of visual scenes has distinct initiative and purpose characteristics. Human beings can quickly extract information of their interest from massive visual input, giving priority to processing. This selective perception is visual attention. With the rise of artificial intelligence, studying the mechanism of human visual attention and establishing the computational model of visual attention has become a new research hotspot. What is the essence of visual attention? What are the basic units of visual attention? What are the factors that affect visual attention? The in-depth analysis of cognitive philosophy in visual attention helps to understand the mechanism of visual attention and to establish an effective artificial intelligence model. The fact shows that the performance of the computer simulation model can be improved by taking full account of the influence of high-level factors such as task, expectation, memory, knowledge and experience in modeling.

Keywords: Visual attention · Visual cognition · AI philosophy
Computational model of visual attention

1 Introduction

Most of human's perception of the world comes from vision. Compared with the massive visual information received by the retina, the neural resources of our visual system are very limited. So there is a contradiction, the total amount of information provided by the perceptual system is far beyond the maximum capacity of the cortex to store information [1]. However, this so-called bottleneck effect does not bring any discomfort to our visual perception. That's because the cognitive process of human vision is an active process of interacting with the outside world. Through eye scanning, it can make the region of interest to the retinal foveal feature to recognize the target priority, using high resolution foveal. The selective perception is visual attention, which can filter visual information, give priority to important information, and control behaviour based on the information.

As far as physiological characteristics are concerned, visual attention is the basic function developed by primates in the long term evolution process for survival needs. Visual attention can help to find the enemy and from the primate class animal food real-time massive visual information, so the visual information is significant and the environment and memory are closely related. As far as the psychological characteristics are concerned, visual attention reflects the importance of human psychological activity initiative and consciousness, so most of the mathematical modeling of visual attention simulate the filtering operation in the human brain in the information process. As far as the characteristics of machine vision are concerned, the construction of machine vision system requires quick response to complex and changeable external environment through attention mechanism, so as to respond in real time. Therefore, visual attention is related to specific tasks. At present, the research on visual attention mechanism is mainly focused on the cognitive study of visual attention and the computer simulation modeling of visual attention. Scholars in the field of cognition mainly study the mechanism of visual attention mechanism and the construction of cognitive models. Scholars in the fields of computer vision, pattern recognition and artificial intelligence are mainly concerned with how to establish a computational model of visual attention by combining the research results of visual physiology and cognition. The effectiveness of the computer simulation model depends largely on the idea of modeling. The idea of modeling often comes from the understanding of the cognitive mechanism of visual attention. Therefore, the in-depth analysis of the cognitive mechanism of visual attention and the analysis of the cognitive philosophical problems have a great guiding role in improving the thinking of modeling and improving the performance of artificial intelligence model.

2   Cognitive Philosophical Problems of Visual Attention

Although we all know visual attention, so far, the cognitive mechanism of visual attention has not been conclusive. In terms of cognitive philosophy, the main philosophical issues discussed include: what is the essence of visual attention? What are the basic units of visual attention? What are the factors that affect the visual attention? Is there a conscious participation in the process of attention? How to understand the relationship between attention and recognition and so on.

2.1   What is the Essence of Visual Attention?

At first, some theories suggest that visual attention function is the front end of the visual system, so it is necessary to find the area of interest first, and then carry on the more elaborate processing of the back end such as the target recognition. They have studied from the point of view of biological evolution [1]. The representative scholar is BroadBent, who emphasizes that the function of visual attention is actually a filter. When the external visual stimulation enters the filter, it passes through several channels to receive preliminary processing first, and then carries out fine processing of object identification and semantic analysis through subsequent processing [2]. Einhauser et al. pointed out that the correct relationship between visual attention and perception has not been solved to a large extent. Visual attention and perceptual models should be integrated, rather than simply taking visual attention as a preprocessing step for target recognition [3]. Grossberg thinks that the basic unit of visual attention is the surface and boundary. He thinks that there is a link between attention and learning, expectation, competition and consciousness [4]. Fazl et al. proposed the ARTSCAN neural model, and proposed the concept of attention coverage. The prediction results of the model have high consistency with the psychological experimental data [5]. It provides a unified explanation for target attention and spatial attention on how to work together and how to learn content in the scene together. Chikkerur et al. believes that attention is part of the reasoning process that solves the visual task of "where there is" [6]. They emphasize that the main role of the visual system is to infer the identity and position of the target in the visual scene. They introduced Bayesian inference theory for visual attention, and integrated the target identity information and target location information of the dorsal pathway.

2.2   What are the Basic Units of Visual Attention?

The basic unit of visual attention has always been a controversial issue. At present, there are several viewpoints on this issue. Based on the discernibility view, the attention system is restricted by the number of discernibility, so the unit of visual attention is related to the system resolved properties [7]. According to the time view, visual attention selects the attributes that occur simultaneously in time [8, 9]. From the viewpoint of feature, visual salience is based on various characteristics of objects, which is the combination of these characteristics in guiding visual attention [10, 11]. The space based view believes that attention is like a "Spotlight" moving in the visual area. Attention needs to be focused on the specific space area in the visual scene, only the visual information falling into the area can be followed by subsequent analysis, and the information that is not in that area is ignored [12]. From the viewpoint of objects, visual attention plays a role in perceiving units or objects that have been well organized in the pre attention stage. Visual attention can directly select discrete objects in the visual area, instead of concentrating attention on a certain spatial area in the visual scene [13–15]. Therefore, when the attention is paid to a certain object, the various components of the object can be parallel processing in time, while other objects can only be processed in time sequence. At present, more and more psychological experimental results provide support for object-based visual attention theory. The process of attention is consciously involved. Attention is related to perception and recognition.

2.3   What are the Factors that Affect the Visual Attention?

At present, a hot topic in the field of visual cognition is to study the main factors that affect visual attention, that is, what affects our visual attention. According to the physiological research results, ventral pathways and dorsal pathways are involved in this process in different ways. Visual attention can be bottom-up and driven by image data. Visual attention can also be top-down and conceptually driven [16]. With the advent of the eye motion recorder, many experimental data sets related to visual attention have been set up. It is possible to obtain more information about attention from the human visual experience.

This paper takes the data set [17, 18] released by Massachusetts Institute of Technology as an example to analyze the key factors that affect visual attention.

The data set contains 1003 images. From the content point of view, the image of the data set covers natural landscape images, human scene images, and images related to portraits, characters, animals, buildings, and so on. The researchers invited 15 people to look at the images and record the focus distribution of each person's gaze when they watched the image through an eye motion recorder. The researchers recorded each of the 15 people watching the first 5 focuses of each image. Then, these focal points in the same image are labeled, and the overall distribution of the focus of each image is gotten, as shown in Fig. 1.

Fig. 1. Eye focus distribution collected from the eye movement data set [17, 18]

By analyzing the focus distribution map of the dataset, we can find [17]: People tend to pay attention to faces, people and texts, followed by body parts, animals and cars. When there are portraits in the image scene, almost all observers observe the faces instinctively, regardless of whether the parts of the face are clear. When text information exists in the scene images, the observers tend to pay attention to these textual features. These data indicate that visual attention is related to the knowledge structure, personal experience and memory in our brain. Figure 2 is a schematic diagram of the relationship between human visual focus and image content provided by data set publishers. It can be seen from the Fig. 2 that objects with high-level semantic features are more likely to arouse the attention of subjects, and this kind of attention is generally unified.

3   The Computational Model of Visual Attention

Human perception of visual scene has a distinct initiative and purpose, and the perceived content is closely related to the specific visual tasks and memory mechanisms. Therefore, visual perception can be regarded as an active process under certain conditions. As the new research direction and hot spot of computer vision, the core of bionic vision is to simulate the active perception ability of the biological vision system

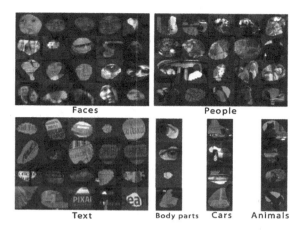

Fig. 2. The image content of the attention focus of people [17, 18]

and accomplish the corresponding task or purpose from the mechanism. The establishment of visual attention computer simulation model is one of the important research contents, which can help us create a bionic visual system which is closer to the human visual cognitive process, in order to accomplish more computer vision tasks.

In the field of computer vision modeling, from the international conferences and periodicals published in the field of computer vision, it can be seen that the computational model of visual attention is one of the current research focus. At first, the research team of California Institute of Technology and the research team of University of Southern California presented their respective representative models respectively. In recent years, with the in-depth research, Massachusetts Institute of Technology and other institutions have carried out related research. The first theoretical framework for visual attention was put forward in 1985 [19]. After several years of exploration and efforts, Itti and Koch put forward the first computer simulation model of visual attention in 1998 [20]. The model has been universally recognized by the academic community, and has become the benchmark of performance evaluation for current visual attention computation models. Walther and others extended the above model, and extended the selected unit to the prototype by hierarchical feedback connection, and realized the continuous multi target recognition in the scene [21]. Navalpakkam and others integrated the bottom-up attention model and top-down statistical knowledge to further optimize the speed of target detection [22]. With the deepening of research, scholars have actively explored the mechanism of attention. Instead of sticking to the original model framework, scholars put forward a series of innovative models from other perspectives, such as: computational models based on spectrum analysis [23–25], computational models based on entropy [26], computational models based on area [27], computational models based on graph theory [28], computational models based on machine learning [17, 29]. Machine learning based modeling is the current research hotspot.

4   Core Issues and Strategies for Constructing Visual Attention Computational Models

Although the human visual system has evolved to a nearly perfect stage, our knowledge of our own visual system is very limited. Thankfully, science always advances and progresses. On the one hand, scientists have developed eye tracking system to obtain the focus distribution of attention when humans are free to watch, and to get "ground truth" through the analysis of the human visual focus. It reflects the response of human visual system to visual scene content, and describes the distribution of saliency in visual scenes. With the appearance of "ground truth", we have real experimental data. On the other hand, with the development of cognitive philosophy, it provides a good theoretical guidance for computer modeling. We can try to think about the core issues of the visual attention computational model from the perspective of cognitive philosophy, and give corresponding coping strategies.

4.1   The Basic Unit Selection Problem of Visual Attention Computational Model

The biggest difference in the basic unit selection of the visual attention computational model is that the visual attention calculation should be based on the basic unit of the object or the space position as the basic unit. In addition, the definition of "object" is also controversial. Does it refer to objects with strict semantic characteristics or some similar target areas? From a psychological point of view, several viewpoints have been supported by experiments. However, from the perspective of cognition, object based visual attention computing is more conducive to subsequent visual analysis. Theoretically, features with more semantic information are more conducive to subsequent identification and analysis. There is no semantic feature in the pure area, and it has no significant effect on the identification and analysis of subsequent dorsal pathways. Therefore, on the basic unit selection problem, the idea of object based modeling is more in line with human visual characteristics.

4.2   Feature Selection of Visual Attention Computational Model

In view of the problem of feature selection in visual attention modeling, different visual attention models are divided mainly in the need to add semantic information on the basis of primary visual features. From a physiological point of view, the computation of visual attention is actually a very complex neural activity, which is not only influenced by the current input signal, but also influenced by the prior knowledge of the memory. Brain science has shown that in the connections between the cerebral cortex and the thalamus, the information transmitted backward is more than an order of magnitude higher than the information transmitted forward. The prior knowledge in memory is a summary of the cerebral cortex for the historical input signal and its response. It has a high level of generality and is a high-level representation of the scene. In addition, cognitive studies show that in the long term memory, the coding of scene images is mainly semantic code. Therefore, from the point of view of physiological structure and cognitive experience, the establishment of visual attention computational model needs to take into account the effects of the primary visual features and the visual features of the semantic information. It is necessary to integrate the integrated effect of the current visual input signal and the historical input signal in memory.

4.3   Selection of Multiple Visual Features Fusion Strategies for Visual Attention Computational Models

In view of the problem of feature selection in visual attention modeling, different visual attention models are divided mainly in the need to add semantic information on the basis of primary visual features. From a physiological point of view, the computation of visual attention is actually a very complex neural activity, which is not only influenced by the current input signal, but also influenced by the prior knowledge of the memory. Brain science has shown that in the connections between the cerebral cortex and the thalamus, the information transmitted backward is more than an order of magnitude higher than the information transmitted forward. The prior knowledge in memory is a summary of the cerebral cortex for the historical input signal and its response. It has a high level of generality and is a high-level representation of the scene. In addition, cognitive studies show that in the long term memory, the coding of scene images is mainly semantic code. Therefore, from the point of view of physiological structure and cognitive experience, the establishment of visual attention computational model needs to take into account the effects of the primary visual features and the visual features of the semantic information. It is necessary to integrate the integrated effect of the current visual input signal and the historical input signal in memory [29].

5   Conclusion

Computer simulation modeling of visual attention has always been a fascinating research topic in the field of artificial intelligence. Therefore, we need to dig into and analyse the cognitive philosophical problems in visual attention, to recognize the basic elements that affect visual attention, to understand the relationship between attention and consciousness, to identify the basic units of visual attention and to understand the mechanism of visual attention. In this way, we can better extract effective visual features in artificial intelligence modeling, build a simulation model based on task and high-level semantic information, and calculate visual attention based on image data to solve more artificial intelligence problems.

Acknowledgement. This work was supported by the National Social Science Fund of China (Visual attention research based on the field of artificial intelligence), Hunan Provincial NSF of China under Grant 2015JJ3018, and Pre research project of National University of Defense Technology under Grant JS17-03-19.

References

1. Baorong, Z.: Applied Psychology. Tsinghua University Press, Beijing (2009)
2. Broadbent, D.: Perception and Communication. Pergamon Press, London (1958)
3. Einhauser, W., Spain, M., Perona, P.: Objects predict fixations better than early saliency. J. Vis. **8**(14), 1–26 (2008)
4. Grossberg, S.: Linking attention to learning, expectation, competition, and consciousness. Neurobiol. Attention, 652–662 (2005)
5. Fazl, A., Grossberg, S., Mingolla, E.: View-invariant object category learning: how spatial and object attention are coordinated using surface-based attentional shrouds. Cogn. Psychol. **58**, 1–48 (2009)
6. Chikkerur, S., Serre, T., Tan, C., et al.: What and where: a Bayesian inference theory of attention. Vis. Res. **50**, 2233–2247 (2010)
7. Isabelle, M., Robert, M.S.: Effects of contrast and size on orientation discrimination. Vis. Res. **44**, 57–67 (2004)
8. Egeth, H.E., Yantis, S.: Visual attention: control, representation, and time course. Annu. Rev. Psychol. **48**, 269–297 (1997)
9. Deco, G., Pollatos, O., Zihl, J.: The time course of selective visual attention: theory and experiments. Vis. Res. **42**, 2925–2945 (2002)
10. Treue, S., Trujillo, J.C.M.: Feature-based attention influences motion processing gain in macaque visual cortex. Nature **399**(6736), 575–579 (1999)
11. Saenz, M., Buracas, G.T., Boynton, G.M.: Global effects of feature-based attention in human visual cortex. Nat. Neurosci. **5**(7), 631–632 (2002)
12. Treisman, A.M., Gelade, G.: A feature integration theory of attention. Cogn. Psychol. **12**(1), 97–136 (1980)
13. Treisman, A.: Feature binding, attention and object perception. Philos. Trans. R. Soc. Lond. B Biol. Sci. **353**, 1295–1306 (1998)
14. Scholl, B.J.: Objects and attention: the state of the art. Cognition **80**, 1–46 (2001)
15. Scholl, B.J., Pylyshynb, Z.W., Feldman, J.: What is a visual object-evidence from target merging in multiple object tracking. Cognition **80**, 159–177 (2001)
16. Koch, C.: The Quest for Consciousness : A Neurobiological Approach. Roberts & Company Publishers (2004)
17. Judd, T., Ehinger, K., Durand, F., Torralba, A.: Learning to predict where humans look. In: IEEE International Conference on Computer Vision (2009)
18. http://people.csail.mit.edu/tjudd/WherePeopleLook/index.html
19. Koch, C., Ullman, S.: Shifts in selective visual attention: towards the underlying neural circuitry. Hum. Neurobiol. **4**(4), 219–227 (1985)
20. Itti, L., Koch, C., Niebur, E.: A model of saliency-based visual attention for rapid scene analysis. IEEE Trans. Pattern Anal. Mach. Intell. **20**(11), 1254–1259 (1998)
21. Walther, D., Koch, C.: Modeling attention to salient proto-objects. Neural Netw. **19**, 1395–1407 (2006)
22. Navalpakkam, V., Itti, L.: An integrated model of top-down and bottom-up attention for optimizing detection speed. In: IEEE Conference on Computer Vision and Pattern Recognition (2006)
23. Hou, X., Zhang, L.: Saliency detection: a spectral residual approach. In: IEEE Conference on Computer Vision and Pattern Recognition (2007)
24. Li, J., Levine, D., An, X., et al.: Visual saliency based on scale space analysis in the frequency domain. IEEE Trans. Pattern Anal. Mach. Intell. **35**, 996–1010 (2012)

25. Guo, C., Zhang, L.: A novel multiresolution spatiotemporal saliency detection model and its applications in image and video compression. IEEE Trans. Image Process. **19**(1), 185–198 (2010)
26. Bruce, N., Tsotsos, J.: Saliency based on information maximization. In: Advances in Neural Information Processing Systems (2006)
27. Gao, D., Han, S., Vasconcelos, N.: Discriminant saliency, the detection of suspicious coincidences, and applications to visual recognition. IEEE Trans. Pattern Anal. Mach. Intell. **31**, 989–1005 (2009)
28. Harel, J., Koch, C., Perona, P.: Graph-based visual saliency. In: Advances in Neural Information Processing Systems (2007)
29. Li, J., Tian, Y., Huang, T., Gao, W.: Probabilistic multi-task learning for visual saliency estimation in video. Int. J. Comput. Vis. **90**(2), 150–165 (2010)

Parallel Dimensionality-Varied Convolutional Neural Network for Hyperspectral Image Classification

Haicheng Qu[✉], Xiu Yin, Xuejian Liang, and Wanjun Liu

Liaoning Technical University, College of Software,
188, longwan south street, XingCheng, Huludao, China
{quhaicheng, liuwanjun}@lntu.edu.cn,
lntuyinxiu@hotmail.com, liangxuejian2015@hotmail.com

Abstract. Many spectral-spatial classification methods of HSI based on convolutional neural network (CNN) are proposed and achieve outstanding performance recently. However, these methods require tremendous computations with complex network and excessively large model. Moreover, single machine is obviously weak when dealing with big data. In this paper, a parallel dimensionality-varied convolutional neural network (DV-CNN) is proposed to address these issues. The dimensionalities of feature maps extracted vary with stages in DV-CNN, and DV-CNN reduces the dimensionalities of feature maps to simplify the computation and the structure of network without information loss. Besides, the parallel architecture of DV-CNN can obviously reduce the training time. The experiments compared with state-of-the-art methods are performed on Indian Pines and Pavia University scene datasets. The results of experiments demonstrate that parallel DV-CNN can obtain better classification performance, reduce the time consuming and improve the training efficiency.

Keywords: Parallel computing · Dimensionality variation
Convolutional neural network · Hyperspectral image classification

1 Introduction

Hyperspectral data contain spectral and spatial information simultaneously [1], which have been used in various fields widely. In hyperspectral image (HSI) processing, dimensionality reduction is one of the most difficult and most necessary problem to be solved. In order to improve the accuracy of HSI classification, numerous methods have been proposed in the past few years.

Traditional HSI classification methods which generally focused on spectral bands included k-nearest neighbours, logistic regression, maximum likelihood, etc. [2]. Spectral-spatial classification could improve the accuracy of classification prominently [3]. And deep learning promoted the development of HSI processing and many methods based on convolutional neural network (CNN) were proposed for classification. Gu et al. proposed a Representative multiple kernel learning (RMKL) method to achieve good classification accuracy and interpretability [4]. Chan et al. proposed PCANet which learned extremely easily and efficiently [5]. Moreover, a stacked

Z. Shi et al. (Eds.): ICIS 2018, IFIP AICT 539, pp. 302–309, 2018.
https://doi.org/10.1007/978-3-030-01313-4_32

autoencoder (SAE) method [6] and a Deep Belief Network (DBN) method [7] were proposed by Chan et al. to deal with HSI classification, which were demonstrated the effectiveness of deep learning in HSI classification. The nonlinear spectral-spatial network (NSSNet), which was based on PCANet and proposed by Pan obtained better performance [8] than other methods. Some CNN-based methods for HSI classification such as 2D-CNN [9] and 3D-CNN [10] were proposed and achieved better performance. However, traditional methods barely made full use of HSI information and CNN-based methods required more tremendous computations and more complex networks because of the increase of kernels for more abundant and accurate features. In addition, the single worker was unable to meet the needs of big data, and parallel computing of cluster can reduce the computing time and improve the efficiency.

In this paper, a parallel dimensionality-varied convolutional neural network (DV-CNN) is proposed. The parallel DV-CNN is designed to solve tremendous amount of HSI data and improve the accuracy of classification. The results of experiments show that the parallel DV-CNN achieves 99.14% of overall accuracy (OA) on Indian Pines and 99.84% of overall accuracy (OA) on Pavia University scene, which is higher than the second best accuracy in CNN-based methods or other state-of-the-art methods. Besides, the parallel architecture of DV-CNN obviously reduces the training time consuming.

The rest of this paper is organized into three sections. In Sect. 2, a detailed description about parallel computing of DV-CNN is presented. Section 3 elaborates the experiments of HSI classification and parallel computing, and the comparisons with state-of-the-art methods are also present. Finally, Sect. 4 presents the conclusions.

2   Proposed Method

This section elaborates the structure of parallel DV-CNN, which include the dimensionality variation of feature maps in four stages of DV-CNN, and the architecture description of parallel computing is elaborated in detail. The structure of parallel DV-CNN is shown in Fig. 1.

2.1   Dimensionality-Varied CNN

Parallel DV-CNN has four stages (shown in Fig. 1) and each of them relates to dimensionality variation. There are two 3D process and three 2D process in the feature extraction of DV-CNN. In stage 1, both 3D feature extraction contain 3D convolution and maxpooling operations. And the size of convolution kernel is related to specific dataset. One of the functions of stage 1 is the spectral-spatial features fusion. In stage 2, some 1D vectors are generated after stage 1 and assembled in to a 2D feature map as the input of the first 2D process. Stage 3 and stage 4 make up a general 2D-CNN for image processing. There are three 2D process for feature extraction which contain 2D convolution, activation and max pooling operations in stage 3. The main components of stage 4 consists of a fully connected layer, a dropout layer and a softmax layer. The output of DV-CNN is also in stage 4. The dimensionality variation of feature map in

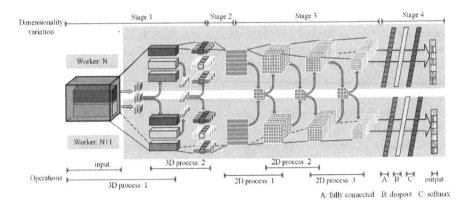

Fig. 1. Structure of parallel DV-CNN

DV-CNN reduces the computations, and the simplified structure of network improves the classification accuracy.

The convolution in first 3D process can be formulated as

$$\text{output}(\boldsymbol{m}_{l,i}) = f\left(conv\left(\text{input}(\boldsymbol{m}_{(l-1)}), \boldsymbol{k}_{l,i}\right) + \boldsymbol{b}_{l,i}\right) \tag{1}$$

where l indicates the layer that is considered, and $m_{(l-1)}$ is the input. $\boldsymbol{k}_{l,i}$ is the convolution kernels with i being the index of feature map generated, and $\boldsymbol{b}_{l,i}$ is the bias corresponded to $\boldsymbol{k}_{l,i}$. The function $conv(\cdot)$ represents convolution operation, and activation function $f(\cdot)$ represents Rectified Linear Unites (ReLU), which is a non-saturation function and faster than other saturating function. $\boldsymbol{m}_{l,i}$ is the result of convolution.

The convolution in second 3D process can be formulated as

$$\text{output}\left(m^h_{(l+1),i,j}\right) = \sum_{x=1}^{L}\sum_{y=1}^{W}\sum_{z=1}^{H} \text{input}\left(n^{g(x),g(y),g(z)}_{l,i}\right) k^{x,y,z}_{(l+1),i,j} \tag{2}$$

$$g(x) = x + s(h-1) \tag{3}$$

In Eqs. (2) and (3), $k^{x,y,z}_{l,i,j}$ is one of the elements in convolution kernel $k_{l,i,j}$ with j being the index of kernel, and $n^{g(x),g(y),g(z)}_{(l-1),i}$ is one of the elements in $n_{(l-1),i}$. (x, y, z) indicates the position of elements in them respectively, and i is the index of input 3D feature map $n_{(l-1),i}$ in $(l-1)$th layer with L, W and H being the length, width and height separately. The $m^h_{(l+1),i,j}$ is one of the elements in 1D vector generated by convolution in the second 3D process with h being the index. Besides, s stands for the stride of convolution in bands.

2.2   Parallel Computing

In order to avoid excessive consumption of time, the DV-CNN is distributed on different workers of a cluster, and each worker typically has a same task for training. Besides, the parallel DV-CNN (show in Fig. 1) has a master server and some worker servers in the cluster which share parameters and datasets. All servers update parameter asynchronously to make the most of computing ability of each worker. The master server is focus on the initialization of parameters, the deployment of parallel network and the test of classification accuracy by sampling every a certain times.

The input of parallel DV-CNN is a cube of HSI which consists of spectral-spatial information. Each worker in this cluster selects different bands of a neighbour region to extract features, and feature maps are generated by 3D/2D processes. After four stages, the parameters in convolution kernels are updated by workers. Thus, every worker contributes to the update of parameters, so that the network is easier to converge and save training time. The time consuming T can be formulated as

$$T = T_d + T_w + T_c \tag{4}$$

where T_d is the time consuming of parameters initialization and tasks deployment, T_w is the computing time consuming of worker in cluster, and T_c is the communication time between workers in the network and I/O time consuming.

3   Experiments

In this paper, we used two real HSI datasets, namely Indian Pines and Pavia University scene for experiments. And the superiority of DV-CNN on accuracy and time consuming was based on these datasets. Besides, the ratio of labelled samples for training and testing was randomly divided into 1:1.

The dropout and learning rate of network were set as 0.8 and 1×10^{-3} respectively, and the value of cross entropy was set as 1×10^{-15} which meant all values in classification results were set as 1 if they were smaller than 1×10^{-15}. The neighbour region selected from HSI cube was 10×10 for Indian Pines dataset and 8×8 for Pavia University scene dataset. Besides, the batch size input into network was 40 for Indian pines and 150 for Pavia University scene. The other parameters are displayed in Tables 1 and 2.

Table 1. Parameters of 3D feature extraction of DV-CNN

Data set	3D feature extraction			
	Kernel number, kernel size		Pooling size	
	Process: 1	Process: 2	Process: 1	Process: 2
Indian Pines	8, $1 \times 5 \times 5$	36, 21 $\times$ 3 $\times$ 3 1 $\times$ 3 $\times$ 3	1 $\times$ 2 $\times$ 2	2 $\times$ 1 $\times$ 1
Pavia University scene	5, 1 $\times$ 3 $\times$ 3	36, 4 $\times$ 3 $\times$ 3	1 $\times$ 2 $\times$ 2	2 $\times$ 1 $\times$ 1

Table 2. Parameters of 2D feature extraction of DV-CNN

Data set	2D feature extraction			
	Kernel number, kernel size			Pooling size
	Process: 1	Process: 2	Process: 3	Process: all
Indian Pines	6, 5 × 5	12, 5 × 5	36, 5 × 5	2 × 2
Pavia University scene	6, 5 × 5	12, 5 × 5	36, 5 × 5	2 × 2

The parallel DV-CNN were based on Tensorflow architecture and implemented on servers with Intel Core i7-6700@3.4 GHz CPU, 16 GB of RAM. And there was a NVIDIA GTX 1080 Ti GPU card, a NVIDIA GTX 1070 GPU card and two NVIDIA GTX 1060 GPU cards in cluster.

3.1   Classification Accuracy of Methods×

Classification Accuracy of Methods Based on CNN. We compared some methods based on CNN (2D-CNN, 3D-CNN) for HSI classification to evaluate the classification performance of parallel DV-CNN, and all of them were based on spectral-spatial feature extraction. The convolution kernels is two dimensions for 2D-CNN and three dimensions for 3D-CNN in all feature extractions, which are different from kernels in DV-CNN. The experiment parameters of DV-CNN were the same as other methods. The results of comparison were shown in Table 3.

Table 3. Results of comparison with CNNs

	2D-CNN	3D-CNN	DV-CNN
Indian Pines			
OA (%)	95.97 ± 0.0938	99.07 ± 0.0345	99.14 ± 0.1175
AA (%)	93.23 ± 0.7629	98.66 ± 0.0345	98.71 ± 0.65606
κ	0.9540 ± 0.0012	0.9893 ± 0.0004	0.9926 ± 0.0010
Pavia University scene			
OA (%)	99.03 ± 0.0142	99.39 ± 0.0098	99.84 ± 0.0343
AA (%)	98.19 ± 0.0268	99.85 ± 0.0609	99.74 ± 0.0693
κ	0.9817 ± 0.0001	0.9920 ± 0.0002	0.9979 ± 0.0005

In Table 3, 3D-CNN and DV-CNN achieved better classification performance than 2D-CNN because of the simultaneous spectral-spatial features extraction, which was different from 2D-CNN. Besides, the performance of DV-CNN was better than 3D-CNN. The OA of DV-CNN was up to 99.14% on Indian Pines and 99.84% on Pavia University scene, which was 0.07% and 0.45% higher than 3D-CNN. Moreover, both average accuracy (AA) and kappa coefficient (κ) achieved similar results as OA.

Classification Accuracy with State-of-the-Art Methods. We also compared some traditional and other deep learning method for HSI classification. RMKL, PCANet,

SAE-LR, DBN-LR, and NSSNet were performed an optimized effect in HSI classification, and introduced to experiments to compare with DV-CNN. The results of comparison were shown in Table 4.

Table 4. Comparison results of classification with other methods

	Indian Pines			Pavia University scene		
	OA (%)	AA (%)	κ	OA (%)	AA (%)	κ
RMKL	95.61	94.20	0.9499	96.06	94.48	0.9443
PCANet	86.58	85.16	0.8471	93.20	91.01	0.8997
SAE-LR	92.58	90.38	0.9152	98.69	98.17	0.9829
DBN-LR	95.95	95.45	0.9539	99.05	98.48	0.9875
NSSNet	96.08	96.40	0.9547	99.50	99.03	0.9910
DV-CNN	99.14	98.71	0.9926	99.84	99.74	0.9979

As shown in Table 4, DV-CNN obtained the best classification results of OA, AA and κ on both Indian Pines and Pavia University scene datasets. And the OA of DV-CNN was 3.18% higher than the second best obtained by NSSNet on Indian Pines dataset and 0.34% higher on Pavia University scene dataset.

3.2 Time Consuming of Methods

In order to verify the efficiency of parallel computing for big data in HSI classification, we conducted some comparison experiments on both Indian Pines dataset and Pavia University dataset. The experiments were divided into four categories based on the number of servers in parallel cluster. We tested the classification efficiency of NVIDIA GTX 1060 GPU cards (1060-1, 1060-2) and NVIDIA GTX 1080 GPU cards (1080Ti) for singe server computing, two NVIDIA GTX 1060 GPU cards for parallel computing (parallel-2), two NVIDIA GTX 1060 GPU cards and a NVIDIA GTX 1080Ti GPU card (as master) for parallel computing (parallel-3), and all of the GPU cards in cluster for parallel computing (parallel-4). Moreover, we also tested the impact of network on computing time. The parallel-3-1 represented that the parallel computing was in a 1000 Mb/s network and parallel-3-2 was in a 100 Mb/s network. We counted the time consuming every 100 iterations. The curve of Indian Pines was shown in Fig. 2 and the curve of Pavia University scene was shown in Fig. 3.

As shown in Figs. 2 and 3, the time consuming of each model in the beginning is much more than in other iteration points, which is resulting from the task deployment and initialization of parallel network. In terms of single computing, 1080Ti is much faster than 1060 s. According to the curves of parallel-2, parallel-3-1 and parallel-4, we can infer that increasing the number of worker in cluster can improve the average computing speed and save lots of time. Besides, parallel-3-1 takes less computing time than parallel-3-2, which demonstrates that the hardware of network can influence the parallel efficiency. Moreover, the fluctuation of the parallel-4 curve is much larger, and the impact of network on the parallel computing is more obvious than others. Thus, the

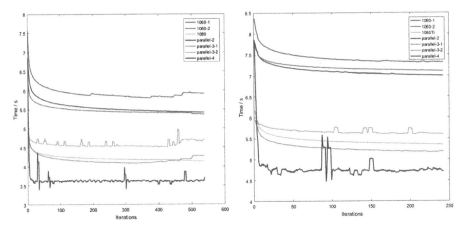

Fig. 2. The curve of time consuming on Indian Pines

Fig. 3. The curve of time consuming on Pavia University scene

parallel efficiency increases when the number of worker in cluster increases, but it decreases when the number of worker is exceeds a certain number.

4   Conclusions

In this paper, we have proposed a parallel DV-CNN framework for HSI classification. The DV-CNN can make full use of spectral-spatial information contained in HSI to extract more accurate features. The experiments compared with the state-of-the-art methods demonstrate that DV-CNN can improve the classification accuracy. DV-CNN also simplifies the complexity and reduces the computations of CNN-based model. Besides, the parallel architecture is feasible for HSI classification methods based on CNNs. It has been shown that parallel DV-CNN is a suitable approach for HSI classification dealing with big data.

However, the communication of workers in cluster is a major factor to influence the time consuming of parallel computing, and the consumption of communication time is obviously influenced by the increased of worker number. In the further research, we mainly research on the specific effect of increasing number of workers in cluster on parallel time consuming.

Acknowledgments. This paper was supported by the National Nature Science Foundation of China (41701479).

References

1. Zhao, W.Z., Du, S.H.: Spectral–spatial feature extraction for hyperspectral image classification: a dimension reduction and deep learning approach. IEEE Trans. Geosci. Remote Sens. **54**(8), 4544–4554 (2016)
2. Qu, H.C., Liang, X.J., Liang, S.C., Liu, W.J.: Dimensionality-varied deep convolutional neural network for spectral–spatial classification of hyperspectral data. J. Appl. Remote Sens. **12**(1), 016007 (2018)
3. Qian, Y.T., Ye, M.C.: Hyperspectral imagery restoration using nonlocal spectral-spatial structured sparse representation with noise estimation. IEEE J. Sel. Top. Appl. Earth Obs. Remote Sens. **6**(2), 499–515 (2013)
4. Gu, Y.F., Wang, C., You, D., et al.: Representative multiple kernel learning for classification in hyperspectral imagery. IEEE Trans. Geosci. Remote Sens. **50**(7), 2852–2865 (2012)
5. Chan, T.H., Jia, K., Gao, S., et al.: PCANet: a simple deep learning baseline for image classification? IEEE Trans. Image Process. **24**(12), 5017–5032 (2015)
6. Chen, Y.S., Lin, Z.H., Zhao, X., et al.: Deep learning-based classification of hyperspectral data. IEEE J. Sel. Top. Appl. Earth Observ. Remote Sens. **7**(6), 2094–2107 (2014)
7. Chen, Y.S., Zhao, X., Jia, X.P.: Spectral–spatial classification of hyperspectral data based on deep belief network. IEEE J. Sel. Top. Appl. Earth Observ. Remote Sens. **8**(6), 2381–2392 (2015)
8. Pan, B., Shi, Z.W., Zhang, N.: Hyperspectral image classification based on nonlinear spectral–spatial network. IEEE Geosci. Remote Sens. Lett. **13**(12), 1782–1786 (2016)
9. Yue, J., Zhao, W., Miao, S.: Spectral–spatial classification of hyperspectral images using deep convolutional neural networks. Remote Sens. Lett. **6**(6), 468–477 (2015)
10. Li, Y., Zhang, H., Shen, Q.: Spectral–spatial classification of hyperspectral imagery with 3d convolutional neural network. Remote Sens. **9**(1), 67 (2017)

Model Selection Prediction
for the Mixture of Gaussian Processes
with RJMCMC

Zhe Qiang[1] and Jinwen Ma[2(✉)]

[1] Department of Mathematics, Northwest University, Xi'an 710069, Shaanxi,
People's Republic of China
zhe.qng@nwu.edu.cn
[2] Department of Information Science, School of Mathematical Sciences & LMAM,
Peking University, Beijing 100871, People's Republic of China
jwma@math.pku.edu.cn

Abstract. Repetition measurements from different sources often occur in data analysis which need to be model and keep track of the original sources. Moreover, data are usually collected as finite vectors which need to be considered as a sample from some certain continuous signal. Actually, these collected finite vectors can be effectively modeled by the mixture of Gaussian processes (MGP) and the key problem is how to make model selection on a given dataset. In fact, model selection prediction of MGP has been investigated by the RJMCMC method. However, the split and merge formula of the RJMCMC method are designed only for the univariables in the past. In this paper, we extend the split and merge formula to the situation of the multivariables. Moreover, we add a Metropolis-Hastings update rule after the RJMCMC process to speed up the convergence. It is demonstrated by simulation experiments that our improved RJMCMC method is feasible and effective.

Keywords: RJMCMC · Mixture of Gaussian processes
Multivariable regression

1 Introduction

In real applications, data often come as measure curves or time series. Although these are gathered as finite points, it is valuable to consider them as sample or trajectories of stochastic processes. In [1], the batch curves are considered from a mixture of Gaussian processes to describe its heterogeneity. However, the problem of well-known model selection for the mixture model comes, which can be solved with BIC criterion [2]. But this method can only make inference about the parameters in a fixed dimension parameter subspace and is weak on exploring the low probability area between the neighbourhood subspaces.

© IFIP International Federation for Information Processing 2018
Published by Springer Nature Switzerland AG 2018. All Rights Reserved
Z. Shi et al. (Eds.): ICIS 2018, IFIP AICT 539, pp. 310–317, 2018.
https://doi.org/10.1007/978-3-030-01313-4_33

Reversible Jump Markov Chain Monte Carlo (RJMCMC) [3] is one of the most important method among Markov chain Monte Carlo simulation methods. It can not only make inference about parameters but also make model selection prediction. Since this simulation method explores the different parameter subspaces with varied dimension, it can prevent the solution to be trapped into local optimum. In [5], we proposed the RJMCMC method for automatic model selection or prediction for mixtures of Gaussian processes and demonstrated its effectiveness. Based on it, we try to extend the split and merge formula of the RJMCMC method from the single-input case to the multi-input case to adapt to more complex or general situation.

Our paper is organized as follows. We introduce the mixture of Gaussian processes and their latent variables in Sect. 2. Section 3 reviews the five type of moves of the parameters and extends the split and formula to the multi-input situation. We demonstrate our improved RJMCMC on simulated data in Sect. 4. Finally, we conclude briefly in Sect. 5.

2   The Mixture of Gaussian Processes and Its Latent Variables

Suppose that M curves are given and the response $y_m(t)$ for the m-th curve is defined as:

$$y_m(t) = \tau_m(t) + \epsilon_m(t) \tag{1}$$

where $\tau_m(t)$ is a Gaussian process and $\epsilon_m(t)$ is a noise term. For the m-th curve, the time points at which we collect data are $(t_{m,1}, \cdots, t_{m,N_m})$ and the target values are $(y_{m,1}, \cdots, y_{m,N_m}) = (y_m(t_{m,1}), \cdots, y_m(t_{m,N_m}))$, where N_m is the total number of observations on this curve. Then this curve can be considered as a sample curve or trajectory of the corresponding Gaussian process [1].

Moreover, data are assumed coming from different source and mixture structure are constructed to model the heterogeneity. For the dataset $\{(x_{m,n}, y_{m,n}), m = 1, \cdots, M, n = 1, \cdots, N_m\}$, the likelihood is as follows

$$L(\boldsymbol{\theta}, \boldsymbol{p}|Y, X) = \prod_{m=1}^{M} \sum_{k=1}^{K} p_k \mathcal{N}(Y; \mu_k(X), \Sigma_k(X)) \tag{2}$$

where K is the total number of the mixtures, μ_k and Σ_k are the mean function and covariance function. Specially, the covariance function is defined as follows:

$$\Sigma_k(x_i, x_j; \theta_k) = v_k \exp\left(-w_k \frac{(x_i - x_j)^2}{2}\right) + \delta_{ij}\sigma_k^2 \tag{3}$$

here $\theta_k = (v_k, w_k, \sigma_k^2)$ and each parameters are positive on $\mathbb{R}$.

As usual, auxiliary variables [6] z_m is the index variable and $z_m = k$ indicate the m-th curve belong to the k-th component. Then the completion of model is:

$$Z_m \sim \mathcal{M}_k(1; \underbrace{\frac{1}{n}, \cdots, \frac{1}{n}}_{K repetitions}), \quad y_m|x_m, z_m, \boldsymbol{\theta} \sim \mathcal{N}(\mu_k(x_m), \Sigma(x_m; \theta_k)|z_m = k) \tag{4}$$

From the Bayesian point of view, the prior of parameters should also be set, and the posterior is proportional to the prior multiples likelihood. For convenience of the derivation, the prior is set to be the conjugate prior:

$$w_k \sim I\Gamma(\frac{1}{2}, \frac{1}{2}), \quad v_k \sim \mathcal{LN}(-1, 1^2), \quad \sigma_k^2 \sim \mathcal{LN}(-3, 3^2) \tag{5}$$

where $I\Gamma$ represents inverse gamma distribution and $\mathcal{LN}$ represents log normal distribution. The mixed proportion coefficient π are from Dirichlet distribution:

$$(\pi_1, \cdots, \pi_K) \sim Dir(1, \cdots, 1) \tag{6}$$

3   Model Selection Prediction with Reversible Jump Markov Chains Monte Carlo

Model selection is an important and difficulty problem on the mixture model. Moreover, the valley among the neighbourhood modes cannot be detected with a general parameter estimation method. Thus, we utilize the Reversible Jump MCMC model to overcome this difficulty.

In fact, our algorithm contains five types of parameter moves:

(a) $\pi = (\pi_1, \cdots, \pi_K)$;
(b) $\theta = (\theta_1, \cdots, \theta_K)$, where $\theta_k = (v_k, w_k, \sigma_k^2)$;
(c) $z = (z_1, \cdots, z_M)$;
(d) split and merge move;
(e) Metropolis-Hastings update.

Actually, the first three belong to the moves with the component number fixed and step (d) is the move with the component number varied, step (e) is also the move with the component number fixed.

3.1   The Moves with the Component Number Fixed for (a)(b)(c)

For the moves with the component bunber fixed, gibbs sampling and hybrid MCMC (one kind of Metropolis-Hastings) are adopt. For π and z, Gibbs sampling are used as follows:

- sample z_m from its posterior $p(z_m = k | X, Y, \theta, \pi) \propto \pi_k p(y_m | \theta_k, x_m)$, $m = 1, \cdots, M$, $k = 1, \cdots, K$
- sample π from the conditional distribution:$(\pi_1, \cdots, \pi_K)|z \sim Dir(1 + n_1, \cdots, 1 + n_K)$

Here, $n_k = \sum_{m=1}^{M} \mathbb{I}_{z_m = k}$. In addition, θ are sampled by hybrid MCMC from its posterior:

$$p(\theta | \mathcal{D}, z) \propto \prod_{k=1}^{K} p(\theta_k | \mathcal{D}_m, z) \tag{7}$$

3.2   The Moves with the Component Number Changed

For the moves with the component number changed, we use the same rules given in [5] but extend the univariables to the multivariables $\boldsymbol{x} = (x_1, \cdots, x_D)$, then the off diagonal entry becomes:

$$\pi_k \left(v_k \exp \left(-\frac{\sum_{d=1}^{D} w_{k,d}(x_{i,d} - x_{j,d})^2}{2} \right) \right) = \pi_{k_1} \left(v_{k_1} \exp \left(-\frac{\sum_{d=1}^{D} w_{k_1,d}(x_{i,d} - x_{j,d})^2}{2} \right) \right)$$
$$+ \pi_{k_2} \left(v_{k_2} \exp \left(-\frac{\sum_{d=1}^{D} w_{k_2,d}(x_{i,d} - x_{j,d})^2}{2} \right) \right) \qquad (8)$$

Since each dimension x_d are independent, we can make the analysis for each variable separately and obtain the new detailed balanced framework:

$$\pi_k = \pi_{k_1} + \pi_{k_2} \qquad (9a)$$
$$\pi_k \sigma_k^2 = \pi_{k_1} \sigma_{k_1}^2 + \pi_{k_2} \sigma_{k_2}^2 \qquad (9b)$$
$$\pi_k v_k = \pi_{k_1} v_{k_1} + \pi_{k_2} v_{k_2} \qquad (9c)$$
$$\pi_k v_k w_{k_d} = \pi_{k_1} v_{k_1} w_{k_1,d} + \pi_{k_2} v_{k_2} w_{k_2,d}, \quad d = 1, \cdots, D \qquad (9d)$$

The merge moves are based on the above Eq. 9, and the split moves are follows: in order to match the dimension before and after split, $(d+3)$-dimension random variable $\boldsymbol{u} = (u_1, u_2, u_3, u_{4,1}, \cdots, u_{4,D})$ need to be generated each are generated from $Beta(2,2)$. Then combining the detailed balanced equation 9, split formula is given as follows:

$$\pi_{k_1} = u_1 \pi_{k^*}, \quad \pi_{k_2} = (1 - u_1)\pi_{k^*}, u_1 \in (0,1) \qquad (10a)$$
$$\sigma_{k_1}^2 = u_2 \sigma_{k^*}^2 \frac{\pi_{k^*}}{\pi_{k_1}}, \quad \sigma_{k_2}^2 = (1 - u_2)\sigma_{k^*}^2 \frac{\pi_{k^*}}{\pi_{k_2}}, u_2 \in (0,1) \qquad (10b)$$
$$v_{k_1} = u_3 v_{k^*} \frac{\pi_{k^*}}{\pi_{k_1}}, \quad v_{k_2} = (1 - u_3)v_0^{k^*} \frac{\pi_{k^*}}{\pi_{k_2}}, u_3 \in (0,1) \qquad (10c)$$
$$w_{k_1,d} = \frac{1 - u_{4,d}}{u_3} w_k, \quad w_{k_2} = \frac{u_{4,d}}{1 - u_3} w_k, u_{4,d} \in (0,1), \quad d = 1, \cdots, D \qquad (10d)$$

In fact, parameters varies under split and merge moves which forms a birth-and-death Markov chain. The birth and death probability for this Markov chain is set to be: $d_1 = b_{k_{max}} = 0$, $d_{k_{max}} = b_1 = 1$, $d_k = b_k = 0.5, \forall k = 2, \cdots, k_{max} - 1$. Moreover, it is simulated by Metropolis-Hastings algorithm which is one kind of Markov chain Monte Carlo simulation methods and is called as Reversible Jump Markov chain Monte Carlo simulation in [3]. Therefore, the acceptance probability need to be calculated:

$$A = \prod_{m=1}^{M} \frac{l(Y_m|\theta_{k+1})}{l(Y|\theta_k)} \times \frac{d_{k+1}}{b_k} \times \frac{p(\theta_{k+1})}{p(\theta_k)} \times \frac{1}{Beta(\boldsymbol{u}|\theta_{k+1}, \theta_k)} \times \left| \frac{\partial \theta_{k+1}}{\partial (\theta_k, \boldsymbol{u})} \right| \qquad (11)$$

here, k, k_1, k_2 are randomly chosen from $\{1, \cdots, k_{max}\}$.

3.3   Metropolis-Hastings Update

For the Metropolis-Hastings update, we adopt a penalized likelihood [4] as follows:

$$\log L^P(\theta, p | \mathcal{D}, k) = \log L(\theta, p | \mathcal{D}, k) - \mathcal{P} \tag{12}$$

where the penalization $\mathcal{P}$ is BIC penalization:

$$\mathcal{P}_{BIC} = \frac{\xi}{2} \log(N) \tag{13}$$

here ξ denote the number of model parameters $(4 * k)$ and N is the number of curve multiples the number of points on each curve. The acceptance probability of this Metropolis-Hastings is:

$$\min\{1, \frac{L^P(\theta_{new} | \mathcal{D})}{L^P(\theta_{old} | \mathcal{D})}\} \tag{14}$$

4   Convergence Diagnosing

In this subsection, we generate 9 curves as the dataset, which are trajectories of three Gaussian processes $GP(\mu_i, \theta_i)$, $i = 1, 2, 3$, respectively. The mean functions are given by

$$\begin{aligned}
\mu_1(x_1, x_2) &= f_1(x_1) + f_2(x_2), \\
\mu_2(x_1, x_2) &= f_2(x_1) + f_3(x_2), \\
\mu_3(x_1, x_2) &= f_3(x_1) + f_1(x_2).
\end{aligned} \tag{15}$$

where

$$\begin{aligned}
f_1(x) &= \exp(x/5) - 1.5 \\
f_2(x) &= \sin(x/4 * \pi) \\
f_3(x) &= -\sin(x/4 * \pi)
\end{aligned} \tag{16}$$

and covariance parameters θ_i are given in Table 1. The convariants are: $x_1 = -4 : 0.08 : 4$ and $x_2 = 0 : 0.08 : 8$. Since the number of points in x_1 and x_2 are the same, there is a one-to-one map between x_1 and x_2 in our experiment. In Fig. 1, we plot the simulated curves. The x-axis is x_1 and y-axis is y. Each 3 curves in one panel are trajectories of the same Gaussian process with common parameters of mean functions and covariance.

Table 1. The covariance parameters for the simulated data.

	w_1	w_2	v	σ^2
θ_1	1	0.2	0.2	0.0025
θ_2	0.5	0.5	1	0.001
θ_3	10	1.0	0.2	0.0005

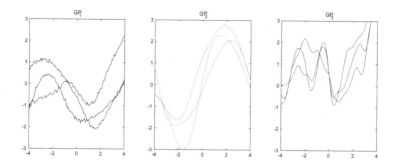

Fig. 1. The simulated dataset. The x-axis is x_1 and y-axis is y. Each 3 curves in one panel are trajectories of the same Gaussian process with common parameters of mean functions and covariance.

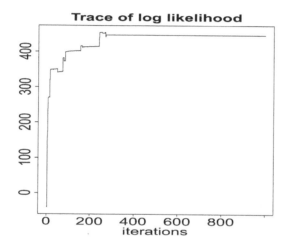

Fig. 2. The trace of log likelihood for 1000 iterations

Figure 2 illustrates the trace of log likelihood for 1000 iterations of Markov chain Monte carlo simulation. It became stable after 400 iterations. In Fig. 3, we plot the trace of K for 1000 iterations and in Fig. 4 we plot the histogram of K for 1000 iterations. It can be seen that K becomes stable after 400 and stays on 3, which is compatible with our setting.

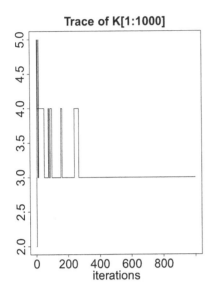

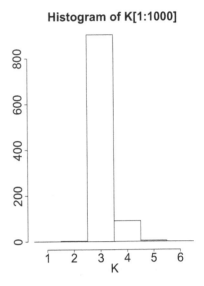

Fig. 3. The trace of K for 1000 iterations

Fig. 4. The histogram of K for 1000 iterations

5   Conclusion

We have improved the RJMCM method for model selection prediction of MGPs on two aspects: first, we extend the split and merge formula for dealing with multi-input regression, which can be used to more complex and interesting data; second, a Metropolis-Hasting update rule is added after the process of the moves, which can remarkably accelerate the convergence. The experimental results on simulated dataset demonstrate that the improved RJMCMC method is feasible and effective.

Acknowledgment. This work is supported by the Natural Science Foundation of China for Grant 61171138.

References

1. Shi, J.Q., Murray-Smith, R., Titterington, D.M.: Hierarchical Gaussian process mixtures for regression. Stat. Comput. **15**(1), 31–41 (2005)
2. Shi, J.Q., Wang, B.: Curve prediction and clustering with mixtures of Gaussian process functional regression models. Stat. Comput. **18**(3), 267–283 (2008)
3. Green, P.J.: Reversible jump Markov chain Monte Carlo computation and Bayesian model determination. Biometrika **82**(4), 711–732 (1995)
4. Andrieu, C., Freitas, N.D., Doucet, A.: Robust full Bayesian learning for radial basis networks. Neural Comput. **13**(10), 2359–2407 (2001)

5. Qiang, Z., Ma, J.: Automatic model selection of the mixtures of Gaussian processes for regression. In: Hu, X., Xia, Y., Zhang, Y., Zhao, D. (eds.) ISNN 2015. LNCS, vol. 9377, pp. 335–344. Springer, Cham (2015). https://doi.org/10.1007/978-3-319-25393-0_37
6. Marin, J.M., Robert, C.P.: Bayesian Core: A Practical Approach to Computational Bayesian Statistics. Springer Texts in Statistics. Springer Science & Business Media, New York (2007). https://doi.org/10.1007/978-0-387-38983-7
7. Marin, J.M., Robert, C.P.: Bayesian Essentials with R. Springer, New York (2014). https://doi.org/10.1007/978-1-4614-8687-9
8. Wu, D., Chen, Z., Ma, J.: An MCMC based EM algorithm for mixtures of Gaussian processes. In: Hu, X., Xia, Y., Zhang, Y., Zhao, D. (eds.) ISNN 2015. LNCS, vol. 9377, pp. 327–334. Springer, Cham (2015). https://doi.org/10.1007/978-3-319-25393-0_36
9. Wu, D., Ma, J.: A two-layer mixture model of Gaussian process functional regressions and its MCMC EM algorithm. IEEE Trans. Neural Netw. Learn. Syst. (2018)
10. Qiang, Z., Luo, J., Ma, J.: Curve clustering via the split learning of mixtures of Gaussian processes. In: 2016 IEEE 13th International Conference on Signal Processing (ICSP), 1089–1094 (2016)

Intelligent Robot

Self-developing Proprioception-Based Robot Internal Models

Tao Zhang, Fan Hu, Yian Deng, Mengxi Nie, Tianlin Liu, Xihong Wu,
and Dingsheng Luo[✉]

Key Lab of Machine Perception (Ministry of Education),
Speech and Hearing Research Center, Department of Machine Intelligence,
School of EECS, Peking University, Beijing 100871, China
{tao_zhang,fan_h,yiandeng,niemengxi,xhwu,dsluo}@pku.edu.cn

Abstract. Research in cognitive science reveals that human central nervous system internally simulates dynamic behavior of the motor system using *internal models* (forward model and inverse model). Being inspired, the question of how a robot develops its internal models for arm motion control is discussed. Considering that human proprioception plays an important role for the development of internal models, we propose to use autoencoder neural networks to establish robot proprioception, and then based on which the robot develops its internal models. All the models are learned in a developmental manner through robot motor babbling like human infants. To evaluate the proprioception-based internal models, we conduct experiments on our PKU-HR6.0 humanoid robots, and the results illustrate the effectiveness of the proposed approach. Additionally, a framework integrating internal models is further proposed for robot arm motion control (reaching, grasping and placing) and its effectiveness is also demonstrated.

Keywords: Development and adaptation of intelligence
Intelligent robots · Robot arm manipulation

1 Introduction

Arm motion skills such as reaching, grasping, placing as well as other manipulations are fundamental for humanoid robots and have been heavily focused for decades. Traditionally, controlling methods like inverse kinematics work well in well-defined environments [10,21]. However, such methods are weak in adaption and lead to poor performance in volatile environments due to the strong reliance on prior knowledge about the kinematic parameters of the robots and the information of environments. As model learning may allow the absence of knowledge about the robot parameters and can work in unstructured environments, it is gaining more interests in robot motion control [8,14,18]. In the work of Nguyen-Tuong and Peters, they pointed out that uncertainty of environments is the key problem in robotics [14]. In this respect, humans usually develop their new skills

© IFIP International Federation for Information Processing 2018
Published by Springer Nature Switzerland AG 2018. All Rights Reserved
Z. Shi et al. (Eds.): ICIS 2018, IFIP AICT 539, pp. 321–332, 2018.
https://doi.org/10.1007/978-3-030-01313-4_34

in a quite short time and the acquired ability is robust in versatile environments. Therefore, the human-inspired learning methods could offer a promising alternative for robot skill acquisition, especially for humanoids which have human-like physical body structures [1,13].

Studies on human cognitive development reveal that the central nervous system (CNS) makes use of *internal models* in planning, controlling and learning of the motor behaviors [23]. Generally, there are two kinds of internal models: forward model (FM) and inverse model (IM). The FM, also called *predictor*, predicts the next sensory state according to the current state and motor commands [9]. While the IM converts desired state into motor commands, which is also known as *controller*. The FM and IM always work in harmony in a feedback manner to fulfill complex arm motion control tasks [15,22]. These views from cognitive science provide a solid inspiration and imply that developing the FM and IM can contribute to gaining the abilities for robots to accomplish various motion tasks.

Actually, many works have focused on how to build internal models for robots. For example, back 2001, D'Souza et al. employed the then-popular Locally Weighted Projection Regression (LWPR) to learn the inverse kinematic mappings [5]. The LWPR performed well on the given task. However, it largely depends on the data distribution and different tasks requires different optimization criteria, and thus the expandability is debased. In 2007, Castellini et al. tried to establish human internal models for motion behaviors using the learning machine [3]. They concentrated on reaching and grasping and collected the biometric data from human users during grasping, then used the data to train a support vector machine (SVM). Yet model is learned for predicting the behaviors of humans or robots, not for motion control. In 2010, Baranes and Oudeyer introduced SAGG-RIAC algorithm to learn the internal models for a simulated robot efficiently and actively [2], while the accuracy of both the inverse model and the forward model seem to be unappealing respectively. There are also relative works on humanoids. In 2012, Schillaci et al. proposed to map the arm and neck joint angles with the estimated 3D position of hand (using a marker) to establish the forward and inverse models for a humanoid robot NAO [19]. However, previous research did not pay much attention to the real human mechanism on how the internal models were developed.

As cognitive psychologists suggested, infants primarily employ proprioceptive information instead of vision to control and direct their hands towards a specific target [17]. Besides, the emergence of human arm motion ability like reaching is the product of a deeply embodied process in which infants firstly learn to direct these arm movements in space using proprioception [4]. The term *proprioception*, firstly introduced by Sherrington in 1907 [20], represents the sense of the relative position of neighboring parts of the body and strength of effort being employed in movement. Intuitively, it originates from the embodied perspective that humans have sense of their own body even when eyes are closed. It has been shown that those findings of human mechanism involving proprioception are beneficial for robots [7,11,12]. In 2010, Huffmann et al. pointed out that the body representations contribute to better adaption in new environments [7]. In 2016 and 2017, Luo et al. simulated the findings from human infants by Corbetta et al. with humanoid robot, and illustrated the effectiveness [11,12].

In this work, inspired by the fact that proprioception plays an important role in human infants developing their internal models, the issue of how a robot establishes its internal models for arm motion control is further discussed. By borrowing the basic idea from Luo et al. [11,12], we propose new proprioception-based internal models for robot arm motion control. Different from other works, the new proposed models in this work not only model the proprioception with autoencoder network and develop the internal models by robot itself just like what human infants do, but also delicately model both FM and IM with deep neural networks where modified cascade-structures are involved. The effectiveness of the new models is illustrated with the improved performance in the experiments, which indicates that human mechanism is deeply captured by the new models.

2   Developing the Internal Models

To develop the internal models is to establish the mapping between *Motions* and *States*. Since the changes of arm joint angles most directly describe arm motions, we choose arm joint angles to represent the motions (for dynamic control it is better to use joint velocity and acceleration to describe the motions). It is studied that when reaching a target, subjects aligned their arms by matching hand position rather than elbow angle [6], so we chose the hand pose (position and orientation) to represent the state of the arm. Therefore, the internal models are essentially the mappings between the arm joint angles and the hand pose.

As is shown in Fig. 1, the learning process originates from the random self-produced arm movements just like human infants' babbling. During the learning process, the proprioception is developed to represent the joint angles, and the two internal models are developed based on the currently learned proprioception. More detailedly, once the robot conducts a self-produced arm movement, proprioception is updated once, and the two internal models are also updated based on the current proprioception. With the capability of deep neural network in describing the mappings between different variables, in this paper, we employ neural networks to formalize both forward models and inverse models.

2.1   Developing the Proprioception

Proprioception is the sense of human own body and is provided by different proprioceptors. From an embodied view, humans have the sense of each part of their own body and such sense of their body is called proprioception. For example, even if we close our eyes, we can feel the position of our arms. What we know is not the exact joint angles but vague sense. For arm motion control we discussed here, the most important proprioception is the sense of arm joints. For humans, the fibrous capsules act as the proprioceptors to help humans get

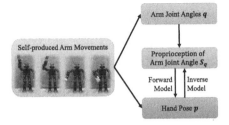

Fig. 1. Illustration of the architecture for learning internal models based on proprioception. The robot learns the proprioception, the forward model and the inverse model during the same process of randomly self-produced arm babbling. Each movement is accompanied by a pair of arm joint angles q and hand pose p which are the raw materials of learning.

the sense of their joints. As for robots, the joint servos sever as the same effect which helps robots learn the proprioception of their joints.

To model the proprioception of robots, we consider the characteristics of human proprioception. First, the exact value of proprioception is unknowable for humans, which means the learning of proprioception is unsupervised. Second, it should be able to represent the body states and as for robots the proprioception should be able to transfer to joint state as the input of joint servos. Thus, the *autoencoder* neural network is considered, which have identical input and output. We take joint state as input of autoencoder. The gotten hidden layer is considered as proprioception, which accords with the fact that the proprioception is invisible to human consciousness and can rebuild the joint state. In this way, the features of joint state which represent proprioception may help to reduce the influence of noise and be more suitable for modeling.

Therefore, autoencoders are employed to learn the proprioception of joints. The joint angles act as the input and the hidden layer as proprioception. The proprioception of each joint is developed respectively. The structures of the models are shown in Fig. 2, where $q_i, i = 1, 2, .., n$ is the joint angle read from the servo and S_{q_i} represents the sense of the i_{th} joint, and n is the total number of joints.

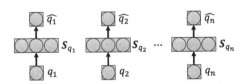

Fig. 2. The structure of the autoencoders employed to develop the proprioception of each joint respectively (the number of circles in the figure do not represent the sizes of the layers).

2.2 Developing the Forward Model

Developing the forward model is essentially learning the mapping from joint angles onto hand states. In this research, the proprioception is introduced and we propose to establish the forward model through the mapping from proprioception of joints S_q onto the hand pose p and we name it proprioception-based neural forward model (PNFM), while the S_q is the splicing of all joint proprioception. However, the most intuitive and common formation of the forward model is the mapping from joint angles q onto the hand pose p, and we name this classical neural forward model NFM. An illustration of the two forward models is shown in Fig. 3. The comparison of NFM's and PNFM's performance may show the effectiveness as proprioception is involved in modeling FM.

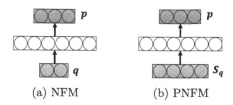

(a) NFM   (b) PNFM

Fig. 3. Two forward model structures.

2.3 Developing the Inverse Model

The inverse model is essentially the mapping from hand pose onto the arm joints. Similar to the forward model, we propose to use the proprioception of joints S_q to replace the joint angles q and the inverse model based on the proprioception is shown in Fig. 4(b) which is named PNIM. The most basic formation of neural inverse model (NIM) is shown in Fig. 4(a) which directly employs the arm joint angles q. However, different from the forward model, the inverse map is more complicated thus is more difficult to learn, especially for the complex-structured redundant robot arms. Inspired by what Riemann et al. mentioned in 2002 that movement of one joint induces movement of another and different joints may play dominant role in separate directions [16], we further propose a cascaded proprioception-based neural inverse model (cPNIM) which is shown in Fig. 4(c). More detailedly, S_{q_1} is directly controlled by p, and S_{q_2} is controlled by p and S_{q_1} jointly. Similarly, $S_{q_i}, i = 2, 3, ..., n$ is controlled by $p, S_{q_1}, ..., S_{q_{i-1}}$ jointly.

3 Experiments

In this section, experiments are conducted to show the effectiveness of our proprioception model, and to compare the two forward models as well as the three inverse models for evaluating the function of proprioception.

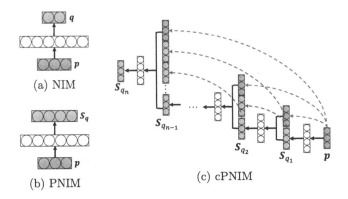

Fig. 4. Three inverse model structures.

We conduct all the experiments on our PKU-HR6.0 robot platform (Fig. 5(a)). The models are trained in the simulation system Gazebo (Fig. 5(b)) where the hand pose can be directly read from the supervisor view and the abrasion of the real robot is avoided. Then the learned models are finally employed in the real robot while the hand pose is calculated by 3D vision system. PKU-HR6.0 is the 6th generation kid-sized humanoid robot (3.93 kg weight and 59.50 cm height) designed by our lab. It has 24 degrees of freedom (DOFs) with four DOFs for moving plus one DOF for grasping in each arm. In this paper, we only consider the right arm (Shoulder Pitch, Shoulder Roll, Elbow Roll and Elbow Yaw). Yet the proposed model is easily applied to more complex tasks such as whole body movements.

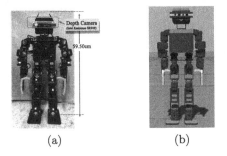

Fig. 5. The appearance of our PKU-HR6.0 robot. (a) the physical platform; and (b) the simulation platform.

3.1 Development of Proprioception

In this experiment, the learning process of proprioception is shown to demonstrate the developmental learning process, and the performance of the learned proprioception are shown to evaluate the effectiveness of autoencoder network in modeling the proprioception.

As described before, the whole learning process is conducted in a developing manner, which means once the robot conducts a self-produced movement, the proprioception model and the two internal models are updated for one time. In the learning process, the angle of each joint is recorded by the joint servo which is a numerical value (in rad).

We conduct *Joint Position Matching* to evaluate the performance of proprioception, which is an established protocol for measuring the joint position proprioception of humans. In the experiments, individuals are blindfolded while a joint is moved to a specific angle q for a given period of time. Then the subjects are asked to replicate the specified angle, and the replicated angle is marked as q'. The difference between q and q' reflects the accuracy of proprioception. For the evaluation of robot's performance, after each iteration, we randomly conduct Num times of Joint Position Matching and calculate the average error (we set $Num = 100$). Figure 6 shows the records of the average error of each iteration during learning the proprioception of joint *Shoulder Pitch*. The three small pictures in Fig. 6 show the results of one same task of *Joint Position Matching* at different learning stages.

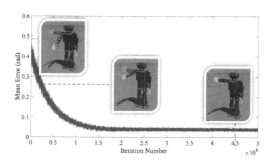

Fig. 6. The learning process of joint *Shoulder Pitch*'s proprioception. The three small pictures visualize the robot's performance in *Joint Position Matching* at iteration 10, 3000, 45000, when the target joint angles are all the same: -1.74 (rad) which is visualized by the shaded arm. The solid arm in each small picture is what the robot re-conducts using the currently developed proprioception.

We can see from the learning curves as well as the robot's performances in the three small pictures that with the learning process going on, the robot is able to replicate the given joint angle more and more accurately which means the proprioception is getting better and better, and this is similar to human infants' development. After learning, the testing errors of the four joint proprioception are 0.0658, 0.0387, 0.0471, 0.0353 (rad) respectively which indicates that the learned proprioception are all quite accurate, and the proposed model is qualified enough to model the proprioception.

3.2   Comparison of the Forward Models

In this experiment, we compare our proprioception-based neural forward model (PNFM) with the classical neural forward map (NFM) to verify the benefits of using proprioception rather than the original joint angles. The hidden layers of the two structures are both configured with 100 units. The learning rate and iteration time are both the same.

To evaluate the learning results of the two models, we randomly create 56118 pairs of joint angles which are within the range limits of robot's arm. We use each pair of data as motor command and send it to the robot, then the robot uses NFM and PNFM to predict the corresponding hand pose $p_{predicted}$ respectively. We calculate the average predicting error after testing all the 56118 pairs of motor commands. Figure 7 shows the comparison of the averaged predicting error of NFM and PNFM in the three axes.

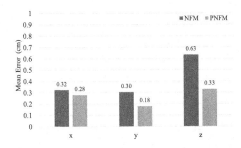

Fig. 7. Comparison of the averaged error of NFM and PNFM in three axes.

We can see from Fig. 7 that PNFM has significantly lower mean errors than NFM in all three directions, which verifies that the proprioception-based forward model is more effective than the classical forward model that uses the raw joint angles. Although this idea is inspired by humans, the reason for the improvement may be that the proprioception (feature of data) is both robust to noises and suitable in format which makes it easier to build a more accurate map. Moreover, the mean errors of our proposed PNFM in three axes are all lower than 0.35 cm which means it is qualified enough to be used in real tasks.

3.3   Comparison of the Inverse Models

Similar to the FM, we compare the proprioception-based neural inverse model (PNIM) with the classical neural inverse model (NIM). Furthermore, we also compare the cascaded proprioception-based neural inverse model (cPNIM) with PNIM to show that the cascaded structure offers a more efficient way to form the IM.

To compare the effectiveness of the three inverse models in controlling the robot arm, the testing inputs are chosen as the 56118 pairs of hand pose which are the expected outputs of forward models. The output of the inverse model is the joint angles which are desired to be able to drive the hand to the expected pose. To evaluate the performance of an inverse model, the robot's arm joints are set according to the output of inverse model, and the difference between the actual hand pose and the expected one is calculated. Figure 8 shows the averaged error of the three inverse models in the three axes.

From the Fig. 8, we can see that PNIM has better performance than NIM which means the proprioception is also effective in improving the accuracy of inverse model. Further, cPNIM has even lower mean errors than PNIM, which means learning the inverse map of each joint in a cascading way do improve the performance. To some extent, the results verify the findings of Riemann's that movement of one joint may induce movement of another as well as that different joints may play dominant role in separate directions.

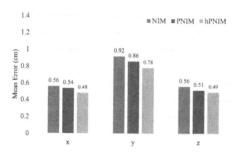

Fig. 8. Comparison of the averaged error of NIM, PNIM and cPNIM models in the three axes.

(a) Reaching   (b) Grasping   (c) Placing

Fig. 9. Examples of successful executions of the three arm motion tasks: (a) reaching, (b) grasping and (c) placing.

3.4 Evaluation of the Integrated Internal Models

To show the effectiveness of the internal models in different arm movements, we integrate them to accomplish reaching, grasping and placing tasks in PKU-HR6.0 robot as shown Fig. 9.

In the integration framework, the input is the 3D pose of the object(s) in robot coordination system. The robot determines the expected hand pose $p_{expected}$ according to the specific task and the pose of the object. After determining the expected hand pose $p_{expected}$, the inverse model (cPNIM in this experiment) outputs the joints proprioception according to $p_{expected}$. Then the forward model (PNFM in this experiment) predict the hand pose $p_{predicted}$ according to the gotten proprioception. If the difference Δp between $p_{expected}$ and $p_{predicted}$ is small enough, the robot executes the command according to the proprioception. While if Δp is too large, the robot chooses not to execute the command and response something like *"Sorry, I can't do it"*.

Table 1 shows the records of the 20 times executions for each task, in which S represents successful trail, F represents failed one and R means the robot responds *"Sorry, I can't do it"*. Both S and R are considered as successful trials. We can see from the records that the successful rates of the three tasks are all quite ideal. However, there are still some failed cases. The reasons may be that the real environment is more unstable and the hand pose calculated by the 3D vision system is not accurate enough.

Table 1. The records of the three different tasks in evaluating the integrated framework of the internal models in PKU-HR6.0. Each task is conducted with two different desk height (1 cm and 3.5 cm) and random cuboid poses.

Task	Desk height (cm)	Times			Successful rate (%)
		S	F	R	
Reaching	3.5	9	0	1	100
	1	10	0	0	
Grasping	3.5	9	0	1	95
	1	8	1	1	
Placing	3.5	8	0	2	95
	1	9	1	0	

4   Conclusion

In this paper, an approach for a robot to develop its internal models is proposed. The approach mimicking the human infants and enables the robot to learn the internal models by itself through self-produced movements. To summarize our approach, firstly the proprioception is developed using the autoencoder neural networks through the robot motor babbling. Then based on the proprioception rather than directly using the motor command, the forward model and inverse model are built using the deep neural network. As for the more complicated inverse map, a cascaded proprioception-based inverse model (cPNIM) is further proposed to learn the inverse map of proprioception in a cascading manner. The learned forward model and inverse model are then integrated in a feedback framework to fulfill different arm movement tasks including reaching, grasping and placing.

With the experiments, we demonstrated the learning process of proprioception, and also compared the two different forward models and three inverse models. The learning process of proprioception shows that the progress of robot's ability is similar to the cognitive development of human infants. The performance of the learned proprioception verifies that the proposed autoencoder model works well in mimicking the proprioception of human. The comparisons between the two forward models as well as the three inverse models both verified the effectiveness and superiority of using the proprioception rather than the original numerical values of body configuration, and that the cascaded model do improve the performance of the inverse model. The integration of forward model and inverse model provides a more confidential mechanism just like human do in accomplishing arm motion tasks. In the future, we will extend our model to more complex tasks where whole body motion control is involved.

Acknowledgment. The work is supported in part by the National Natural Science Foundation of China (No. U1713217, No. 11590773), the Key Program of National Social Science Foundation of China (No. 12 & ZD119) and the National Basic Research Program of China (973 Program; No. 2013CB329304).

References

1. Asada, M., MacDorman, K.F., Ishiguro, H., Kuniyoshi, Y.: Cognitive developmental robotics as a new paradigm for the design of humanoid robots. Robot. Auton. Syst. **37**(2), 185–193 (2001)
2. Baranes, A., Oudeyer, P.-Y.: Intrinsically motivated goal exploration for active motor learning in robots: a case study. In: 2010 IEEE/RSJ International Conference on Intelligent Robots and Systems (IROS), pp. 1766–1773. IEEE (2010)
3. Castellini, C., Orabona, F., Metta, G., Sandini, G.: Internal models of reaching and grasping. Adv. Robot. **21**(13), 1545–1564 (2007)
4. Corbetta, D., Thurman, S.L., Wiener, R.F., Guan, Y., Williams, J.L.: Mapping the feel of the arm with the sight of the object: on the embodied origins of infant reaching. Front. Psychol. **5**, 576 (2014)
5. D'Souza, A., Vijayakumar, S., Schaal, S.: Learning inverse kinematics. In: IEEE/RSJ International Conference on Intelligent Robots and Systems, vol. 1, pp. 298–303. IEEE (2001)
6. Gooey, K., Bradfield, O., Talbot, J., Morgan, D.L., Proske, U.: Effects of body orientation, load and vibration on sensing position and movement at the human elbow joint. Exp. Brain Res. **133**(3), 340–348 (2000)
7. Hoffmann, M., Marques, H., Arieta, A., Sumioka, H., Lungarella, M., Pfeifer, R.: Body schema in robotics: a review. IEEE Trans. Auton. Ment. Dev. **2**(4), 304–324 (2010)
8. Jiang, H., et al.: A two-level approach for solving the inverse kinematics of an extensible soft arm considering viscoelastic behavior. In: 2017 IEEE International Conference on Robotics and Automation (ICRA), pp. 6127–6133. IEEE (2017)
9. Jordan, M.I., Rumelhart, D.E.: Forward models: supervised learning with a distal teacher. Cogn. Sci. **16**(3), 307–354 (1992)

10. Kofinas, N., Orfanoudakis, E., Lagoudakis, M.G.: Complete analytical forward and inverse kinematics for the NAO humanoid robot. J. Intell. Robot. Syst. **77**(2), 251–264 (2015)

11. Luo, D., Hu, F., Deng, Y., Liu, W., Wu, X.: An infant-inspired model for robot developing its reaching ability. In: 2016 Joint IEEE International Conference on Development and Learning and Epigenetic Robotics (ICDL-EpiRob), pp. 310–317. IEEE (2016)

12. Luo, D., Hu, F., Zhang, T., Deng, Y., Nie, M., Wu, X.: Human-inspired internal models for robot arm motions. In: 2017 IEEE/RSJ International Conference on Intelligent Robots and Systems (IROS), p. 5469. IEEE (2017)

13. Metta, G.: Babybot: a study into sensorimotor development. Ph.D. thesis, LIRA-Lab (DIST) (2000)

14. Nguyen-Tuong, D., Peters, J.: Model learning for robot control: a survey. Cogn. Process. **12**(4), 319–340 (2011)

15. Pickering, M.J., Clark, A.: Getting ahead: forward models and their place in cognitive architecture. Trends Cogn. Sci. **18**(9), 451–456 (2014)

16. Riemann, B.L., Lephart, S.M.: The sensorimotor system, part II: the role of proprioception in motor control and functional joint stability. J. Athl. Train. **37**(1), 80 (2002)

17. Robin, D.J., Berthier, N.E., Clifton, R.K.: Infants' predictive reaching for moving objects in the dark. Dev. Psychol. **32**(5), 824 (1996)

18. Rolf, M., Steil, J.J.: Efficient exploratory learning of inverse kinematics on a bionic elephant trunk. IEEE Trans. Neural Netw. Learn. Syst. **25**(6), 1147–1160 (2014)

19. Schillaci, G., Hafner, V.V., Lara, B.: Coupled inverse-forward models for action execution leading to tool-use in a humanoid robot. In: ACM/IEEE International Conference on Human-Robot Interaction, pp. 231–232. ACM (2012)

20. Sherrington, C.S.: On the proprio-ceptive system, especially in its reflex aspect. Brain **29**(4), 467–482 (1907)

21. Tolani, D., Goswami, A., Badler, N.I.: Real-time inverse kinematics techniques for anthropomorphic limbs. Graph. Model. **62**(5), 353–388 (2000)

22. Wolpert, D.M.: Computational approaches to motor control. Trends Cogn. Sci. **1**(6), 209–216 (1997)

23. Wolpert, D.M., Ghahramani, Z., Jordan, M.I.: An internal model for sensorimotor integration. Science **269**(5232), 1880 (1995)

Artificial Unintelligence: Anti-intelligence of Intelligent Algorithms

Yuhong Zhang[1,2(✉)] and Umer Nauman[1]

[1] Information Science and Engineering College, Henan University of Technology,
Zhengzhou 450001, China
zhangyuhong001@gmail.com, stormy.umer@gmail.com
[2] Network and Data Security Key Laboratory of Sichuan, University of Electronic
Science and Technology, Chengdu 610054, China

Abstract. In the age of big data, artificial intelligence (AI) algorithms play an important role in demonstrating the value of data, but its negative effects are increasingly prominent. Researchers focus more on the obvious issues like privacy protection these days, and we assume that technologies like AI will make people wiser. But due to the not-so-obvious limits of AI, it may push us to the opposite side–artificial unintelligence. In this paper, a detailed analysis of the process of value creation from data is presented first. Then, based on the above process, the role played by AI algorithms is analyzed. Next, the anti-intelligence features of AI algorithms, such as the tendency of *'fatuous King'*, the generalization of quick thinking, a faster transformation of master/slave at the individual level, and retribalization at the group level, are deeply discussed. Finally, some possilbe ways of circumventing the aforementioned problems are offered.

Keywords: Big data · Artificial intelligence algorithm
Quick thinking · Master-slave transformation

1 Introduction

We are in the era of big data. The value of big data has been confirmed to a large extent. Obviously, the value of it is not the data. Data itself is not the solution. It is just part of the path to that solution (say, the valuable insights hidden from the data) [1]. However, finding insights from the massive data is not easy. People are incapable of dealing with this issue directly due to the deficiencies in terms of memory and computational power. As a result, we often mechanize

This paper is partly supported by Henan Provincial Key Scientific and Technological Plan (no. 152102210261, no. 162300410056), Plan of Nature Science Fundamental Research in Henan University of Technology (no. 2015QNJH17) and National Natural and Science Foundation of China (no. 61602154).

and engineer mental activities that humans are not good at, and then outsource them to machines. In simple terms, this kind of mental activity process in which machines replace humans can be collectively referred to as artificial intelligence (AI) algorithms.

With the help of AI algorithms, the value of big data is shining. However, value comes at its fair price, because there is no free lunch in the world. Therefore, we need to check carefully what kind of role AI algorithms play in the production chain of big data value.

Some ongoing researches have shown that the technologies including AI may not make us smarter. We have to pay the price for the 'lunch' provided by AI algorithms. The price includes the tendency to be a 'fatuous king' in digital age, the generalization for quick thinking, the faster transformation between master and slavery at the individual level, whose idea proposed by Hegel, and retribalization at a group level. All of the secondary products of AI algorithms may create a dangerous situation of artificial unintelligence. We will discuss these topics separately in the rest of the paper.

2   Value Creation From Big Data

2.1   Core Value of Big Data

The core value of big data is rooted in its ability to prediction [2]. When it comes to the value of big data, many people think of it as scientific value. But in fact, the scientific value of big data is only a by-product during realizing its commercial value. This point can be seen from how domestic and foreign famous big data companies (such as Alibaba, Google, Amazon, etc.) monetize their investments.

There are many Internet giants described above, which are typically big data companies. In terms of business model, they may vary widely. But in the way of monetizing big data, they all come together in the same way. Most of them can be classified as advertising technology companies since their key businesses are mostly based on computational advertising (CA).

In CA, the role of prediction is evident. There is a need to predict which user is interested in which advertisement, and then the corresponding advertisement is pushed. Herein we use a simple case to illustrate how the value of data is produced.

2.2   Demonstration of Data Value

In order to monetize their assets, Internet companies (namely, big data companies) can achieve it in two ways. One is the use of traffic. This approach means that, if people visit your website, or use your App, in addition to serving content that the user is interested in (mostly free), the website (or App) can also carry some sponsored contents (i.e., advertisements). The sponsored contents are carried in free ones, which is the basis for the monetization of traffic.

Obviously, the monetization of traditional traffic-based advertising is inefficient. This is because all of people will see the same advertisement. Accordingly, the ad conversion rate is expected to be too low since most of users have no feelings of these ads or even disgusted with them. For Internet companies, it is a big deal to boost the conversion rate. Actually, these companies have a very important magic weapon, which is data [3].

For example, suppose we have a website with $100,000$ visits per day. Then we can place an ad space on it (for instance, razor advertising). This ad space can have a quotation, say $5,000$\$, which is the value of the traffic monetization. Clearly, it has flaws in this way, since the effective audiences for razor advertising are basically men. In other words, about half of the audiences (female) are wasted. Now the question remains: how to rationally use the other half of traffic?

In fact, in computational advertising system, we can do exactly that, leaving only half of the traffic to this razor advertiser for men audience. For the advertiser, the reach of effective audience is not compromised. But because he only uses half of the traffic, search-ad platform can give him a discount, say $3,000$\$.

As a result, Internet companies can sell the remaining half of the women's traffic to another advertiser who sells cosmetics at a price of $3,000$\$. For advertiser, he/she should be satisfied with the result because he/she spends less money to get same effective audience ($5,000$\$ $- 3,000$\$ $= 2000$\$). As for the Internet companies, they earn more money than before ($3,000$\$ $+ 3,000$\$ $= 6,000$\$).

There is no doubt that it is a win-win business. However, the win-win (or multi-win) situation is, to some extent, inconsistent with the principle of interests conservation. To ahead for a win-win situation, new resources must be injected into this process of profit distribution. What is this extra resource? It is data. In fact, Internet companies are able to get a premium of $1,000$\$ just because they have got the gender data of each user. This is the value that data create.

2.3   No Free Lunch Theorem

As previously analyzed, data indeed can create value. If we continue to cross-examine, why do Internet companies have such data? In fact, the reason is not complicated. To a large extent, it is because Internet users make use of their products for free. This is the price of the so-called *"free"*. As we all know, *"There is no free lunch in the world."*

Many Internet companies offer a variety of Internet products (such as search engines, social networks, and e-commerce APPs). Most of these products are free. When users use these products, they leave behind clues so regardlessly. Internet companies collect this information (i.e., electronic imprinting) and form a user profile. A user profile is a collection of information associated with a user. It is the explicit digital representation of the identity of the user. The user profile helps in ascertaining the interactive behavior of the user along with preferences.

After obtaining the user profiles, Internet companies can use them to deliver related paid content or products. Naturally, such *"targeted"* or *"catering to"* promotion is much better than blind promotion without any information guidelines.

2.4   Role of Artificial Intelligence Algorithms

The case listed above is relatively simple, but in a real business environment, the number of users, the complexity of data and the sorts of advertisement rise far above that of the simple case.

Computational advertising companies (i.e. big data companies) want to find the *"best match"* among particular users and the corresponding advertisement in a context. The biggest challenge facing this "best match" is the large-scale optimization and search problem under complex constraints. In such scenarios, if there is no AI algorithm heavily involved, it is impossible to efficiently complete the massive matching task.

Therefore, it can be said that AI algorithms play a very important role in extracting useful knowledge (namely, insights) from large amount of data. In the process of mining value from big data, researchers and the general public often show great concern about the fall of personal privacy.

But is it so simple? Of course not. Besides that, AI algorithms have aggravated its potentials of "anti-intellectual" in the digital age [4]. Next Let us put this issue below the fold.

3   Anti-intelligence Features of AI Algorithms

3.1   Tendency of 'fatuous King'

With the help of AI algorithms, even if you are in a minority in any small scale, your preferences or opinions can be realized to some extent. Even if your viewpoint is wrong or your behavior is not in line with mainstream social values, it is not difficult to find your followers by collaborative filtering recommendation algorithm. Clearly, it will worsen the cognitive prejudice of the world around us, which gradually results in a modern version of the "fatuous king" in digital age. As we know, in ancient history, we usually called a ruler as *"fatuous king"*, who only listened to one-sided story to make wrong decision.

This argument is supported by facts. For example, China's emerging news client giant, Jinri Toutiao (i.e., today's headlines), which is a Beijing-based news and information content platform, short video giants, Tik Tok (also known as Douyin), which is a Chinese music video platform and social network, and the US social giant Facebook, the users number of these companies are hundreds of millions, and their AI models will generate a tailored feed list of content (including news, videos, friends and even political ads) for each user, by analyzing the features of content and user profiles.

The well-known big data scholar Zhipei Tu called this type of data a "data pill", "once you continue to take it, you subconsciously change your judgment of something." In this situation, it is very easy for us to fall into a narrow view. It seems that every choice is made by us. However, those things we do not like or prone to reject are all filtered by AI algorithms. Gradually, we will hold a misguided belief that the world we see is the entire world, and naturally become considerable arrogance.

Therefore, one will face a disadvantageous scenario in which all the personalized services (including information) are provided by AI algorithms. There is no doubt that it will contribute us to be a *"fatuous king"* in AI age.

3.2 Generalization of Quick Thinking

From the analysis above, we can see that, due to information overload, one has to rely on the recommendation of personalized content. These seemingly good recommendations will strengthen our ability of quick thinking and weaken our ability of thinking slow.

The concepts of "thinking fast" and "thinking slow" are put forward by Daniel Kahneman in his book *"Thinking, Fast and Slow"* [5]. In this book, Kahneman explains the dichotomy between two modes of thought "System 1" that is fast, instinctive and emotional; "System 2" that is slower, more deliberative, and more logical.

As for human decision-making, due to the laziness of the slow-thinking system (it will consume more energy), brain does not start it easily. As a result, most of the time, brain is dominated by quick thinking.

But quick thinking is flawed because it produces more cognitive bias. It means that, to achieve rapid decision-making, we have to repeatedly adopt a psychological shortcut - heuristics (i.e, quick thinking).

In the era of AI, AI algorithms can provide us all kinds of decisions that look good. Due to highly engineered solutions provided by intelligent algorithms, even if our mind can switch to "slow thinking" model, finally we find that decisions which are made by machine are usually better than ours. Overtime, people tend to quit the thankless job, which results in weakening our capacity of "slow thinking" gradually. The serious consequence that followed is to generalize our ability of quick thinking.

If a further deduction is made, we will see that, the *"free-will"* of which human beings have been proud may become *"pseudo-free"*, since all of choices seem to be our own, but in fact, they just are the derivative results that are recommended by AI algorithms.

3.3 Retribalization of the Digital Age

The famous media theorist and thinker Marshall McLuhan made a three-phase division of the development of human society: *"tribalization - detribalization - retribalization."* In the age of AI, the evolution of the three phases will be accelerated.

In simple terms, *"tribalization"* means that, starting from the primitive society, dispersed individuals need to build a network of collaborative relationships to form a relatively stable and small alliance, which is a natural community of interests and discourse. The tribes are isolated and closed from each other.

In Web 1.0, the essence of the Internet is connection. It realizes the online expression of the physical world and generates convergent information. Through the online connection, the *"tribalization"* of the online community is realized.

In Web 2.0, the information began to flow interactively, and Internet flattened the world. In other words, each point can be easily interconnected. Even if users are separated from each other by thousands of miles, and the media still can push similar news, social media, which allows us to "get closer to the world (namely global village)". The closed tribe gradually begin to disintegrate. This process is called "*detribalization*."

Internet forecasters even believed that Internet would lead to a massive increase in "*detribalization*". However, decades have passed and the theory has not been applied yet. Instead, with the rise of Web 3.0, human beings do not realize the Great Unity. On the contrary, everyone seems to enter a small circle and is barely able to scrape himself off the floor.

The most significant feature of Web 3.0 is that information begin to "actively (intelligently)" find people. Through the personalized recommendation of AI algorithms, even if a very small group of "individuals" can become "groups" again. Interconnecting technologies make it easier for us to find our own memes, so that, once again, they can hide in "the call of the tribe", whose concept is proposed by Karl Popper. This phenomenon is called "*retribalization*".

In other words, in the age of the Internet, we should have been more interconnected, but on the recommendation of AI algorithms, it may make people more closed from each other, not get closer to one another. Since the adaptive recommend framework used by Jinri Toutiao (a news App) and Facebook, usually uphold the principle of Matthew: "*For unto everyone that hath shall be given and he shall have abundance.*" But is this really the "Gospel" for general public? This is an important issue that is worthy of deep consideration.

4   Countermeasures and Conclusions

From the above analysis, we can see that in the era of big data, AI algorithms do indeed make us see the world of data more clearly. However, everything has two sides, because it may also bring about potential negative effects. The former Amazon Chief scientist Andreas Weigend presented us with a general principle: "Our lives should not be driven by data. They should be empowered by it." [6]

Part of the solutions to address many of those concerns have actually been written into the wisdom of our sages. For example, in order to relieve tendency of being a *fatuous king* in digital age, the ancient Chinese philosophy helps, "Listen to both sides and you will be enlightened; heed only one side and you will be benighted." More "listening" to several dimensions of "facts" independently, will bring a clearer fact.

For another example, the solution to both the generalization of quick thinking and Hegel's master-slave identity transformation are similar, they exit both due to over-reliance on AI algorithm. Therefore, independence, self-reliance is not only applicable to individuals and countries, but also applies to individuals in the era of big data. In this scenario, Occam's razor also works. "If not necessary, not by entity." This is because that, once one entity (such as AI algorithm) is involved in our life or work, then this object may give us an uncontrollable result.

In addition, with the increasing maturity of blockchain technology, it also provides a technical perspective for us to solve *retribalization*. Blockchain is not an extension of the Internet, but a great subversion of the Internet. At present, each of us lives in a block (similar to a tribe) and is constrained by this small environment, and it is increasingly difficult to escape. If we can make innovations in the *"chain"* of the blockchain, break the barriers of the blocks and make the blocks link with each other efficiently, this will yield a huge dividend in the future.

AI will become IA (Intelligence Augmentation) if it is used properly. However, if used improperly, artificial unintelligence will happen and AI may become an elusive Leviathan (a primeval monster that symbolizes evil in Bible). In this regard, we are expected to have a cautious optimism, embrace the uncertainties of AI era [7], and form a cross-border harmony between artificial intelligence and human intelligence.

References

1. Hammond, K.J.: The Value of Big Data Isn't the Data. Harvard Business Review. https://hbr.org/2013/05/the-value-of-big-data-isnt-the. Accessed 7 Aug 2018
2. Victor, M.-S., Keneth, C.: Big Data: A Revolution That Will Transform How We Live, Work, and Think. Houghton Mifflin Harcourt, New York (2013)
3. Wang, J., Yuan, S.: Real-time bidding: a new frontier of computational advertising research. In: Proceedings of the Eighth ACM International Conference on Web Search and Data Mining, pp. 415–416. ACM (2015)
4. Meredith, B.: Artificial Unintelligence: How Computers Misunderstand the World. The MIT Press (2018)
5. Kahneman, D., Egan, P.: Thinking, Fast and Slow. Farrar, Straus and Giroux New York (2011)
6. Weigend, A.S.: Data for the People: How to Make Our Post-Privacy Economy Work for You. Basic Books (2017)
7. Sam, B., Chris, W., Simon, W.: Embrace the uncertainty of AI, McKinsey & Company. https://www.mckinsey.com/business-functions/organization/our-insights/the-organization-blog/embrace-the-uncertainty-of-ai. Accessed 7 Aug 2018

XiaoA: A Robot Editor for Popularity Prediction of Online News Based on Ensemble Learning

Fei Long[1(✉)], Meixia Xu[1], Yulei Li[1], Zhihua Wu[1], and Qiang Ling[2]

[1] Chinaso Inc., Beijing 100077, China
`{longfei,xumeixia,liyulei,wuzhihua}@chinaso.com`
[2] Department of Automation, University of Science and Technology of China, Hefei 230027, China
`qling@ustc.edu.cn`
`http://www.chinaso.com`

Abstract. In this paper, we propose a robot editor called XiaoA to predict the popularity of online news. A method for predicting the popularity of online news based on ensemble learning is proposed with the component learners such as support vector machine, random forest, and neural network. The page view (PV) of news article is selected as the surrogate of popularity. A document embedding method Doc2vec is used as the basic analysis tool and the topic of the news is modeled by Latent Dirichlet Allocation (LDA). Experimental results demonstrate that our robot outperforms the state of the art method on popularity prediction.

Keywords: Robot · Popularity prediction · Ensemble learning · LDA

1 Introduction

Online news articles are attractive to a large amount of Internet users for the short length and rich content. However, the popularity of those articles are not evenly distributed. Only a small fraction of news attract the public attention successfully and become the so called hot news. For the popularity of online content is always related to the revenue, it is important to predict it beforehand.

The Blossom bot built by New York Times can solve this problem well. It is a chat bot within the messaging app Slack, which utilizes machine learning in its backend. It helps decide which story to post to social media, which got 380 percent more clicks than a typical post. In view of this, we develop a robot editor similar to Blossom named XiaoA to help editors improve there works. The main task of XiaoA is to predict the popularity of a large amount of news.

Predicting the popularity of online news is a great challenge for it is affected by several factors. According to the previous researches, quality of the content, article topic and the title influence the popularity of the article a lot, so these

Z. Shi et al. (Eds.): ICIS 2018, IFIP AICT 539, pp. 340–350, 2018.
https://doi.org/10.1007/978-3-030-01313-4_36

three factors are considered in our following prediction. In order to predict the popularity of the online news, the popularity itself needs to be quantified. Here, page view is selected as the only surrogate of popularity.

As mentioned in many former research papers, the lives of most online news articles are very short. Hence it is more valuable to predict the early popularity of a news. Fortunately, these data are available from kinds of news rankings. Moreover, the prediction of online content was modeled as regression or classification problem in preceding researches. So we present a method for popularity prediction of online news based on ensemble learning. The news ranking data of 163 (http://news.163.com/rank/) is used as our training and testing set.

The contributions of our paper are:

(1) We propose a popularity prediction method for online news based on ensemble learning which outperforms the state of the art method.
(2) We evaluate the performance of several classifiers on popularity prediction and get some meaningful conclusions.
(3) We find the relationship between popularity and the news features.

2 Related Work

Linear regression was used to predict the views of Digg and Youtube [11]. In [11], the long term popularity after 30 days can be predicted based on the early popularity within one hour. However, the popularity news needs to be predicted before promotion. Zhang [15] presented a rationale augmented Convolutional Neural Network (CNN) model for text classification. The RA-CNN model outperforms some baseline CNN models. Although the popularity prediction of online news is also modeled as a classification problem, the unique features of online news make the application scenario a little different from this paper.

Several papers [1,2,7,9,12,13] were published to predict the popularity of online news. Keneshloo et al. [7] defined the popularity of an article as page views within the first day. They casted popularity prediction as a regression problem. Various features were used in the prediction while some of which such as social media features are not available for news. [12] and [2] both used regression models to predict the popularity of news articles. However, those classifiers may not perform well among unpopular articles [1]. [13] addressed the prediction problem as a two stage classification. This kind of two stage classifier was demonstrated to perform below the average of other classifiers under the dataset in our former experiments. [9] used the number of votes to present the popularity of a story. However, it does not reflect the relationship between the article itself and its popularity. According to the works mentioned above, several classifiers are used comprehensively here to predict the popularity. We will show that our method can substantially improve the prediction accuracy of online news.

3   Preliminaries

3.1   Problem Statement

We seek to predict whether a given news will be popular given its content and title. If a dataset of news and their corresponding PVs is available, the prediction can be formulated as a classification problem. An article with its PVs above a certain threshold is deemed as popular and vice versa. The contents and titles of the articles are transformed into vectors, besides, the topics of the articles are also transformed into vectors through LDA. These three vectors of an article are combined as a comprehensive vector $x = [x_1, x_2, ..., x_n]$, as the input of the classifier. The output of the classifier is whether the article is popular or not.

3.2   Original Dataset

Our dataset is crawled from 163 news. As shown in Fig. 1, the category of the news is labelled in the top red box. Here the characters in the red box means 'news'. The red box in the second line means the 'PV rank within 24 h'. The page views of a single news article is marked in the column of the right red box.

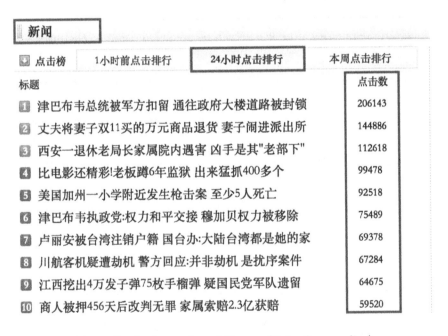

Fig. 1. Rank of news from 163.com. (Color figure online)

The dataset contains 25733 pieces of news in 20 weeks. We use this dataset to train and test our popularity model.

3.3   Data Preprocessing

As mentioned above, the article should be transformed into vector. The title, content and topic of an article are transformed into vectors respectively, and combined as a comprehensive vector. Doc2Vec [8] is used to vectorize the title and content, and LDA [4] is used to represent the topic. Before vectorization, the stop words should be deleted. The stop words are some function words such as 'is', 'and', 'but' and 'or' etc., and some meaningless symbols. For we know the existing categories of 163 news is 10, the topic of LDA is chosen to 10 accordingly.

4   Proposed Methodology

According to previous literatures, there is no single classifier that can overwhelm all the others. Therefore, we bring ensemble learning for popularity classification. Several classifiers are chosen as our component learners such as Random forest (RF), Neural network (NN), Support vector machine (SVM), Logistic regression (LR), Nearest centroid (NC) and Restricted Boltzmann machine (RBM). We will give a brief introduction of three main learners in this paper.

4.1   Component Learners

Support Vector Machine (SVM): As mentioned above, the content, title and topic of an article will be transformed into vectors and combined together as the input. If the dimension of the combined vector is n, the input of SVM is $x^{(i)} = [x_1^{(i)}, x_2^{(i)}, ..., x_n^{(i)}]$. The key idea of SVMs is to find a maximum margin hyperplane that separates two sets of points in a higher dimensional space [5]. Each article in the training set is deemed as a point. The combined vector $x^{(i)} = [x_1^{(i)}, x_2^{(i)}, ..., x_n^{(i)}]$ is a n-dimensional vector represents the article i itself, the value $y^{(i)}$ is the PVs of article i. Hence the article can be deemed as a $n+1$-dimensional point $\mathbf{x}^{(i)} = (x_1^{(i)}, x_2^{(i)}, ..., x_n^{(i)}, y^{(i)})$. What we want is to find a 'line' as the boundary of the two sets of points that represents popular and unpopular points respectively. The boundary of the two sets is defined as:

$$\mathbf{y}(\mathbf{x}) = w^T \mathbf{x} + b \tag{1}$$

The problem can be represented as:

$$min_{w,b,\xi} \frac{1}{2}\|w\|^2 + \frac{\gamma}{2}\sum_{i=1}^{m}\xi_i^2 \tag{2}$$

subject to
$$\mathbf{y}_i(w\mathbf{x}_i + b) \geq 1 - \xi_i, i = 1, 2, ..., m$$
$$\xi_i \geq 0, i = 1, 2, ..., m$$
where γ is the penalty parameter, and ξ_i is the slack variable [3].

Neural Network (NN): Artificial neural networks were proved to be good classifier. if the weight is w_{ij}, where i represents the start node and j represents the end node. Node i has s input nodes, the output of this node can be represented as:

$$a_i = f(\sum_{j=1}^{s} w_{ij} + b_i) \qquad (3)$$

where b_i is the bias of node i, and $f(\cdot)$ is the activation function. In this paper, the combined vector $x^{(i)} = [x_1^{(i)}, x_2^{(i)}, ..., x_n^{(i)}]$ of article i is the input of the neural network. If $y^{(i)}$ is the PVs of article i, $(x^{(i)}, y^{(i)})$ is a training point accordingly. The network can be solved by a back propagation (BP) algorithm [10].

Random Forest (RF): Random forest is an ensemble learning method based on decision trees [6]. A news article can be represented as a vector $x = [x_1, x_2, ..., x_n]$, and y represents whether the article is popular, $y = \{0, 1\}$. We know that the training set $\{(x^{(1)}, y^{(1)}), ..., (x^{(m)}, y^{(m)})\}$ can form a decision tree based on information gain. However, the decision tree may overfit the training set. Random forest utilizes a multitude of decision trees to make classification by voting of these decision trees. The training set of each tree is a subset of the whole training set, and is created by a bootstrap manner.

4.2   Ensemble Learning

Ensemble learning accomplish learning task through constructing and combining various learners. Our component learners are already introduced above, ensemble learning integrate all the component learners to achieve a better performance. Before introducing the algorithm, we should answer some questions:

(1) How to quantify the influence of topic, content and title of the article on its popularity?
(2) How to determine the threshold τ of the popularity of these news articles?
(3) How to integrate the component learners to get a strong learner?

Title and content are coupled tightly. Although same content can have totally different titles, the title should be related to the content finally. On the other hand, the topic is not so tightly coupled with the content. The contents of articles under the same topic can be of a wide variety. If the topic vector o is p-dimensional, $o = [o_1, o_2, ..., o_p]$, the content vector c is q-dimensional, $c = [c_1, c_2, ..., c_q]$, and the title vector t is also q-dimensional, $t = [t_1, t_2, ..., t_q]$, the input vector x can be defined as $x = [o, c, t]$. In order to test the contributions of the content and title to the popularity, we use $x = [o, c', t']$ instead of $[o, c, t]$, in which $c' = \alpha c$, $t' = (1 - \alpha)t$, where α is the contribution weight.

The threshold τ can be estimated by the dataset and parameter α will be tested by experiments. We use voting policy to determine the classification result. Assume the whole set is: $D = \{d_1, d_2, ..., d_n\}$, where d_i is a news contains title and content. $U = \{d_1, d_2, ..., d_m\}$ is the training set, $m < n$. Accordingly, the test set is $V = D - U$. The N component leaners can be deemed as functions. Each function F_i can be trained using the training set U by SGD or other methods given the parameters τ and α. The input of function F_i is the vector x of an article mentioned above, and the output is the classification result y, $y \in \{0, 1\}$. The algorithm of the ensemble learning is as follows.

Algorithm 1. Popularity prediction based on ensemble learning

Input: U, x, τ, α
Output: y
1: Transform each news article in D into vector x using LDA and Doc2Vec with certain α
2: **for all** $i \in [1, N]$ **do**
3:     Train the function F_i using training set U with certain threshold τ
4: **end for**
5: **for all** $F_i \in \{F_1, F_2, ..., F_N\}$ **do**
6:     **if** $F_i(x) == 1$ **then**
7:         $T + +$
8:     **else**
9:         $F + +$
10:     **end if**
11: **end for**
12: **if** $T > F$ **then**
13:     **return** $y = 1$
14: **else**
15:     **return** $y = 0$
16: **end if**

---

5   Experiments

We traced the 163 news rank for 20 weeks and got 25733 pieces of news with their PVs in 24 h. Since the measurement of popularity differs in each field (for example, the average PVs of entertainment is apparently higher than other fields), the threshold of popularity should be considered comprehensively. According to our statistics, the percentage of the popular news is about 10% of the whole dataset, the threshold of all the fields is about 15000 PVs, that is, $\tau = 15000$. To balance the positive and negative samples, the popular news and unpopular news are chosen with a proportion of 1:1. Finally, 8000 articles are chosen as our training set with 4000 popular news and 4000 unpopular news, 1600 news are chosen as the test set similarly.

In order to find out the influence between title and content on the popularity of an article, we tune the parameter α to 0, 0.5 and 1 respectively. $\alpha = 0$ means that we only consider the influence of the title. $\alpha = 1$ means that we only consider the influence of the content. $\alpha = 0.5$ means that we consider both the title and the content equally. The dimension of the topic vector is 4, the dimension of the content and title vector both are 200, hence the dimension of vector x is 404.

We evaluate the accuracy, precision, recall and F1 score of each base learner, the parameter α is tuned to 0, 0.5 and 1. Since random forest (RF) is already an ensemble learning method, the rest learners are combined to form the ensemble learning in our first experiment. The results are shown in the following tables.

We use F1 score to measure the performance of all the leaners mentioned in Tables 1, 2 and 3. Three conclusions can be drawn from the results:

(1) Some base learners perform obviously better than others on our dataset. Seen from the results, RF, NN, and SVM are much better than the other three in F1 score about 10% on average whenever α is 0, 0.5 or 1.
(2) The ensemble learning can not always perform better than each of the base learner especially when the performances of the base learners differ a lot. As shown in the results, the ensemble learning method is better than LR, NC and RBM in all situations. However, it can not always perform better than NN and SVM. The poor performances of LR, NC and RBM may degrade the performance of the ensemble learning.
(3) The content is more important than title for we find that the average performance of the base learners and ensemble learning increases when α increases. Although some clickbait title may attract many clicks at the beginning, the netizens will soon be familiar with this kind of title and ignore them finally.

Table 1. Popularity prediction results with $\alpha = 0.5$

Base learner	Accuracy	Precision	Recall	F1
RF	0.778	0.791	0.759	0.775
NN	0.772	0.739	0.843	0.788
SVM	0.784	0.938	0.610	0.740
LR	0.608	0.608	0.622	0.615
NC	0.648	0.674	0.582	0.625
RBM	0.545	0.528	0.924	0.672
Ensemble (without RF)	0.752	0.725	0.815	0.767

Table 2. Popularity prediction results with $\alpha = 0$

Base learner	Accuracy	Precision	Recall	F1
RF	0.756	0.764	0.743	0.754
NN	0.707	0.678	0.795	0.732
SVM	0.792	0.956	0.614	0.748
LR	0.519	0.512	0.964	0.669
NC	0.574	0.577	0.570	0.574
RBM	0.523	0.514	0.968	0.671
Ensemble (without RF)	0.636	0.593	0.880	0.709

Table 3. Popularity prediction results with $\alpha = 1$

Base learner	Accuracy	Precision	Recall	F1
RF	0.784	0.771	0.811	0.791
NN	0.766	0.744	0.815	0.778
SVM	0.774	0.810	0.719	0.762
LR	0.667	0.679	0.639	0.658
NC	0.653	0.676	0.594	0.632
RBM	0.507	0.505	0.984	0.668
Ensemble (without RF)	0.766	0.724	0.863	0.788

Notice that the performance of RF, NN and SVM is similar and obviously better than others, we use them as the base learners in our second experiment. The ensemble learnings in the first and second experiment are marked as En1 and En2. The experiment results are shown in Tables 4, 5 and 6.

As shown in Tables 4, 5 and 6, the performance of En2 improves a lot. The increasing of F1 score of En2 when α is 0, 0.5 and 1 can also demonstrate conclusion (3). The good performance of En2 demonstrate conclusion (2) from the opposite side that the ensemble learning can improve the performance when the performances of the base learner are similar. The comparison between base learners and En1, En2 can be seen from Fig. 2, which represents the trend that the content improving the performance with the increasing of weight α.

Table 4. Improved popularity prediction results with $\alpha = 0.5$

Base learner	Accuracy	Precision	Recall	F1
RF	0.778	0.791	0.759	0.775
NN	0.772	0.739	0.843	0.788
SVM	0.784	0.938	0.610	0.740
En1	0.752	0.725	0.815	0.767
En2	0.800	0.861	0.719	0.783

Table 5. Improved popularity prediction results with $\alpha = 0$

Base learner	Accuracy	Precision	Recall	F1
RF	0.756	0.764	0.743	0.754
NN	0.707	0.678	0.795	0.732
SVM	0.792	0.956	0.614	0.748
En1	0.636	0.593	0.880	0.709
En2	0.776	0.835	0.691	0.756

Table 6. Improved popularity prediction results with $\alpha = 1$

Base learner	Accuracy	Precision	Recall	F1
RF	0.784	0.771	0.811	0.791
NN	0.766	0.744	0.815	0.778
SVM	0.774	0.810	0.719	0.762
En1	0.766	0.724	0.863	0.788
En2	0.804	0.814	0.791	0.802

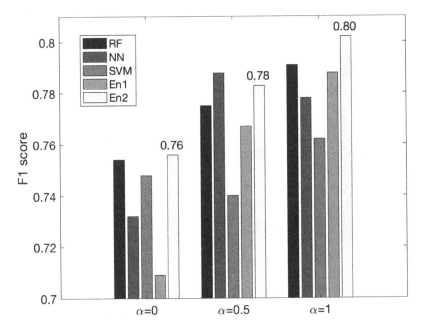

Fig. 2. F1 score of the learners

Our work is also compared with a state of the art news popularity prediction method [14]. Three different algorithms such as LPBoost, Random Forest and AdaBoost were implemented in this paper to predict the popularity of news articles. The dataset of 39797 news were collected from UCI machine learning repository. According to this paper, the best popularity prediction model was adaptive Boosting on the MCI dataset, which had achieved F1 score of 73% and accuracy of 69%. Our En2 method can achieve F1 score of 80.2% and accuracy of 80.4% when $\alpha = 1$ under a much less dataset of 8000 news articles.

6   Conclusion

This paper presents a robot editor called XiaoA to predict the popularity of online news based on ensemble learning. Our ensemble learning method outperforms the state of the art prediction method. Besides, we find that the content of an article plays the most important role in determining the popularity. We also find that learners with similar performance will have a better performance using ensemble learning. Although ensemble learning achieves good performance in popularity prediction, a lot of factors will also affect the popularity of online news, which will be discussed in out future researches.

References

1. Arapakis, I., Cambazoglu, B.B., Lalmas, M.: On the feasibility of predicting news popularity at cold start. In: Aiello, L.M., McFarland, D. (eds.) SocInfo 2014. LNCS, vol. 8851, pp. 290–299. Springer, Cham (2014). https://doi.org/10.1007/978-3-319-13734-6_21

2. Bandari, R., Asur, S., Huberman, B.A.: The pulse of news in social media: forecasting popularity. In: Proceedings of the Sixth International AAAI Conference on Weblogs and Social Media, pp. 26–33 (2012)

3. Bishop, C.M.: Pattern Recognition and Machine Learning. Information Science and Statistics. Springer, New York (2006)

4. Blei, D.M., Ng, A.Y., Jordan, M.I.: Latent Dirichlet allocation. J. Mach. Learn. Res. **3**, 993–1022 (2003). http://dl.acm.org/citation.cfm?id=944919.944937

5. Cortes, C., Vapnik, V.: Support-vector networks. Mach. Learn. **20**(3), 273–297 (1995). https://doi.org/10.1007/BF00994018

6. Ho, T.K.: Random decision forests. In: International Conference on Document Analysis and Recognition, pp. 278–283 (1995)

7. Keneshloo, Y., Wang, S., Han, E.H., Ramakrishnan, N.: Predicting the popularity of news articles. In: Siam International Conference on Data Mining, pp. 441–449 (2016)

8. Le, Q., Mikolov, T.: Distributed representations of sentences and documents. In: Proceedings of the 31st International Conference on International Conference on Machine Learning - Volume 32, ICML 2014, pp. II-1188–II-1196. JMLR.org (2014). http://dl.acm.org/citation.cfm?id=3044805.3045025

9. Lerman, K., Hogg, T.: Using a model of social dynamics to predict popularity of news. In: Proceedings of the 19th International Conference on World Wide Web, WWW 2010, pp. 621–630. ACM, New York (2010). https://doi.org/10.1145/1772690.1772754

10. Rumelhart, D.E., Hinton, G.E., Williams, R.J.: Learning representations by back-propagating errors. In: Neurocomputing: Foundations of Research, pp. 696–699. MIT Press, Cambridge (1988). http://dl.acm.org/citation.cfm?id=65669.104451

11. Szabo, G., Huberman, B.A.: Predicting the popularity of online content. Commun. ACM **53**(8), 80–88 (2008)

12. Tatar, A., Antoniadis, P., Amorim, M.D.D., Fdida, S.: Ranking news articles based on popularity prediction. In: International Conference on Advances in Social Networks Analysis and Mining, pp. 106–110 (2012)

13. Tsagkias, M., Weerkamp, W., Rijke, M.D.: Predicting the volume of comments on online news stories. In: ACM Conference on Information and Knowledge Management, pp. 1765–1768 (2009)
14. Vilas, T., Dhanashree, D.: Analysis of online news popularity prediction and evaluation using machine intelligence. Int. J. Math. Comput. Methods **2**(2), 120–131 (2017)
15. Zhang, Y., Marshall, I., Wallace, B.C.: Rationale-augmented convolutional neural networks for text classification. In: Proceedings of the Conference on Empirical Methods in Natural Language Processing, pp. 795–804 (2016)

Design and Implementation of Location Analysis System for Mobile Devices

Yu Rao[✉], Shunxiang Wu, Bin Xi, Huan Li, and Jingchun Jiang

Department of Automation, Xiamen University,
No. 422, Siming South Road, Xiamen, Fujian, China
q360584748@qq.com, wsxl009@163.com, bxi@xmu.edu.cn,
1748468754@qq.com, 945366381@qq.com

Abstract. With the rapid development of Internet technology and the rapid spread of mobile devices, mobile devices have become an indispensable tool for people today and store a large amount of user's private information. The use of mobile devices for fraud, illegal transactions, dissemination of rumors, and the spread of Trojan horse crimes and other criminal activities are increasing, which has a serious impact on social stability. Therefore, how to locate the suspect's geographical position and narrow the scope of arrest become particularly important. Based on the principles of mobile digital forensics, this paper studies and designs a complete location analysis system for mobile terminal device from the perspective of data visualization.

Keywords: Mobile device · Digital forensics · Geolocation
Data visualization

1 Introduction

Foreign digital forensic research began in the 1970s and continued to develop in the 1990s and then became mature. Relative to the progress of foreign research, the research on domestic mobile digital forensics technology started late and is still in the development stage. For the time being, there are no standardized forensic procedures and scientific forensics systems, and there are no comprehensive laws and regulations on digital evidence forensics. In recent years, with the continuous improvement in the status of electronic evidence in the judiciary, some domestic government research departments, university research institutes, and corporate research departments have begun research on smart mobile terminal forensics-related technologies, and the number of people investing in research has also rapidly increased.

With the memory capacity of smart mobile terminals is getting larger, more and more applications are installed, and the types of data formats are also increasing. This causes an increase in the workload of traditionally relying on manual search for digital evidence. Therefore, it is very important to develop effective mobile digital forensics tools. Mobile terminal equipment will generate a large amount of geographical location information in daily use. In digital forensics, the locational information can well reflect the trajectory of the behavior of suspects, and play a very important role in determining and narrowing the range of activities of suspects and can provide important clues for forensic personnel.

2 System Related Technologies

2.1 NodeJS

NodeJS is an open source, cross-platform JavaScript runtime environment that can run JavaScript on the server side. As we know, JavaScript is the only standard for front-end development. NodeJS brings JavaScript to the development of back-end servers. Its greatest advantage is the use of JavaScript's inherent event-driven mechanism plus the V8 high-performance engine, making write a high-performance Web service easily. With the increasing popularity of NodeJS, NPM was born. NPM is a Node package management tool. It is the largest software registry in the world, with approximately 3 billion downloads per week and over 600,000 packages. Allowing users to download third-party packages written by others from the NPM server for local use, therefore NodeJS can greatly accelerates the efficiency of development.

2.2 Bootstrap

Bootstrap is an open source toolkit for front-end development designed by Twitter. It can help us quickly build the interface. Providing rich font icons, drop-down menus, buttons, navigation, carousels, and other various components designed to make the development of dynamic web pages and web applications easier. Its unique grid system allows Web pages to adapt to different resolution devices.

2.3 Tencent Map API

The Tencent Map open platform provides locational services and solutions for various developers. It provides components such as location display, route planning, and map selection points, markers can be displayed on the map by latitude and longitude information. By marking on the map, it is possible to display the suspect's geographic information.

3 Design and Implementation of Geographical Analysis System of Mobile Terminal Equipment

The geographical analysis system of the mobile device can be divided into two parts: data collection and data analysis. The collection client can collect data of mainstream mobile devices on the market, and the data analysis client can analysis data, filter data and visual data. Through the data acquisition equipment, the positioning data in each application can be obtained. The system only needs to visualize the position, latitude and longitude, and time stamp data in the positioning information table in the analysis system interface.

The system does not display all the data of the database directly, but needs to filter and classify the data in the database, such as the scope of activities for a specific time period, the visual display of geographical information of a specific App, and then visually shows the target suspect's detailed information for locating or navigating with an App during a certain period of time. The system uses Tencent map to display the suspect's mark. The overall process design of the system is shown in Fig. 1.

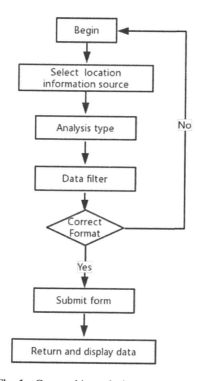

Fig. 1. Geographic analysis system process

The geographic location analysis system is mainly composed of client and server. The client needs to provide the selection of the geographical location information data source, statistical types, and data screening conditions to filter the data as follows:

– Choose geographical location information data source: Include some apps that can generate geographic location information. Baidu map, Gaode map, Tencent map, location information in Taobao, location information of photos are included in this system.
– Statistics Type: Contains navigation type and location type. The navigation type is displayed in the Tencent map as a line segment containing the start and end points. The location type is displayed as a marked point on the Tencent map.
– Start and end time: You can filter geographical location information according to start time and end time, and then display it through Tencent map.

The data submitted by the client is sent to the server via ajax. The server processes the data in the database and then returns data to the client for displaying. System function module diagram shown in Fig. 2:

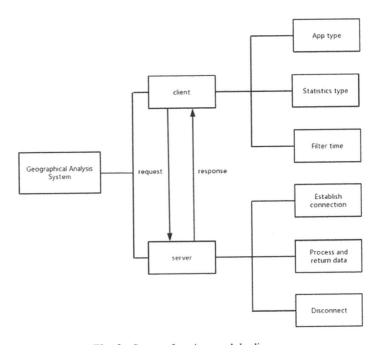

Fig. 2. System function module diagram

3.1   System Front-End Interface Design and Implementation

The web interface design of this system mainly uses the bootstrap framework. First, jQuery must be introduced before the introduction of bootstrap, because all javascript plugins of bootstrap rely on jQuery, and then introduce bootstrap through the free CDN acceleration service. According to the above functional analysis, the entire page is designed as a two-column layout: the bootstrap grid system divides the entire browser viewport into 12 columns, and it can scale the devices with different rates, which can be well performed on different devices. Run, the system uses the left 3 columns of the browser viewport as a conditional filter layout, from top to bottom as follows:

- Select tag creates a selection list with options. The selection list includes the data source and the type of statistics that generated the geographic location information.
- The DateTimepicker time selection control of bootstrap filters the data generated by the geographic location information in time.
- Button tag is used to submit form data to the server.

A set of buttons for manipulating the suspect's mark: including location query, display location, hide and clear tags.

The remaining 9 columns on the right are used to display detailed geographic information and Tencent map markers as follows:

- Datatables plugin of bootstrap is used to store the data returned by the filtered server. The table plugin contains the search box and sorting functions.
- The page loads the Tencent map by embedding the iframe and marks the suspect's geographical information.

The operational effect of the system is shown in Fig. 3:

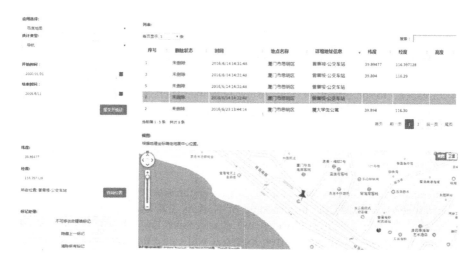

Fig. 3. System operation result graph

3.2 Design and Implementation of System Front-End Logical Functions

The front-end logic functions of this system mainly include.

When the form is submitted, the selected conditions are recorded and the data is submitted to the server via ajax: The value of the select list can be obtained by setting the value, obtaining the start and end time of the DateTimepicker control, and submitting the data to the server. Then submit to the server through jQuery's $.ajax method and request the processed data, and convert the data into JSON format.

After the data submitted to the server is filtered, the data matching the conditions is returned to the front end, and the data is inserted into the Datatables table plugin. Then the data binding event for each row in the table, when the lines in table are clicked, the position information can be obtained such as latitude and longitude, the latitude and longitude information is passed as an argument to the Tencent map api, and the marker or the route is displayed on the map. In addition, even if there is no latitude and longitude information can use the place name to search for places. Here are the Tencent map APIs used by this system as follows:

- CityService: According to the city name, latitude and longitude, IP address, telephone area code to obtain city information. This function sends the request parameters to the server asynchronously and returns the result through a customized callback function.

- Geocoder: The address resolution class is used to translate between addresses and latitude and longitude. This function sends the retrieval condition asynchronously to the server and returns the result via a custom callback function.
- Add custom controls: Add custom controls to the map by generating a qq.maps. Control instance. This control requires initialization of the options parameter, where the content attribute specifies the DOM node element or HTML string that the custom control presents in the page.
- Navigation route display: The line segments connecting the start and end points are displayed on the map by obtaining the values of the start and end latitude and longitude. Concrete implementation: Add a polyline to the map by generating an instance of the class maps.polyline. This class requires initialization of the options parameter, where path is a set of coordinates on the map and the coordinates correspond to the inflection point of the polyline. Then add polyline to the corresponding map object by setting the map property.

3.3 The Backend Logic Function of This System is Realized Through NodeJS

- Create a server: The Http protocol is a generic application layer protocol that creates a http server through the http module of Node.js and listens for client requests. Because Node.js is single-threaded, we must ensure the stability of the system. Otherwise, once the program error occurs, it will cause the entire server to crash. Therefore, the system uses a third-party package of NPM: Log4js log module, which has a log, daemons and exception capture and other functions.
- Listen to the request sent by the client and operate the database according to the parameters passed to the server. Node.js and MySQL database connection there are many ways, this system is selected in the NPM installation mysql package, and then in the server require ('mysql'), and then the parameters passed to the client, and finally execute MySQL Query the statement and return the data to the front end.

4 Conclusion

This paper introduces the design and implementation of a mobile terminal device geographic location analysis system. The advantage of this system lies in the display of geographical location information through data visualization technology. At the same time, different applications and positioning time can be screened to narrow the scope of activities of criminal suspects under certain circumstances, providing new ideas for the geographical location of mobile devices.

References

1. Petraityte, M., Dehghantanha, A., Epiphaniou, G.: Mobile phone forensics: an investigative framework based on user impulsivity and secure collaboration errors, pp. 79–89 (2017)
2. Xiao, C.W., Jun, L.U., Li-Geng, Y.U.: Application and optimization of algorithm in mobile phone forensics. Electron. Des. Eng. (2017)
3. Yang, L.I., Xue-Hui, G.U., Jia, Y.J.: Forensics of android mobile phone SkyDrive client. J. Crim. Invest. Police Univ. China (2017)
4. Ran, M.O., Dai, Z.C., Law, D.O.: Research on mobile phone forensics standardization program. J. Guangxi Police Acad. (2016)
5. Wang, G.Q.: Mobile phone forensics and its application in crime investigation. Forensic Sci. Technol. (2006)
6. Mislan, R.P.: Creating laboratories for undergraduate courses in mobile phone forensics. In: ACM Conference on Information Technology Education, pp. 111–116. ACM (2010)
7. Ding, H.: The study of mobile phone forensics normalization. Netinfo Secur. (2012)
8. Induruwa, A.: Mobile phone forensics: an overview of technical and legal aspects (2009). Inderscience Publishers

Control Information Acquisition
and Processing of the AMT System Based
in LabVIEW and MATLAB

Zhisen Zhang[✉] and Chengfu Yang

Sichuan University of Arts and Science,
519, Tashi Road, Dazhou City, Sichuan, China
thrywood@163.com, 443563624@qq.com

Abstract. In this paper, the information acquisition and processing system is proposed for the automated mechanical transmission (AMT) control by LabVIEW and MATLAB. Firstly, an interface of the data acquisition was designed with the automated mechanical transmission system based in LabVIEW. Secondly, the data acquisition hardware was built by employing the single chip system incorporating with necessary sensors. Thirdly, the data acquired was transmitted into the data processing system built by MATLAB. By using the data processing system, the transmission status and driving parameters was presented in MATLAB which can be viewed easily and ready to be utilized further more.

Keywords: Control information · Acquisition · Processing · AMT
LabVIEW · MATLAB

1 Introduction

The automated mechanical transmission (AMT) system is widely used in vehicles. Basically, the AMT system is composed of two parts, one is the mechanical transmission and the other is the shift mechanism which functions as an experienced personnel driver. By the shift mechanism, the mechanical transmission can be shifted from one gear to another according to the control strategies. For the purpose of obtaining efficient control strategies and implementing the control operation, the control information such as the driving motor (or engine) speed, clutch control data and mechanism motors driving currents data etc. have to be acquired and processed. In this paper, an information acquisition and processing system of the AMT based in LabVIEW and MATLAB was fabricated. By the LabVIEW, the acquired data can be presented in the Man-Machine interface and stored simultaneously. By MATLAB, the data acquired from different AMT locations which carries coupled information can be viewed easily and available for further utilized.

2   Overview of Data Acquisition and Process Techniques

Data acquisition (DAQ) is the process in which the physical signals such as voltage, current, temperature, pressure and voice etc. can be measured, transmitted and stored in computer. Usually, a DAQ system is composed of sensors, DAQ measurement hardware and computer armed by programmable software. The schematic diagram of a typical DAQ system is shown in Fig. 1. Compared with a traditional system, the computer DAQ system can provide more powerful, flexible and cheap resolutions by using standardized processing, displaying and communicating.

Sensors   DAQ Device   Computer

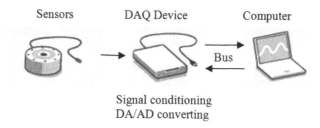

Bus

Signal conditioning
DA/AD converting

Fig. 1. Schematic diagram of DAQ system

A data processing system is a combination of machines, people, and processes that for a set of inputs produces a defined set of outputs. Data Processing ensures that the data is presented in a clean and systematic manner and is easy to understand and be used for further purposes. Usually, a data process involves steps like editing, coding, classification, tabulation and analysis which can be shown in Fig. 2.

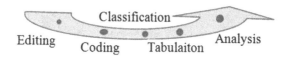

Classification

Editing   Coding   Tabulaiton   Analysis

Fig. 2. Data processing steps

3   Interface of the DAQ System

In this research, the DAQ system includes sensors, data bus, and signal conditioning devices, AD/DA converters and man-machine interfaces. The transmission control unit (TCU) is powered by a single chip produced by Freescale (MC9S12DJ64). The shift mechanism has two driving motors. One of which selects the proper gear position and the other changes the gear by pushing the handle to the target position. During the changing gear process, the clutch must be interrupted for the protection of the transmission and shift mechanism. Thus the clutch openness was controlled as well. LabVIEW is systems engineering software for applications that require test, measurement, and control with rapid access to hardware and data insights. The gear control and clutch control interfaces are developed in the LabVIEW environment which are illustrated in Figs. 3 and 4.

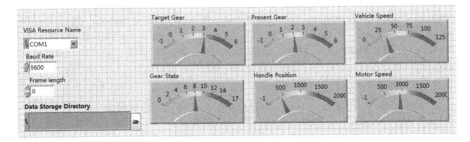

Fig. 3. Gear control interface

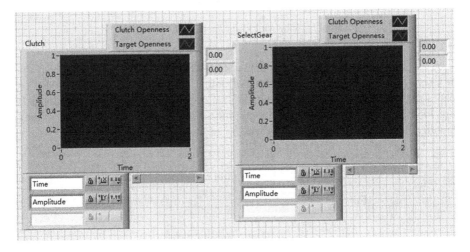

Fig. 4. Clutch control interface

4   Data Processing and Results Presentation

MATLAB (matrix laboratory) is a multi-paradigm numerical computing environment developed by MathWorks. MATLAB allows matrix manipulations, plotting of functions and data, implementation of algorithms, creation of user interfaces, and interfacing with programs written in other software languages. The visualization of data in MATLAB environment by programming is user-friendly and can be completed easily. The acquired data such as gear status, engine speed, mechanism motors driving currents, handle position and transmission input/output speeds etc. are presented in Fig. 5.

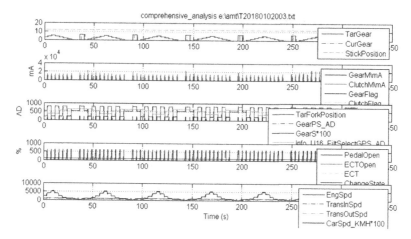

Fig. 5. Data processing by MATLAB programming

5 Conclusions

By using the data acquisition and processing system, the transmission status and driving parameters was presented in MATLAB which can be viewed easily and ready to be utilized further more in control strategy design and analysis. The method developed here can be used by analogous systems as well and is of practice.

Acknowledgement. This work was financially supported by the Education Department of Sichuan (No. 15ZA0318) and Sichuan University of Arts and Sciences.

References

Xi, J., Wang, L., Fu, W., Liang, W.: Shifting control technology on automated mechanical transmission of pure electric buses. Trans. Beijing Inst. Technol. **30**(1), 42–45 (2010)

Taguchi, Y., et al.: Development of an automated manual transmission system based on robust design. In: Transmission & Driveline Systems Symposium, SP-1760, pp. 79–85 (2003)

Junzhi, Z., Qingchun, L.: Automated mechanical transmission in China. In: Seoul 2000 FISITA World Automotive Congress, F2000A151, pp. 1–7 (2000)

Yang, L., et al.: LabVIEW Programming and Application, 2nd edn. Publishing House of Electronic Industry, Beijing (2005). (in Chinese)

National Instruments Corp: LabVIEW Simulation Interface Toolkit User Guide. National Instruments, Inc. (2003)

Vitale, V., et al.: A Matlab based framework for the real-time environment at FTU. Fusion Eng. Des. **82**, 1089–1093 (2007)

Multi-robot Distributed Cooperative Monitoring of Mobile Targets

J. Q. Jiang[1(✉)], B. Xin[2], L. H. Dou[3], and Y. L. Ding[1]

[1] School of Automation, Beijing Institute of Technology, Beijing 100081, China
976798059@qq.com, dingyvlong@163.com
[2] State Key Laboratory of Intelligent Control and Decision of Complex Systems, Beijing Institute of Technology, Beijing 100081, China
brucebin@bit.edu.cn
[3] Beijing Advanced Innovation Center for Intelligent Robots and Systems, Beijing Institute of Technology, Beijing 100081, China
doulihua@bit.edu.cn

Abstract. Cooperative monitoring targets of mobile robots is of great importance in military, civil, and medical applications. In order to achieve multi-robot coordinated monitoring, this paper proposes a new distributed path planning algorithm. In this algorithm, a method of dynamic sequential decision making is used firstly to determine which robots should be moved to the neighbourhoods of the target path. A distributed cooperative path planning algorithm is then proposed in which the robots can only utilize their local information to make the joint monitoring areas of robots cover the target path. To reduce the chance of detection and ensure enough time to implement monitoring handover, our distributed algorithm can guarantee enough overlapped monitoring areas between adjacent robots. The effectiveness of the algorithm was verified by simulation experiments and a comparison experiment in typical scenarios.

Keywords: Cooperative monitoring · Dynamic sequential decision making Distributed coverage

1 Introduction

Target monitoring has a wide range of application prospects and potential economic value, including search and rescue program [9], path planning [6], decision problems of different types of unmanned aerial and ground vehicles systems (UAGVS) [2], Dubins Traveling Salesman Problem (DTSPN) [8] and Cooperative multi-area coverage (CMAC) [7]. Khan et al. reviews control techniques for cooperative mobile robots monitoring multiple targets [5]. In recent years, the problem has been studied extensively in the centralized setting. Chakrabarty et al. [1] proposed a distributed method which is used to monitor the targets using a robot network. They adopted a grid coverage strategy. Although the targets can be monitored in a wider range, they used too many robots and the cost is too high. Fu et al. [4], distributed methods are used to enhance the scalability of the network, reduce the communication costs, Fu et al. [4]

Z. Shi et al. (Eds.): ICIS 2018, IFIP AICT 539, pp. 362–372, 2018.
https://doi.org/10.1007/978-3-030-01313-4_39

establish an integer programming model, and obtains the optimal value. But there is no monitoring handover in their research, robots are easily found by the targets.

This paper will study the monitoring problem of multiple mobile robots for moving objects. Each robot needs to monitors an area. After the target (see the red car) moves, the monitoring handover takes place and the next robot continues to monitor. As shown in Fig. 1.

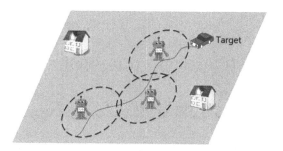

Fig. 1. Monitoring of moving targets by multiple mobile robots, the dotted circular area represents the monitoring range of the robot. The curve represents the path of the moving target (Color figure online)

This paper propose a novel distributed target monitoring algorithm. When solving this monitoring problem, we apply a decoupling strategy to decompose the problem into two subproblems: one is to select suitable robots and quickly move near the target by a distributed algorithm based on dynamic sequential decision making, and the other is to achieve the full coverage of the target trajectory by a distributed path planning algorithm. In addition, the distributed path planning algorithm can generate enough overlaps between the monitoring ranges of robots to effectively reduce the probability that robots is monitored and ensures enough time for robots to implement monitoring handover. As far as the author knows, target monitoring with monitoring handover has not been studied.

The paper is structured as follows. Section 1 introduces the research background and related work of cooperative target monitoring. Section 2 describes the problem and modelling. Section 3 introduces a algorithm based on sequential decision making. Section 3.1 proposes the distributed cooperative path planning algorithm of the robots. Section 4 gives the experimental results and analysis. Section 5 concludes the paper.

2   Problem Description

We consider a set of n robots $\mathbf{S} = \{s_1, s_2, \cdots, s_n\}$. Let s_i^x, s_i^y and R_i be the x-coordinate, y-coordinate and the monitoring radius of the robots s_i, respectively. We consider a set of m targets $T = \{T_1, T_2, \cdots T_m\}$. Let T_j^x, T_j^y, V_j and θ be the x-coordinate, y-coordinate, the speed of the target T_i and its horizontal angle, respectively.

The target path is modeled as multiple polyline segments. Note that if the target trajectory is an arbitrary curve, multiple polyline segments can approximate the curve. The objective function g of the problem is the shortest movement distance for all robots, that is,

$$g = \min\left\{\sum_{j=1}^{n} V_j \cdot t_j\right\} \tag{1}$$

subject to:

$$\forall t, \left(T_i^x(t) - s_i^x(t)\right)^2 + \left(T_i^y(t) - s_i^y(t)\right)^2 \leq R_i^2 \tag{2}$$

$$\sum_{i=0}^{N_k} R_i \cdot \cos \theta_k \geq len_{r_k} + f, \quad s_i \in T_k \tag{3}$$

where T_k is the set of robots for selecting the polygonal segment r_k, there are N_k elements in T_k, and f is the overlapped area between the all robots.

The constraint (2) indicates that the target is within the monitoring range of the robots, and can be monitored by the robots. Constraint (3) indicates that the length of the monitoring range projection of all the platforms which select a polygonal segment, is greater than the length of the polygonal segment. This problem confronts us with the following challenges:

(i) The problem is a mixed-variable optimization problem. Whether the platform or line segment is selected is an integer variable. The guiding position of the platform is a continuous variable.

(ii) There is a coupling relationship between selecting polyline segments and selecting monitoring points near the polyline segments for each robot.

3   Distributed Cooperative Path Planning Algorithm

In this section, a two-stage method is designed. The first stage is a distributed robot selection method based on sequential decision making, which can make the robots move towards the target path. The second stage is a robot cooperative path planning algorithm. The multi-robot monitoring areas fully cover target path.

3.1   Distributed Robot Selection Method

The distance from each robot s_i to each polygonal segment r_k is denoted by $D_{i,k}$, and $D_{i,k}$ will be sorted by polyline segments. The distance D_i from each robot s_i to the target path is

$$D_i = \min_k D_{ik}. \tag{4}$$

Each robot broadcasts its own distance D_i to the target path to other robots. Each robot receives information from other robots and sorts them. In this way, each robot will get the same sequence of the robots selection polyline segments.

The ith robot calculates the objective function for each polyline segment r_k when selecting the appropriate polyline segment

$$g_k = \frac{slen_{r_k}}{R_i} \cdot d. \tag{5}$$

In the above formula, R_i is the radius of the monitoring range of the ith robot. Then the kth polyline segment is selected by the ith robot, and

$$k = \arg\min\{g_k\}. \tag{6}$$

After each robot selects a polyline segment, the remaining length of the polyline segment is updated to its original value minus the projection of the robot on the polyline, that is

$$slen_{r_k} \leftarrow slen_{r_k} - R_i \cdot \cos\theta_k. \tag{7}$$

In the above formula, θ_k is the angle with the horizontal direction when the target moves on the kth polyline segment.

At the same time, it is ensured that the length $R_i \cdot \cos\theta_k$ of the projection of the robots for selecting each polyline segment r_k is larger than the length len_{r_k} of the polyline segment. Also there is enough monitoring overlapping area between the robots, that is

$$\sum_{i=0}^{N_k} R_i \cdot \cos\theta_k \geq len_{r_k} + f, s_i \in T_k. \tag{8}$$

In the above formula, T_k is the set of robots for selecting the polygonal segment r_k, there are N_k elements in T_k, and f is the overlapped area between the all robots.

In order to shorten the robots' movement distance, the idea of dynamic sequential decision-making was introduced. When the length of a polyline segment is no more than 0 according to (7), that is $slen_{r_k} \leq 0$, all robots need to redetermine the selected sequence. At this time, when calculating the distance $D_{i,k}$ of the polyline segments, all robots should remove the polyline segment whose length is reduced to 0 or less than 0.

Near the target motion path, two safety lines parallel to the path are generated, and randomly N_k points are generated on two safety lines as the candidate monitoring points of the robots. Every robot needs to select the closest monitoring point and moves to it.

3.2   Multi-robot Distributed Cooperative Path Planning Algorithm

The previous section completed selection of the polyline segments and the initial monitoring points on the safety lines near the polyline segments. In this section, we mainly design a distributed cooperative path planning algorithm [3], which will allow robots that have moved to the initial monitoring point to achieve full coverage of the target path through distributed cooperative path planning, and at the same time ensure that there is sufficient monitoring handover area. This paper models the problem of coverage of the target path by the robots' monitoring area as a barrier coverage problem.

A line segment of length L is used to represent the target path $[0, L]$ on the x axis. There are n robots, each of which is equivalent to one node S_i. In order to avoid separate considerations for the first robot and the last robot, we set two virtual robots s_{-1} and s_{n+1} on the left of $x = 0$ and right of $x = L$, respectively. We denote L_1 as one of the multiple polyline segments of the target path, L_2 as the safety line nearby L_1, and D as the vertical distance from the point on L_2 to L_1, θ as the angle between L_1 and the horizontal direction, R is the radius of the robot guidance area, and r as the radius of the robot monitoring area when projected onto a one-dimensional straight line, as shown in Fig. 2.

$$r = \sqrt{R^2 - D^2} \cos \theta \tag{9}$$

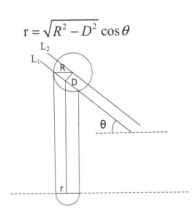

Fig. 2. Coordinate transformation diagram

At this time, all robots' sensing range is mapped to one dimension. Eftekhari et al. [3] studied the problem of covering a straight line with nodes with a length on a one-dimensional line when robots have an identical monitoring radius. In this study, we expand their previous work to make robots with different monitoring radius cover the straight line (the monitoring radius of robot s_i is denoted as r_i). In addition, the robot s_i ($i \in [-1, n+1]$) needs to judge whether the overlap between its monitoring areas and that of its neighbourhood satisfies Formula (3). Therefore, every robot needs to make decisions on whether and where to move according to the following conditions:

If $s_{i+1}^x - s_i^x \geq r_i + r_{i+1} + f$ and $s_i^x - s_{i-1}^x \leq r_i + r_{i-1} + f$, then $d = -1$, $s_i^x = s_i^x + vd$

If $s_{i+1}^x - s_i^x \leq r_i + r_{i+1} + f$ and $s_i^x - s_{i-1}^x \geq r_i + r_{i-1} + f$, then $d = 1$, $s_i^x = s_i^x + vd$

If $s_{i+1}^x - s_i^x \geq r_i + r_{i+1} + f$ and $s_i^x - s_{i-1}^x \geq r_i + r_{i-1} + f$, then $d = -1$, $s_i^x = s_i^x + vd$

where d indicates the movement direction of the robot after each movement, $d = -1$ indicates that the robot moves to the left, $d = 1$ indicates that the robot moves to the right, v is the movement speed of the robot, and f is the projection of the overlapping area between the robots in T_k on one dimension.

According to the above algorithm, full coverage of the target path can be achieved on each polyline segment, and sufficient monitoring of the handover area is ensured. However, at the turning point of each of the two polyline segments, the robots located at both ends of the two polyline segments do not have any interaction with each other, and there is no guarantee that there will be enough areas for monitoring handover. Therefore, The turning points of two polyline segments are also required to be analyzed. The schematic diagram at the turning point is shown in Fig. 3, judging whether | $AB| + |BC| \geq f$ is satisfied or not.

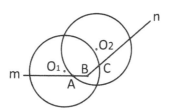

Fig. 3. Inflection point

If $|AB| + |BC| \geq f$, two robots' positions do not need to be adjusted. If $|AB| + |BC| < f$, s_1 will move to the turning point B until $|AB| + |BC| \geq f$; If during the movement, the intersection with the guidance junction is no longer satisfied between the robot on the left side and s_1, s_2 should move to the turning point B until $|AB| + |BC| \geq f$. According to the above ideas, we proposed Algorithm 1.

Algorithm 1 Improved barrier coverage algorithm based on mobile robots

---

1: for $k = 1 : q$

    Set two virtual robots s_{-1} and s_{n+1}

    for $i = 1 : N_k$

        if s_i moved last stage

            if there is a gap in the direction of movement before

                s_i move one unit toward the gap

            else

                s_i do not move

            end if

        else if

            if $s_{i+1}^x - s_i^x \geq r_i + r_{i+1} + f$ and $s_i^x - s_{i-1}^x \leq r_i + r_{i-1} + f$

                s_i move one unit toward the left gap

            else if $s_{i+1}^x - s_i^x \leq r_i + r_{i+1} + f$ and $s_i^x - s_{i-1}^x \geq r_i + r_{i-1} + f$

                s_i move one unit toward the right gap

            end if

        else

            if $s_{i+1}^x - s_i^x \geq r_i + r_{i+1} + f$ and $s_i^x - s_{i-1}^x \geq r_i + r_{i-1} + f$

                s_i move one unit toward the left gap

            else

                s_i do not move

            end if

        end if

    end for

    end for

2: if the projection target path is fully covered

    Convert the motion of the robots into a motion on a two-dimensional plane

  else

    Goto step1

  end if

3: if the requirements for the guidance handover area at the turning point satisfy

    end

  else

    Adjust the positions of the robots at the turning points, goto step3

  end if

---

4 Comparative Experiment

In this section, we simulate the program, and obtain experimental results, then analyze the experimental results. Through comparison experiment, it is verified that we can shorten the movement distance of the robots by dynamic sequential decision making.

4.1 Experimental Results

The experimental results are shown in Fig. 4a. The robots move to the safety line near the target path from their initial positions. The red line indicates the target path, the green line indicates the safety line near the target path, and the blue dots indicates the robot's initial positions, and the connections between the robots and the initial points on the safety line (grey line) represent the paths taken by each robot to reach its proper initial monitoring point.

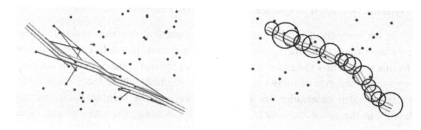

Fig. 4. a The results of the robots have selected the polyline segments. **b** The target trajectory obtained after the robots has moved through distributed path planning algorithms (Color figure online)

From Fig. 4b, after the selection of the polyline segments based on the sequential decision making is completed, the robot most suitable for the initial monitoring point can be selected and moves towards the target path. Distributed and coordinated movement between the robots achieves full coverage of the target path. At the same time, it can ensure that two adjacent robots meet the requirements for monitoring overlapping areas. The experimental results is shown in Table 1.

Percentage per robot in Table 1 is the ratio of average moving distance per robot and total moving distance of all robots. We expanded the number of targets from 1 to 3 for testing. From the above table, it can be seen that when the number of targets is 1, the robot has a shorter movement distance, but as the number of targets increases, the robot moving distance gradually increases because the initial position of the robot and the total number of robots in the space are fixed. When the number of targets increases, the available robots decrease, and the available robots' positions gradually move away from the target path, so the robot needs to move longer distance to the path.

Table 1. Experimental result

Target's number	Example	Robot's number	Total moving distance	Average distance per robot	Percentage per robot
1	Example 1	12	1714.21	142.851	8.33%
	Example 2	14	1621.84	115.846	7.14%
	Example 3	13	1677.95	129.073	7.69%
2	Example 1	25	3676.01	147.0404	4%
	Example 2	27	3800.16	140.747	3.7%
	Example 3	27	3709.03	137.371	3.7%
3	Example 1	40	6443.7	161.0925	2.5%
	Example 2	41	6018.23	146.786	2.44%
	Example 3	42	6230.53	148.346	2.38%

4.2 Dynamic Decision Sequence Versus Fixed Decision Sequence

In Sect. 3, we mentioned the idea of dynamic sequential decision-making. If there is update of the order of platform selection in the solution, just a fixed order method is used. In this way, when the order is determined in the next phase, there is no practical meaning for the distance of robots to the polyline segment whose length is 0. Although the robots can still determine the order of selecting, the original task can still be completed, but the robot movement distance will increase, the value of the objective function will increase greatly. Table 2 shows the experimental results when a fixed sequential decision making is used.

Table 2. Experimental results during fixed sequence decision making

Target's number	Example	Robot's number	Movement distance		Percentage
			Fixed sequence	Dynamic sequence	
1	Example 1	15	3618.21	1714.21	47.38%
	Example 2	14	3414.95	1621.84	47.49%
	Example 3	15	4464.26	1677.95	37.59%
2	Example 1	28	5903.12	3676.01	62.27%
	Example 2	29	7163.65	3800.16	53.05%
	Example 3	30	7369.85	3709.03	50.33%
3	Example 1	46	9575.96	6443.7	67.29%
	Example 2	43	9121.53	6018.23	65.98%
	Example 3	44	10756.4	6230.53	57.92%

The percentage in Table 2 refers to the ratio of the robot's movement distance in the fixed sequential decision making to the robot's movement distance in the dynamic sequential decision making. By comparison with Table 1, it can be seen that when the fixed sequential decision is made, the robot's movement distance is always longer than making the dynamic sequential decision making (see Table 2), no matter how many the target is. Because the number of robots in space is increasing, when the number of targets increases, the distance of dynamic sequential decision-making moves closer to a fixed sequential decision making. But it is always shorter than fixed-order decision-making. Therefore, it can be concluded that dynamic sequential decision making is beneficial to shorten the robot's movement distance.

5 Conclusion

This paper studies the cooperative monitoring problem for multiple moving robots, and proposes a strategy for multiple robots to achieve full coverage of the target path, so that the intrusion targets can be monitored by the robots in real time. An in-depth analysis is performed on the comparison of the time of the robots moving to the monitoring points and the time when the targets moved to the same monitoring points.

Acknowledgment. This work was supported in part by the National Natural Science Foundation of China under Grant 61673058, in part by NSFC-Zhejiang Joint Fund for the Integration of Industrialization and Informatization under Grant U1609214, in part by the Foundation for Innovative Research Groups of the National Natural Science Foundation of China under Grant 61621063, in part by the Projects of Major International (Regional) Joint Research Program NSFC under Grant 61720106011, and in part by International Graduate Exchange Program of Beijing Institute of Technology.

References

1. Chakrabarty, K., Iyengar, S.S., Qi, H., et al.: Grid coverage for monitoring and target location in distributed sensor networks. IEEE Trans. Comput. **51**(12), 1448–1453 (2002)
2. Chen, J., Zhang, X., Xin, B., Fang, H.: Coordination between unmanned aerial and ground vehicles: a taxonomy and optimization perspective. IEEE Trans. Cybern. **46**(4), 959–972 (2016)
3. Eftekhari, M., Kranakis, E., Krizanc, D., et al.: Distributed algorithms for barrier coverage using relocatable sensors. Distrib. Comput. **29**(5), 361–376 (2013)
4. Fu, Y., Ling, Q., Tian, Z.: Distributed sensor allocation for multi-target pathing in wireless sensor networks. IEEE Trans. Aerosp. Electron. Syst. **48**(4), 3538–3553 (2012)
5. Khan, A., Rinner, B., Cavallaro, A.: Cooperative robots to observe moving targets: review. IEEE Trans. Cybern. **48**(1), 187–198 (2016)
6. Qi, M.F., Dou, L.H., Xin, B., et al.: Optimal path planning for an unmanned aerial vehicle under navigation relayed by multiple stations to intercept a moving target. In: 56th IEEE Conference on Decision and Control, pp. 2744–2749 (2017)
7. Xin, B., Gao, G.Q., Ding, Y.L., et al.: Distributed multi-robot motion planning for cooperative multi-area coverage. In: IEEE International Conference on Control and Automation, pp. 361–366, August 2017

8. Zhang, X., Chen, J., Xin, B., et al.: A memetic algorithm for path planning of curvature-constrained UAVs performing surveillance of multiple ground targets. Chin. J. Aeronaut. **27**(3), 622–633 (2014)
9. Zhao, W., Meng, Q., Chung, P.W.H.: A heuristic distributed task allocation method for multivehicle multitask problems and its application to search and rescue scenario. IEEE Trans. Cybern. **46**(4), 902–915 (2016)

Research on the Micro-blog User Behavior Model Based on Behavior Matrix

Zhongbao Liu$^{(\boxtimes)}$, Changfeng Fu, and Chia-Cheng Hu

Quanzhou University of Information Engineering, Quanzhou 362000, China
liu_zhongbao@hotmail.com

Abstract. The micro-blog user behavior model based on behavior matrix is proposed based on the analysis of the current related researches. In the proposed method, two behavior matrices named original behavior matrix and individual behavior matrix are constructed; And then, the analysis method of behavior matrix is utilized to mine the representative behaviors, which can reflect the disciplines of the user; Finally, the experiments on the 2043 blogs of 9 users from June 15, 2016 to August 1, 2016 reflect the individual behavior disciplines.

Keywords: Micro-blog user · Behavior model · Behavior matrix

1 Introduction

With the development of information technology, micro-blog has attracted much more attention because of its instantaneity and interactivity. We can use micro-blog freely to express our own opinions and easily obtain interested information. More and more micro-blog platforms have appeared since 2016, such as Twitter, Sina, Tencent, Netease. The number of micro-blog users has grown exponentially and its number has arrived at 200 million.

The user behavior in micro-blog has attracted many researchers' interest, and a lot of achievements related the above research have been made. Kshay et al. draws the conclusions that the purpose of the micro-blog user is to discuss their activities and obtain their interested information. Brian points out that Twitter plays an important role in the communication between users in the disaster relief. Kaye analyzes the relationship between motivation, leadership and social media from the perspective of the user. Schwab et al. obtained the user's interests by analyzing the keywords in micro-blog. Maloof et al. discusses the problem of user interest drift in micro-blog with the help of forgetting mechanism. Xia discusses the features of user's behavior based on analyzing the topics, comments and other information the user visits in the micro-blog. Ping et al. analyzes the relationship of micro-blog network based on the topology relationship between micro-blog users. Wang et al. analyzes the number of concerns and fans, and the relationship between the number of blog articles based on the analysis of the structure and propagation of micro-blog. Zhao et al. conducts statistical analysis on the number of concerns and blog articles.

It can be seen that the researches on micro-blog users are gradually popular, and the achievements greatly improve the micro-blog service. However, the problem of

Z. Shi et al. (Eds.): ICIS 2018, IFIP AICT 539, pp. 373–377, 2018.
https://doi.org/10.1007/978-3-030-01313-4_40

micro-blog authenticity has emerged with the popularity of micro-blog. In view of this, the micro-blog user behavior model based on behavior matrix is proposed in this paper. In the proposed method, two behavior matrices named original behavior matrix and individual behavior matrix are constructed; And then, the behavior vector space is generated by user behavior matrix, and the most representative characteristic behavior is mined by Principal Component Analysis (PCA), which reflected the disciplines of user behavior; Finally, simulation experiment results on the Sina micro-blog show the effectiveness of the proposed method.

2 Micro-blog User Behavior Model Based on Behavior Matrix

We try to establish the micro-blog user behavior model by analysis of blog articles and the user behavior on the micro-blog. The micro-blog user behavior refers to the micro-blog user's behaviors, such as release, transmit and praise blog articles, concerns users, login and withdraw, and make comments, when they communicate with each others on the micro-blog platform in a certain period of time.

2.1 Behavior Matrix Model

In order to accurately describe the user behavior, two behavior matrices, the original behavior matrix and the individual behavior matrix, are introduced. The original behavior matrix is used to store user behavior records, which describes the user behavior disciplines as the user activities in a certain period of time. The individual behavior matrix is used to analyze the behavior disciplines of different individuals, which aims to record the fragmented and irregular micro-blog behaviors, and represent the micro-blog user behavior with appropriate rules and algorithms.

(1) Original behavior matrix
In order to describe the user behavior and analyze the behavior disciplines, the original behavior matrix is proposed. The basic idea of the original behavior matrix is to describe the user behavior disciplines as the user activities in a period of time. The original behavior matrix describes the behavior of user k in n time steps within m observation days. The original behavior matrix is used to store user behavior records, which is the fundamental model of a series of behavioral matrices.

(2) Individual behavior matrix
The individual behavior matrix is used to analyze the behavior disciplines of different individuals. In order to further analyze the user behavior, the individual behavior matrix is proposed based on original behavior matrix. The individual behavior matrix describes the behavior of user k in n time steps, and the behavior of each time step is the summarization of m observation days.

2.2   The Analysis Method of Behavior Matrix

It can be seen from the statistical analysis, the observation samples of user behaviors reflect the stochastic characteristics, however, their behavior disciplines can be still identified. The main idea of the analysis method of behavior matrix is to mine the representative behaviors based on PCA, which can reflect the user disciplines, by constructing the user behavior vector space model based on user behavior matrices.

(1) Behavior vector space model

The main idea of the behavior vector space model is to characterize the behavior disciplines of different individuals. The behavior disciplines are reflected by the distribution of the behavior vectors in a period of time. The vector representing the characteristic of user behaviors is defined as the behavior vector, and the n-dimensional space whose behavior vector belongs to is named as behavior vector space. The behavior discipline similarity of different individuals can be obtained by calculating the correlation coefficient of their behavior vectors.

(2) Characteristic behavior analysis method

The main idea of the characteristic behavior analysis method is to mine the representative behaviors in the behavior vector by PCA. The eigenvectors (characteristic behavior) of correlation coefficient matrix can be obtained by the characteristic behavior analysis method, and let the disciplines of high-frequency behavior with high weights. The correlation coefficient matrix is used to describe the relationship between variables of the behavior matrix. The process of characteristic behavior analysis method is as follows.

Input: the user behavior matrix
Output: eigenvectors and their eigenvalues
Step1: calculate the correlation coefficient matrix R of the user behavior matrix;
Step2: calculate the eigenvectors of R and its eigenvalues;
Step3: descending sort according to the eigenvalues;
Step4: calculate the proportion and the cumulative proportion of each eigenvalue;
Step5: obtain the eigenvectors and eigenvalues of the behavior matrix and the proportional and cumulative proportions of the eigenvalues.

3   Experimental Analysis

The 2043 micro-blogs released by 9 Sina micro-blog users from June 15, 2016 to August 1, 2016 were selected in our experiment, which was obtained by Sina micro-blog Application Programming Interface (API). In our experiment, a day was equally divided into 24 intervals and each time step was 60 min. The behavior matrix is analyzed by the characteristic behavior analysis method. The experiment results are recorded in Table 1, in which U_i (i = 1, 2, 9) stands for the micro-blog users.

It can be seen from Table 1 that the correlation coefficient between U8 and other users is quite small, which indicates his activity disciplines are irrelevance to other users, and it can be inferred that his micro-blog behaviors is quite different from other users. The correlation coefficient U1 and U4 is largest, which indicates these two users

Table 1. The correlation coefficients of 9 micro-blog users' individual behavior matrix

	U_1	U_2	U_3	U_4	U_5	U_6	U_7	U_8	U_9
U_1	1								
U_2	0.24	1							
U_3	0.52	0.51	1						
U_4	0.82	0.54	0.70	1					
U_5	0.64	0.64	0.79	0.70	1				
U_6	0.44	0.28	0.43	0.78	0.34	1			
U_7	0.77	0.45	0.72	0.77	0.81	0.70	1		
U_8	0.02	0.03	0.02	0.01	0.15	0.29	0.20	1	
U_9	0.67	0.30	0.73	0.71	0.62	0.66	0.75	0.03	1

have similar activity disciplines, and it can be inferred that their work environment and habits and customs maybe the same.

4   Conclusions

With the popularity of micro-blog, the number of micro-blog user grows exponentially. The problem of authenticity restricts the improvement of micro-blog services. In view of this, the micro-blog user behavior model based on behavior matrix is proposed based on the analysis of micro-blog user behaviors. Two behavior matrices named the original behavior matrix and the individual behavior matrix are introduced in our method, the former matrix is used to store the user original behavior records, and the latter is used to analyze the behavior disciplines of different individuals. The characteristic behavior analysis method is to mine the characteristic behaviors in the behavior vector by PCA. The simulation experiment results on Sina micro-blog show the effectiveness of the proposed method.

References

Java, A., Song, X., Finin, T., et al.: Why we twitter: understanding micro-blogging usage and communities. In: Proceedings of the 9th Web KDD and 1st SNA-KDD 2007 Workshop on Web Mining and Social Network Analysis, pp. 56–65 (2007)

Brian, B.G.: Socially distributing public relations: Twitter, Haiti, and interactivity in social media. Publ. Relat. Rev. **36**(4), 329–335 (2010)

Sweetser, K.D., Kelleher, T.: A survey of social media use, motivation and leadership among public relations practitioners. Publ. Relat. Rev. **37**(4), 425–428 (2011)

Schwab, I., Pohl, W.: Learning user profiles from positive examples. In: Proceedings of the ACAI 1999 Workshop on Machine Learning in User Modeling (1970)

Maloof, M., Michalski, S.: Selecting examples for partial memory learning. Mach. Learn. **41**, 27–52 (2000)

Xia, Y.H.: The structure and mechanism of microblog interaction-based on the empirical study of Sina Weibo. J. Commun. **4**, 60–69 (2010)

Ping, L., Zong, Y.L.: Research on microblog information dissemination based on SNA centrality analysis-a case study with Sina microblog. Doc. Inf. Knowl. **6**, 92–97 (2010)

Wang, X.G.: Empirical analysis on behavior characteristics and relation characteristics of micro-blogging users-take "Sina micro-blog" for example. Libr. Inf. Serv. **54**(14), 66–70 (2010)

Zhao, W.B., Zhu, Q.H., Wu, K.W., et al.: Analysis of micro-blogging user character and motivation-take micro-blogging of Hexun.com as an example. New Technol. Libr. Inf. Serv. **2**, 69–75 (2011)

Probe Machine Based Consecutive Route Filtering Approach to Symmetric Travelling Salesman Problem

Md. Azizur Rahman and Jinwen Ma[⊠]

Department of Information Science, School of Mathematical Sciences and LMAM,
Peking University, Beijing 100871, People's Republic of China
mdazizur201171@pku.edu.cn, jwma@math.pku.edu.cn

Abstract. The travelling salesman problem (TSP) is one of the NPC combinatorial optimization problems and still now it remains as an interesting and challenging problem in the field of combinatorial optimization. In this paper, we propose a consecutive route filtering approach to solving the symmetric TSP with the help of probe concept such that the worse routes are filtered out step by step by using a rigorous predesigned step proportion. In this way, it is important to set up a reasonable value of the step proportion which is needed in each step during the filtering process. Actually, our proposed algorithm is implemented on the set of symmetric TSP benchmarks with both small and large numbers of cities from the TSPLIB dataset. It is demonstrated by the experimental results that our proposed algorithm can obtain the best results in some cases and generally get the approximation results close to the best known solutions.

Keywords: Probe machine · Travelling salesman problem
Filtering proportion · Discrete optimization

1 Introduction

In the field of combinatorial optimization, TSP is arguably the most prominent, popular, and widely studied problem. It belongs to the class of NP-complete problems [1], i.e., the most difficult problems without any exact algorithm to effectively solve it in polynomial time. In fact, with the increase of number of cities, the executive time of any existing algorithm increases super-polynomially or even almost exponentially [2]. So, its improvement has been drawn much attention to researchers due to the growing demands from vast practical applications in different areas relevant to real life such as vehicle routing, drilling holes in a circuit board, overhauling gas turbine engine, X-ray diffraction, storage and picking of stock in warehousing, computer wiring, interview scheduling, crew scheduling, mission planning, DNA sequencing, data association, image processing, pattern recognition and so on [3,4].

© IFIP International Federation for Information Processing 2018
Published by Springer Nature Switzerland AG 2018. All Rights Reserved
Z. Shi et al. (Eds.): ICIS 2018, IFIP AICT 539, pp. 378–387, 2018.
https://doi.org/10.1007/978-3-030-01313-4_41

Originally, this problem comes into being with a salesman who wants to visit every city exactly once from a list of cities to sell his products and finally return to the starting city where his purpose is to minimize the total tour cost. In graph theory, the symmetric TSP is defined by a complete undirected graph $G = (V, E)$, where $V = \{v_1, v_2, \cdots, v_N\}$ is the set of nodes and $E = \{(v_i, v_j) : v_i, v_j \in V, i \neq j\}$ represent the set of edges [3]. Moreover, a symmetric cost matrix $D_{N \times N}$ is assigned on E for representing the weight of edges. Also, the TSP can be stated as a permutation problem [24,25] with the motive of finding a permutation ψ among N cities that minimize the following objective function:

$$f(\psi) = \sum_{i=1}^{N-1} d_{\psi(i),\psi(i+1)} + d_{\psi(N),\psi(1)} \tag{1}$$

where $\psi(i)$ indicate the city which is visited at step i, $i = 1, 2, \cdots, N$ and $f(\psi)$ is the cost of a permutation ψ. The Euclidean distance $d_{i,j}$, between any two cities $i(x_1, y_1)$ and $i(x_2, y_2)$ is calculated by:

$$d_{i,j} = \sqrt{(x_1 - x_2)^2 + (y_1 - y_2)^2} \tag{2}$$

In the case of symmetric TSP, the distance from city i to city j is the same as the distance from city j to city i, i.e., $d_{i,j} = d_{j,i}$.

Although the TSP problem is quite simple, the main complexity comes from the large number of possible solutions. To solve an N-city symmetric TSP problem, $(N-1)!/2$ different possible routes arise so that the direct optimization procedure cannot accomplish in a polynomial time. For this reason, researchers have been trying to solve this problem in two alternative ways instead of finding the exact solution. The first way is to develop an optimization algorithm to ensure the optimal solution with longer running time, while the second way is to develop a heuristical algorithm which can reduce the computational time significantly and provide only near-optimal solution.

The objective of this work is to propose a new algorithm to solve the symmetric TSP with the help of probe machine. The proposed approach is able to filter out worse routes and keep potential routes step by step by using an appropriate choice of proportion value. The rest of the paper is organized as follows. We review some related works in Sect. 2. Section 3 presents our propose framework. The experimental results are summarized in Sect. 4 and the final section gives the conclusion and future research.

2   Related Work

During the past decades, various TSP algorithms have been proposed for finding an optimal or near-optimal solution. Branch and bound algorithm [5], dynamic programming algorithm [6] and cutting plane algorithms [7] are most popular and well known. However, these algorithms are limited to small size instances and unable to implement in large-scale problems. To get rid of this limitation, researchers have developed the heuristic methods. In fact, the nearest neighbour algorithm (NNA) [8] is a simple and easily implementable heuristic algorithm. It starts with a randomly chosen city and adds the nearest unvisited city step by step until all the cities are contained in the tour. Rosenkrantz et al. [9] developed a repetitive nearest neighbour algorithm (RNNA) and insertion algorithm (IA) to extend the NNA.

Several algorithms were reported to launch with a complete route and improve it iteratively through a simple modification. 2-opt [10] and 3-opt [11] are exoteric methods in this category, where a few edges (2 for 2-opt and 3 for 3-opt) are firstly removed from the current tour and then replace them by the corresponding number of different edges to obtain a shorter tour. Lin and Kernighan [12] enlarged this concept to k-opt where k is chosen in a reasonable way. The variable neighbourhood search (VNS) [13, 14] was also based on the neighbours that are obtained iteratively by a systematic change of the node of the initial tour. Most recently, Hore et al. [15] made a significant improvement on the VNS algorithm.

In this context, population-based metaheuristic algorithms draw much more attention among the researchers in the long run. Simulated annealing (SA) [16] can be successfully applied to get a global optimal solution, but it requires longer computational time. Several types of Genetic algorithms (GAs) [17] are also introduced to rely on the principles of natural selection and genetics, and there are some developments using the assorted operator and selection methods [18]. Ant colony optimization (ACO) [19] is an alternative technique which depends on the foraging behaviour of ant colony and its modifications are also reported in some works [20–22]. Recent population-based algorithm referred to as Symbiotic organisms search (SOS) proposed by Cheng and Prayogo [23] employs the optimization scheme through the mutualism, commensalism and parasitism phase, and an extension of this algorithm developed by Ezugwu et al. [24] strengthens the use of the mutation operator in the local search process.

However, population-based methods can solve the TSP problem quickly, but might be easily trapped into a local optimum solution because there are a group of random numbers and fine-tuning parameters in the process. Certainly, some hybrid algorithms can be established to enhance the overall algorithm performance. Recently, Ezugwu et al. [25] proposed a hybrid optimization algorithm which fuses the SOS and SA algorithm together. On the other hand, Ozden et al. [26] demonstrated how the parallel computing techniques can be used for TSP and significantly decrease the overall computational time with the increase of CPU utilization.

3   Proposed Algorithm

In the proposed framework, we take an attempt to solve the symmetric TSP with the help of probe concept by using a rigorous step proportion value which filters out worse route gradually and finally reaches to an optimal route. For an N-city problem, the process needs to complete $\lceil N/2 \rceil$ steps, where

$$[N/2] = \begin{cases} \frac{N}{2} & \text{if } N \text{ is even} \\ \frac{N-1}{2} & \text{if } N \text{ is odd} \end{cases} \tag{3}$$

The concept of probe, working steps of our proposed framework and the filtering proportion value in each step are given in details in the following subsection consecutively.

3.1 Probe Concept

Actually, the probe is a tool which is used to detect a certain substance accurately. The hypothetical probe machine was introduced by Xu [27], in which there were two types of probes: connective and transitive probes. The connective probes were used to connect two data, while the transitive probes were used to pass information from one data to another data through the data fibers. The probe machine was used to solve two types of NP-complete problems: Hamilton and the graph colouring problem. By the inspiration of his work, we try to present a new type of probe and also present a new filtering mechanism for solving the symmetric TSP using a proportion value. Here, each of the sub-route treats as a probe which is able to automatically find out two unvisited cities and connect them through their wings. In this way, each probe gradually enhances in every step and continue until all cities are included in the route. A sample of 3 cities probe and 5, 7, 9 cities probes generated from 3 cities probe in first four steps are shown in Fig. 1. In the Fig. 1, the sample probes are denoted by x_{ijk}, x_{ijklm}, $x_{ijklmnp}$, $x_{ijklmnpst}$ and their wings are as follows:

$$\omega(x_{ijk}) = \{x_{ijk}^j, x_{ijk}^k\}, \omega(x_{ijklm}) = \{x_{ijklm}^l, x_{ijklm}^m\}$$

$$\omega(x_{ijklmnp}) = \{x_{ijklmnp}^n, x_{ijklmnp}^p\}, \omega(x_{ijklmnpst}) = \{x_{ijklmnpst}^s, x_{ijklmnpst}^t\}$$

3.2 Working Steps

The proposed algorithm accomplishes two tasks in each step. That is, it firstly generates the possible probes and then filters out the worse probes. The whole process consists of several steps which are described in details in the followings:

Step-1: In the first step, the process generates three city probes by considering two paths for each city position. The two path is constructed with an internal city having other two cities that are adjacent to the internal city. More formally, consider c_i, c_j and c_k are three different cities, where c_i is the internal city with c_j and c_k are two different cities both are adjacent to c_i, then the set of all two paths with internal city c_i is denoted by $F^2(c_i)$ and is defined by Eq. (4) [27]:

$$F^2(c_i) = \{c_j c_i c_k \triangleq x_{ijk} : c_j, c_k \in I(c_i); i, j, k \text{ are mutually different}\} \tag{4}$$

where $I(c_i)$ is the set of cities adjacent to c_i and x_{ijk} represent the probe that covers three cities c_i, c_j and c_k. Thus, we construct all possible three cities probe for an N-city problem as follows:

$$X_3 = \cup_{i=1}^N F^2(c_i) = \cup_{i=1}^N \{x_{ijk} : c_j, c_k \in I(c_i); i \neq j, k; k \neq j\} \tag{5}$$

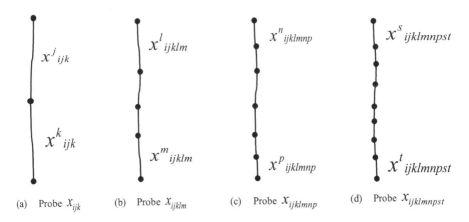

(a) Probe X_{ijk} (b) Probe X_{ijklm} (c) Probe $X_{ijklmnp}$ (d) Probe $X_{ijklmnpst}$

Fig. 1. A sample of 3, 5, 7 and 9 cities probes with their wings.

So there are $N_3 =\mid X_3 \mid= \frac{N(N-1)(N-2)}{2}$ such types of 3 city probes produced in the first step for an N-city symmetric problem. Each probe x_{ijk} has exactly two types of wings x_{ijk}^j and x_{ijk}^k, and this wing includes two other unvisited cities automatically on the probe in next step.

The filtering process starts after the probes have been created. In this phase, the system keeps some potential probes by using a proportion value defined in Subsect. 3.3 and we consider these probes as good probes. The algorithm adopts Euclidean distance (defined in Eq. (2)) to calculate the cost of the probes, and based on the cost it keeps the best N_{G_3} probes. Consequently, those probes are good whose best order lies within the range $[1, N_{G_3}]$, where N_{G_3} indicates the total number of good probes with 3 cities. If we choose α_3 as a proportion value in the first step, then the total number of good probes is as follows:

$$N_{G_3} =\mid G_3 \mid= \alpha_3 N_3, G_3 \subset X_3 \tag{6}$$

where G_3 is the set of all good probes after completing the filtering task in first step.

Step-2: Based on good probes obtained from the first step, the algorithm produces 5 cities probe in the second step. Each good probes enlarge the route through the wings by adding two unvisited cities. In this step, the system needs to add exactly two cities in the probe from the remaining $(N - 3)$ unvisited cities. So based on G_3, the probe can be built by the following ways:

$$X_5 = \{c_l c_j c_i c_k c_m \triangleq x_{ijklm} : c_j c_i c_k \in G_3; c_l, c_m \in R_{N-3}(c_j c_i c_k); i \neq j \neq k \neq l \neq m\} \tag{7}$$

where $R_{N-3}(c_j c_i c_k)$ is the set of remaining unvisited cities corresponding to good probe $c_j c_i c_k$, i.e., $R_{N-3}(c_j c_i c_k) = \{c_1, c_2, \cdots, c_N\} - \{c_i, c_j, c_k\}$. It is noted that the two city c_l and c_m are chosen in both order from $R_{N-3}(c_j c_i c_k)$. Hence, the total produced probe is estimated based on Eq. (8):

$$N_5 =\mid X_5 \mid= 2N_{G_3}\binom{N-3}{2} = N_{G_3}(N-3)(N-4) \tag{8}$$

After generating 5 cities probe the process also takes a filtering proportion value in this step which gives some good probes and this probe will be used in the next step to create 7 cities probe. Therefore, the number of good probes at the end of this stage can be calculated by the following equation:

$$N_{G_5} = \mid G_5 \mid = \alpha_5 N_5, G_5 \subset X_5 \tag{9}$$

where α_5 represent the filtering proportion value in the step and G_5 is the set of all good 5 cities probe, respectively.

Step-(k+1): The $(k+1)^{th}$ step of the procedure starts with the good probes (G_{2k+1}) and their corresponding cost which are obtained from k^{th} step, where

$$N_{G_{2k+1}} = \mid G_{2k+1} \mid = \alpha_{2k+1} N_{2k+1} \tag{10}$$

In the above Eq. (10), $N_{G_{2k+1}}$ indicates the number of total good probes, α_{2k+1} is the filtering proportion value and N_{2k+1} represent the set of possible generated probes in k^{th} step. There are $(2k+1)$ cities in each probe of the above mentioned good probes. In fact, when the algorithm arrive in this step, it needs to include $(N - 2k - 1)$ more cities in the probe. By adding two cities from $(N - 2k - 1)$, each probe becomes $(2k+3)$ cities probe. The set of creates probe in the current step based on good probes (G_{2k+1}) is denoted by X_{2k+3} and is defined by:

$$X_{2k+3} = \{c_s \underbrace{\overbrace{c_l c_j \dots c_k c_m}^{(2k+1)\text{cities good probe}} c_t}_{(2k+3)\text{cities probe}} : c_l \dots c_m \in G_{2k+1}; c_s, c_t \in R_{N-2k-1}(c_l \dots c_m), s \neq t\} \tag{11}$$

The two cities c_s and c_t are taken in both ways, i.e., (c_s, c_t) and (c_t, c_s) both are included and $R_{N-2k-1}(c_l \dots c_m) = \{c_1, c_2, \cdots, c_N\} - \{c_l, c_j, \dots, c_k, c_m\}$. The total produced probes in this step as shown in Eq. (12):

$$N_{2k+3} = \mid X_{2k+3} \mid = 2N_{G_{2k+1}} \binom{N - 2k - 1}{2} = N_{G_{2k+1}}(N - 2k - 1)(N - 2k - 2) \tag{12}$$

Now, the algorithm adopts a filtering proportion value α_{2k+3} that filled out the potential probe. So the number of good probes with $(2k + 3)$ cities obtained as:

$$N_{G_{2k+3}} = \mid G_{2k+3} \mid = \alpha_{2k+3} N_{2k+3}, G_{2k+3} \subset X_{2k+3} \tag{13}$$

where G_{2k+3} is the set of all good probes at the end of this step.

In this way, the probe construction and filtering mechanism continue until all the cities are contained in the probe. When the algorithm reaches in the last step, it generates probes of N cities and finds out the best probes from all generated probes. It is important to mention that in the last step of the even number of problems, there is one city remaining to visit, in this case the probe uses any one of its wings to include the city properly.

3.3 Filtering Proportion in Each Step

The filtering proportion plays a vital role in our proposed framework. Usually, an inappropriate choice of proportion value leads to trap the whole process and yields a worse solution as well as take longer running time. So it is very challenging task to design an efficient filtering proportion value that keeps potential probes from produced probes in each step. In this work, we address a filtering proportion value of kth step through the experiments by trial and error method that is denoted as α_{2k+1} and defined as:

$$\alpha_{2k+1} = \frac{C}{N + \sqrt{k}}; k = 1, 2, 3, \cdots \cdots \tag{14}$$

where N represents the number of cities and C is the constant. The above proportion value has been chosen in such a way that it decreases slowly step by step, i.e., it maintains the relation $\alpha_3 > \alpha_5 > \cdots > \alpha_{2k+1} > \alpha_{2k+3} \cdots$. During the experiment, we found that the number of good probes increases with the gradually decreases of proportion value in each step, but after a certain step later it also starts to decrease. In some cases, we use some strategies in the remaining steps such as trade off to decrease the proportion value and sometimes it is increased by $\frac{1}{(N-2k-1)(N-2k-2)}$. By the experiment, we estimated that the value of C lie within the interval $[\frac{2(N+1)}{N(N-1)(N-2)}, \frac{800(N+1)}{N(N-1)(N-2)}]$ for our computational results.

4 Experimental Results

In this section, we conduct a number of experiments for measuring the capability of our proposed framework based on several TSPLIB [28,29] datasets ranging from 14 up to 1432 cities. Actually, TSPLIB is a publicly available library that contains the sample of TSP instances and their corresponding optimum solutions. The technical computations in this study are performed in MATLAB R2016b software by using 4 core GPU system. The average required time is measured by running each instance ten (10) consecutive times. The experimental results from our experiments have been displayed in Table 1. In the left part of the table, the first column represents the serial number (S/N) of the instances, the second column contains the name of each instance, in the third column "Scale" denote the total number of cities in each instance, the fourth column "Results" stands for the length of obtained solution from the experiment, the fifth one represent average required times (in seconds) of each instance, the sixth column available for the best known results (BKR) obtained from the TSPLIB and the right part of the table brings similar indicator as the left part.

Table 1. The computational results of the proposed algorithm for the symmetric TSP

S/N	Instances	Scale	Results	Time (Se.)	BKR	S/N	Instances	Scale	Results	Time (Se.)	BKR
1	burma14	14	3323	2.1148	3323	38	u159	159	51624	137.3	42080
2	p01	15	291	0.0101	291	39	si175	175	22051	166.0453	21407
3	ulysses16	16	6859	4.1738	6859	40	brg180	180	1960	347.3403	1950
4	gr17	17	2085	0.0181	2085	41	rat195	195	2512.8	6.1452	2323
5	gr21	21	2707	0.3412	2707	42	d198	198	17366	2.2832	15780
6	gr24	24	1272	31.278	1272	43	kroA200	200	35420	2.1057	29368
7	fri26	26	937	0.4982	937	44	kroB200	200	35895	2.3493	29437
8	bays29	29	2093	613.2	2020	45	ts225	225	135000	958.7	126643
9	bayg29	29	1667	0.0282	1610	46	tsp225	225	4550.5	16.8851	3916
10	dantzig42	42	762	93.6	699	47	pr226	226	94850	863.8	80369
11	swiss42	42	1376	11.8	1273	48	gil262	262	2736.7	12.0015	2378
12	att48	48	10671	1265.2	10628	49	a280	280	3111.8	8.1597	2579
13	hk48	48	11461	206.8	11461	50	pr299	299	57931	43.1237	48191
14	eil51	51	455.86	17.48	426	51	lin318	318	52547	65.6026	42029
15	berlin52	52	7982.2	3.3017	7542	52	rd400	400	18581	101.6123	15281
16	brazil58	58	25649	2.2556	25395	53	fl417	417	15406	44.7756	11861
17	st70	70	762.34	0.1033	675	54	pcb442	442	61346	55.0617	50778
18	eil76	76	601.46	39.8	538	55	d493	493	42408	136.3764	35002
19	pr76	76	124740	63.4	108159	56	att532	532	34239	258.6158	27686
20	gr96	96	61741	2119.6	55209	57	si535	535	50286	322.1954	48450
21	rat99	99	1443.6	0.9600	1211	58	pa561	561	3313	248.8954	2763
22	rd100	100	9283	21.5	7910	59	u574	574	46191	290.0283	36905
23	kroA100	100	24511	0.2025	21282	60	rat575	575	8066.2	272.50	6773
24	kroB100	100	23568	0.3084	22141	61	p654	654	48380	1489.9	34643
25	kroD100	100	25767	45.3334	21294	62	d657	657	63099	282.42	48912
26	kroE100	100	24571	0.4640	22068	63	u724	724	50391	399.1771	41910
27	eil101	101	729.2	83.4142	629	64	rat783	783	11140	558.2935	8806
28	lin105	105	16638	68.2125	14379	65	pr1002	1002	314850	1566.7	259045
29	pr107	107	50448	36.26	44303	66	si1032	1032	96145	1737.3	92650
30	gr120	120	8255	59.3458	6942	67	u1060	1060	287620	2584.8	224094
31	pr124	124	70605	80.9855	59030	68	vm1084	1084	305270	2816.4	239297
32	bier127	127	132320	383.1	118282	69	pcb1173	1173	73093	2408.1	56892
33	ch130	130	7002	1.2577	6110	70	d1291	1291	64050.8	4280.1	50801
34	pr136	136	116580	2.2622	96772	71	rl1304	1304	323060	4228.5	252948
35	pr144	144	62370	356.24	58537	72	rl1323	1323	357230	4386.1	270199
36	ch150	150	7195.8	0.800	6528	73	nrw1379	1379	68157	4541.2	56638
37	kroB150	150	29319	5.7321	26130	74	u1432	1432	190400	5494.4	152970

5   Conclusion and Future Research

We have established a consecutive route filtering approach with the help of probe concept by using an appropriate filtering proportion value in each step as a new approach to solving the symmetric TSP. From our experimental results, it is found that the proposed approach can effectively reach at the optimum point to all of the tested datasets that contain up to 26 cities and also for the 48 cities dataset (hk48, Instance no. 13 in Table 1). It is also noted that our obtained solutions are close to the best known solution of some other datasets whereas, the performance of our implemented algorithm is also satisfactory on big scale datasets. In the future, we try to investigate more effective and efficient filtering approach to implement this framework to solve the existing asymmetric travelling salesman problems (aTSPs) and the multi travelling salesman problems (mTSPs). In addition, it is also a good direction to fuse this algorithm with the other existing heuristic or meta-heuristic algorithms together in our further study.

Acknowledgment. This work is supported by the Natural Science Foundation of China for Grant 61171138.

References

1. Papadimitriou, C.H.: The Euclidean traveling salesman problem is NP-complete. Theor. Comput. Sci. **4**(3), 237–244 (1977)
2. Garey, M.R., Johnson, D.S.: Computers and Intractability: A Guide to the Theoryof NP-Completeness. W. H. Freeman, New York (1979)
3. Matai, R., Singh, S.P., Mittal, M.L.: Traveling salesman problem: an overview of applications, formulations, and solution approaches. In: Davendra, D. (ed.) Traveling Salesman Problem, Theory and Applications, pp. 1–24. InTech, Croatia (2010)
4. MacGregor, J.N., Chu, Y.: Human performance on the traveling salesman and related problems: a review. J. Probl. Solving **3**(2), article 2 (2011)
5. Finke, G., Claus, A., Gunn, E.: A two-commodity network flow approach to the traveling salesman problem. Congr. Numer. **41**, 167–178 (1984)
6. Held, M., Karp, R.M.: A dynamic programming approach to sequencing problems. J. Soc. Ind. Appl. Math. **10**(1), 196–210 (1962)
7. Fleischmann, B.: A cutting plane procedure for the travelling salesman problem on road networks. Eur. J. Oper. Res. **21**(3), 307–317 (1985)
8. Bellmore, M., Nemhauser, G.L.: The traveling salesman problem: a survey. Oper. Res. **16**(3), 538–558 (1968)
9. Rosenkrantz, D.J., Stearns, R.E., Philip, M.L.I.: An analysis of several heuristics for the traveling salesman problem. SIAM J. Comput. **6**(3), 563–581 (1977)
10. Croes, G.A.: A method for solving traveling-salesman problems. Oper. Res. **6**(6), 791–812 (1958)
11. Lin, S.: Computer solutions of the travelling salesman problem. Bell Syst. Tech. J. **44**(10), 2245–2269 (1965)
12. Lin, S., Kernighan, B.W.: An effective heuristic algorithm for the traveling salesman problem. Oper. Res. **21**(2), 498–516 (1973)
13. Hansen, P., Mladenović, N.: An introduction to variable neighborhood search. In: Voss, S., Martello, S., Osman, I., Roucairol, C.C. (eds.) Meta-Heuristics: Advances and Trends in Local Search Paradigms for Optimization, pp. 433–458. Kluwer Academic Publishers, Boston (1999)
14. Hansen, P., Mladenović, N.: Variable neighborhood search: principles and applications. Eur. J. Oper. Res. **130**(3), 449–467 (2001)
15. Hore, S., Chatterjee, A., Dewanji, A.: Improving variable neighborhood search to solve the traveling salesman problem. Appl. Soft Comput. **68**, 83–91 (2018)
16. Kirkpatrick, S., Gelatt, C.D., Vecchi, M.: Optimization by simulated annealing. Sci. New Ser. **220**(4598), 671–680 (1983)
17. Whitley, D.: A genetic algorithm tutorial. Stat. Comput. **4**(2), 65–85 (1994)

18. Guo, D., Chen, H., Wang, B.: An improved genetic algorithm with decision function for solving travelling salesman problem. In: Proceeding of 12th ISKE. IEEE Conferences, Nanjing, China, pp. 1–7 (2017). https://doi.org/10.1109/ISKE.2017.8258774

19. Colorni, A., Dorigo, M., Maniezzo, V.: Distributed optimization by ant colonies. In: Varela, F., Bourgine, P. (eds.) Proceedings of European Conference on Artificial Life, Paris, France, pp. 134–142 (1991)

20. Shufen, L., Huang, L., Lu, H.: Pheromone model selection in ant colony optimization for the travelling salesman problem. Chin. J. Electron. **26**(2), 223–229 (2017)

21. Xiong, N., Wu, W., Wu, C.: An improved routing optimization algorithm based on travelling salesman problem for social networks. Auton. Sustain. Comput. Prep. Internet Things Environ. **9**(6), 985 (2017)

22. Ratanavilisagul, C.: Modified ant colony optimization with pheromone mutation for travelling salesman problem. In: Proceeding of 14th ECTI-CON. IEEE Conferences, Phuket, Thailand, pp. 411–414 (2017). https://doi.org/10.1109/ECTICon.2017.8096261

23. Cheng, M.-Y., Prayogo, D.: Symbiotic organisms search: a new metaheuristic optimization algorithm. Comput. Struct. **139**, 98–112 (2014)

24. Ezugwu, A.E.-S., Adewumi, A.O.: Discrete symbiotic organisms search algorithm for travelling salesman problem. Expert. Syst. Appl. **87**, 70–78 (2017)

25. Ezugwu, A.E.-S., Adewumi, A.O., Frîncu, M.E.: Simulated annealing based symbiotic organisms search optimization algorithm for traveling salesman problem. Expert. Syst. Appl. **77**(1), 189–210 (2017)

26. Ozden, S.G., Smith, A.E., Gue, K.R.: Solving large batches of travelling salesman problems with parallel and distributed computing. Comput. Oper. Res. **85**, 87–96 (2017)

27. Xu, J.: Probe machine. IEEE Trans. Neural Netw. Learn. Syst. **27**(7), 1405–1416 (2016)

28. TSPLIB. http://elib.zib.de/pub/mp-testdata/tsp/tsplib/tsplib.html. Accessed 16 June 2018

29. https://people.sc.fsu.edu/~jburkardt/datasets/tsp/tsp.html. Accessed 24 June 2018

Fault Diagnosis

Automatic Fault Detection for 2D Seismic Data Based on the Seismic Coherence of Mutative Scale Analysis Window

Wenli Zheng[1,2]($\boxtimes$) and Jinwen Ma[2]

[1] School of Science, Xi'an Shiyou University,
Xi'an 710065, Shaanxi, People's Republic of China
wlzheng@xsyu.edu.cn
[2] School of Mathematical Sciences & LMAM, Peking University,
Beijing 100871, People's Republic of China
jwma@math.pku.edu.cn

Abstract. Fault detection is a very challenging problem on seismic interpretation. In fact, the process of fault detection contains seismic attribute extraction, seismic attribute enhancement and fault line detection. It is clear that the extracted seismic attributes are key to the fault detection process. Traditionally, as an important seismic attribute, the seismic coherence is generally employed in the fault detection process, but the size of the analysis window often effects the value of the seismic coherence, and there is a trade-off between the vertical resolution and the lateral resolution. In order to overcome this problem, we propose a new kind of seismic coherence with a mutative scale analysis window, and utilize it to locate the fault lines. It is demonstrated by the experimental results that our proposed seismic coherence is more suitable for the fault line detection in comparison with the traditional seismic coherence.

Keywords: Fault detection · Seismic attribute · Seismic coherence
Mutative scale · Analysis window

1 Introduction

For detecting the positions of energy resources under a specific ground, it is often necessary to collect and analyze the seismic data generated from a seismic exploration work which contains seismic data collection, seismic data processing and seismic interpretation. After collected and calibrated, a seismic data volume can be formed corresponding to the space under the ground. Finally we will analyze the seismic data by some theory and technology to obtain underground information.

Fault detection is a very challenging problem in the seismic interpretation within a potentially enormous seismic volume. Geological faults are important since they are often associated with the formation of subsurface traps in which

© IFIP International Federation for Information Processing 2018
Published by Springer Nature Switzerland AG 2018. All Rights Reserved
Z. Shi et al. (Eds.): ICIS 2018, IFIP AICT 539, pp. 391–400, 2018.
https://doi.org/10.1007/978-3-030-01313-4_42

petroleum might accumulate. The process of automatic fault detection contains seismic attribute extraction, seismic attribute enhancement, and fault detection, where the seismic attributes extraction is a major step, and the seismic coherence is often used as one of seismic attributes in the fault detection process. Bahorich and Farmer [1] firstly introduced a classical normalized crosscorrelation to measure the continuity between neighboring windowed seismic traces, which was referred to as an attribute of coherence. Although this original coherence is computed efficiently, but lacks the robustness for noisy seismic data. Then Marfurt et al. [2] proposed a multi-trace semblance coherency, which estimates the coherency over multi seismic traces. In comparison with the original coherence, this one becomes stable in the emergence of noise, and improves the vertical resolution. However, by increasing seismic traces on computing the semblance coherence, the lateral resolution will decrease while the computational cost will increase. Then, Gersztenkorn and Marfurt [3] introduced an eigenstructure-based coherence which provides more stable seismic coherence but requires the computation of eigenvalues. In order to overcome the drawbacks of the eigenstructure-based coherence, Cohen et al. proposed two local discontinuity measure named seismic local structural entropy (LFE) [4] and normalized differential entropy (NDE) [5], respectively, being used to extract the faults. But these seismic attributes are dependent on the size of analysis windows which are selected by experience. In fact, the size of analysis window is rather difficult to be selected in practical fault detection. In this paper, we propose a new kind of seismic coherency measure with mutative scale analysis window to avoid the shortness of the above seismic attributes. In fact, this new seismic coherence can be effectively used for fault detection.

The rest of this paper is organized as follows. We begin to introduce the semblance coherence and the eigenstrcture-based coherence in Sect. 2. In Sect. 3, we present a new seismic coherence. Section 4 summarizes the experimental results on a real seismic dataset. Finally, we conclude briefly in Sect. 5.

2   Conventional Seismic Coherence

Before we introduce the semblance coherence and the eigenstrcture-based coherence, we firstly present a seismic slice and define an analysis window. The value of every points on a seismic slice is the seismic reflection amplitude of the seismic wave, shown in Fig. 1. The horizontal direction is the inline or crossline, and the vertical direction is the time axis. When we estimate the seismic coherency value at a point, we should firstly design an analysis window at the center of this point, of which the size is $2t + 1 \times 2K + 1$, where t and K is set according to experts' experience. After that we present the semblance coherence and the eigenstrcture-based coherence, respectively.

2.1 Semblance Coherence

Let $C2(i, j, \theta)$ be the semblance coherency value at the point (i, j), and θ is the direction of the analysis window. The semblance coherency equation takes the following form.

$$C2(i,j,\theta) = \frac{\sum_{\tau=-t}^{t}\{(\sum_{k=-K}^{K} D(i+\tau+pk, j+k))^2 + (\sum_{k=-K}^{K} D^H(i+\tau+pk, j+k))^2\}}{(2K+1)\sum_{\tau=-t}^{t}\sum_{k=-K}^{K}(D(i+\tau, j+k))^2 + (D^H(i+\tau, j+k))^2}$$

(1)

where D is the seismic slice and $D(i, j)$ represents the amplitude at the point (i, j). Generally, θ is 0.

2.2 Eigenstrcture-Based Coherence

After introducing the semblance coherence, we show the eigenstrcture-based coherency calculation procedure. First, we extract all seismic data enclosed by the analysis window at the center of the point (i, j), and get a matrix SD_{ij} which is the following form.

$$SD_{ij} = \begin{bmatrix} d_{11} & d_{12} & \cdots & d_{1J} \\ d_{21} & d_{22} & \cdots & d_{2J} \\ \vdots & \vdots & \ddots & \vdots \\ d_{N1} & d_{N2} & \cdots & d_{NJ} \end{bmatrix}$$

(2)

where $N = 2t + 1$ and $J = 2K + 1$. Then construct a covariance matrix, denoted C, shown in Eq. 3

$$C = SD_{ij}^T SD_{ij} = \begin{bmatrix} \sum_{n=1}^{N} d_{n1}^2 & \sum_{n=1}^{N} d_{n1}d_{n2} & \cdots & \sum_{n=1}^{N} d_{n1}d_{nJ} \\ \sum_{n=1}^{N} d_{n1}d_{n2} & \sum_{n=1}^{N} d_{n2}^2 & \cdots & \sum_{n=1}^{N} d_{n2}d_{nJ} \\ \vdots & \vdots & \ddots & \vdots \\ \sum_{n=1}^{N} d_{n1}d_{nJ} & \sum_{n=1}^{N} d_{n2}d_{nJ} & \cdots & \sum_{n=1}^{N} d_{nJ}^2 \end{bmatrix}$$

(3)

Finally, compute the eigenvalue of C, denoted as $(\lambda_1, \lambda_2, \cdots, \lambda_J)$, where $\lambda_1 \geq \lambda_2 \geq \cdots \geq \lambda_J$. Therefore, The eigenstrcture-based coherency value takes the following form.

$$C3(i,j,\theta) = \frac{\lambda_1}{\sum_{n=1}^{J} \lambda_n},$$

(4)

where $C3(i, j, \theta)$ ranges form 0 to 1.

Those two seismic coherence attributes are strongly affected by the size of the analysis window. If the length of the analysis window is larger, the more seismic traces are enclosed by the analysis window, and the lateral resolution will decrease. If the width of the analysis window is larger, the vertical resolution will decrease. There is a trade-off between the length and width of the analysis window, so we will propose a new attribute to estimate the seismic coherence.

3 Proposed Seismic Coherence of Mutative Scale Analysis Window

Here we introduce a new seismic coherence based on the semblance coherence and the eigenstrcture-based coherence. The traditional seismic coherency value changes as the size of the analysis window changes, so it is important to choose the size of analysis window. In order to overcome this problem, we propose a new attribute to estimate the seismic coherence with a series of analysis windows in different size. Next we take the eigenstrcture-based coherence as an example to present the variable scale eigenstrcture-based coherence.

Let D be the seismic slice and $p = (i, j)$ be a point, then we compute the variable scale eigenstrcture-based coherency value at the point p. We initialize an analysis window, of which the size is $N \times K$ and the direction is θ, and extract a dataset enclosed by the initial analysis window, then we estimates the eigenstrcture-based coherency value of p by Eq. 4 and record this value. Next let $N = N + \Delta n$ and $K = K + \Delta w$ where Δn and Δw are constants, and repeat the above process. Finally, we obtain a series of the eigenstrcture-based coherency values by varying the size of the analysis window, and record these values in a matrix, denoted G. Because the eigenstrcture-based coherency value ranges from 0 to 1, we name G as a eigenstrcture-based coherency gray image of the point p. If this point p located in different geological structure, its eigenstrcture-based coherency gray image has different texture. In order to measure those different texture, we introduce the Gray-level co-occurrence matrix (GLCM) of G and use the average of the vertical and horizontal angular second moment (ASM) to measure the coherence of the point p, denoted $SC3(i, j, \theta)$. Using the same method, we estimates the variable scale semblance coherency value of the point, p, denoted $SC2(i, j, \theta)$. In order to take advantages of both variable scale coherence, we compute the average of $SC2(i, j, \theta)$ and $SC3(i, j, \theta)$, shown in Eq. 6

$$SC(i, j, \theta) = \frac{SC2(i, j, \theta) + SC3(i, j, \theta)}{2}.$$ (5)

Then let $\theta = \theta + \Delta\theta$, and get a sequence of $SC(i, j, \theta)$. Finally we define the new seismic coherence of the point p, which takes the following form.

$$SC(i, j, \theta^*) = \arg\max_{\theta_{min} \leq \theta \leq \theta_{max}} SC(i, j, \theta).$$ (6)

So far, we have introduced our proposed attribute to estimate the new seismic coherence, and then we use this new seismic coherence as the seismic attribute to locate faults on a seismic slice.

4 Fault Detection Through the Proposed Coherence

After compute the new seismic coherence, we apply the Log-Gabor filter to enhance the new seismic coherence [7]. Finally we locate the fault lines on the seismic slice by the curve detection algorithm proposed by Carsten Streger [9].

The Log-Gabor filter is popular in the image processing [6,8], because it considers the orientation information to save the image detail. The 2D Log-Gabor filter takes following form,

$$G(f, \theta) = exp(\frac{-(log(f/f_0))^2}{2log(\sigma_f/f_0)})exp(\frac{-(log(\|\theta - \theta_0\|)^2}{2\sigma_\theta}) \tag{7}$$

where f and θ are parameters of the Log-Gabor filter. f_0 is the center frequency, and σ_f is the width parameter for the of frequency. θ_0 is the center orientation, and σ_θ is the width parameter for the orientation.

After the new coherence enhancement, we do the fault detection operation by the curve detection algorithm, named Steger's curve detection algorithm [9]. The Steger's curve detection algorithm is wildly used to detect the curves of satellite images and computed tomography images. Then we apply this new attribute on the new enhanced seismic coherence to detect the fault lines of seismic slices.

5   Experimental Results

We have presented the new coherence which is the average of the variable scale semblance coherence and the variable scale eigenstrcture-based coherence, and then we apply the Log-Gabor filter to improve the quality of our proposed seismic coherence. Finally the Steger curve detection algorithm is used to detect faults. In order to test the performance our proposed coherence, we estimate the coherence of a real data, named Qikou, of which the inline number is 2654, shown in Fig. 1, and compare with the traditional semblance coherence, the traditional eigenstrcture-based coherence, and the local structure entropy (LSE) proposed by Cohen. Furthermore, we do the fault detection operation by the Steger curve algorithm on our proposed coherence, and compare with the manual labeled fault

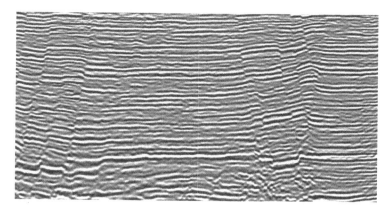

s_plot

23-Mar-2017 15:53:32

Fig. 1. The Qikou seismic slice is in the gray scale and the inline number is 2654

lines. Let the initial window size be $N \times K$, where N ranges from 21 to 61 with the step 4, and K ranges from 3 to 7 with the step 2. The analysis window orientation varies from $-70°$ to $70°$ with the common difference $25°$. Then we apply our propose seismic coherence on the Qikou seismic slice. Firstly we compare the traditional semblance coherence with the variable scale semblance coherence and show in Fig. 2. On Fig. 2, The positions pointed by red arrows are on the fault lines, but the traditional semblance coherency can not make them clear. The areas enclosed by the green circle are special geological structures, and the variable scale semblance coherency can distinct them. So the variable scale semblance coherency not only improve the vertical resolution but also increase the horizontal resolution, and it overcome weaknesses of the traditional semblance coherence. Then we compare the traditional eigenstructure-based coherence with the variable scale eigenstrcture-based coherence, shown in Fig. 3, and we obtain that the variable scale eigenstrcture-based coherence is better than the traditional scale eigenstrcture-based coherence.

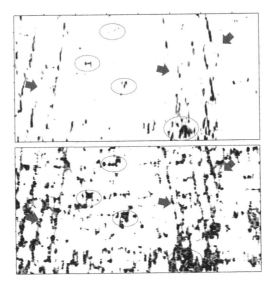

Fig. 2. The traditional semblance coherence (above) and the variable scale semblance coherence (below) (Color figure online)

Following we compare the variable scale semblance coherence, the variable scale semblance coherence and the average of those two variable scale seismic coherence, shown in Fig. 4. Although the variable scale semblance coherence and eigenstructure-based coherence are better than the traditional coherence, there are still details missing pointed by the blue and red arrow in Fig. 4(above) and Fig. 4(middle). Figure 4(below) is the average of those two variable scale seismic coherence, and we can see that the missing details in Fig. 4(above) and Fig. 4(middle) appear again. Additionally we compare the average of those two

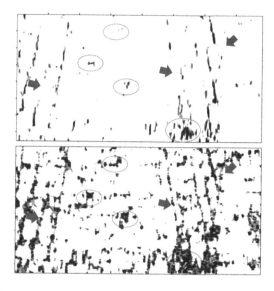

Fig. 3. The traditional eigenstructure-based coherence (above) and the variable scale eigenstructure-based coherence (below) (Color figure online)

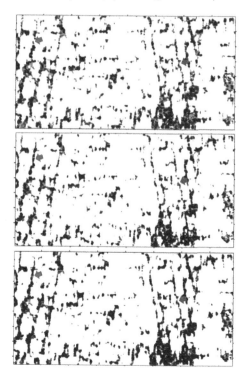

Fig. 4. The variable scale semblance coherence (above), the variable scale eigenstructure-based coherence (middle) and the average of the two variable scale coherence (below) (Color figure online)

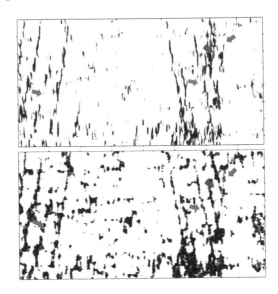

Fig. 5. LSE (above) and the average of the two types of the variable scale seismic coherence (below) (Color figure online)

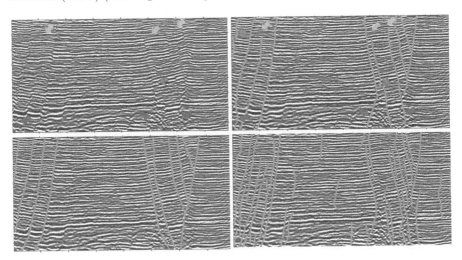

Fig. 6. From left to right and from up to down, the original seismic data of Qikou (a), the fault lines detected by the Steger's curve detection algorithm (b), the fault lines labeled manually (c), and the fault lines detected by the Canny operator (d) (Color figure online)

variable scale seismic coherence with LSE, shown in Fig. 5. Therefore, we take the average of those two variable scale seismic coherence as a new seismic coherence. Finally apply the Log-Gabor filter on the new seismic coherence and use the Steger curve detection algorithm to locate fault lines of the Qikou seismic slice. The result is compared with the manual labeled fault lines, shown in Fig. 6.

Compared with the fault lines labeled manually, the fault lines detected by our proposed method is not totally same to the fault lines labeled manually, and the differences are pointed by yellow, red and blue arrows. But the performance of the Steger's curve detection algorithm is better than the Canny operator.

Further, we use the Steger curve detection algorithm on the semblance coherence, the eigenstructure-based coherence, the variable scale semblance coherence, the variable scale eigenstructure-based coherence, the average of those two variable scale seismic coherence and the LSE, respectively, shown in Fig. 7. From the six images, we obtain that our proposed new seismic coherence is more suitable to be used as the seismic attributes making fault information clear.

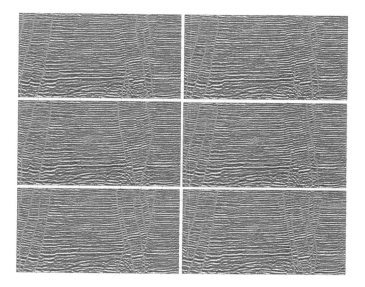

Fig. 7. From left to right and from up to down, the result of detection of the semblance coherence, the eigenstructure-based coherence, the variable scale semblance coherence, the variable scale eigenstructure-based coherence, the average of the two types of the variable scale seismic coherence and the LSE (Color figure online)

6   Conclusion

We have proposed two types of seismic coherence, i.e., the semblance coherence and the eigenstructure-based coherence with mutative scale analysis window and fused the two seismic coherence attributes to form a new seismic coherence. Log-Gabor filter is further implemented on the new seismic coherence to improve its quality, and Steger's curve detection algorithm is used to locate the fault lines. It is demonstrated by the experimental results on the real seismic data that our proposed seismic coherence is better than the traditional seismic coherence and more suitable for fault detection.

Acknowledgment. This work was supported by the BGP Inc., China National Petroleum Corporation.

References

1. Bahorich, M., Farmer, S.: 3-D seismic discontinuity for faults and stratigraphic features: the coherence cube Seg Technical Program Expanded, pp. 93–96 (1995)
2. Marfurt, K.J., Farmer, S.L., Bahorich, M.S.: 3-D seismic attributes using a semblance-based coherency algorithm. Geophysics **63**(4), 1150 (1998)
3. Gersztenkorn, A., Marfurt, K.J.: Eigen structure-based coherence computations as an aid to 3-D structural and stratigraphic mapping. Geophysics **64**(5), 1468 (1999)
4. Cohen, I., Coifman, R.R.: Local discontinuity measures for 3-D seismic data. Geophysics **67**(6), 1933–1945 (2002)
5. Cohen, I., Coult, N., Vassiliou, A.A.: Detection and extraction of fault surfaces in 3D seismic data. Geophysics **71**(4), P21–P27 (2006)
6. Field, D.J.: Relations between the statistics of natural images and the response properties of cortical cells. J. Optical Soc. America A-optics Image Sci. Vis. **4**(12), 2379–2394 (1987)
7. Yu, Y.: Fault enhancement and visualization with 3D log-Gabor filter array. In: Seg Technical Program Expanded, pp. 1960–1965 (2016)
8. Wang, W., Li, J., Huang, F.: Design and implementation of Log-Gabor filter in fingerprint image enhancement. Pattern Recogn. Lett. **29**(3), 301–308 (2008)
9. Steger, C.: An Unbiased Detector of Curvilinear Structures. IEEE Computer Society (1998)

UAV Assisted Bridge Defect Inspection System

Shuzhan Yang[1,3], Zhen Shen[1,2], Xiao Wang[1,2], Tianxiang Bai[1,4], Yingliang Ji[2],
Yuyi Jiang[2], Xiwei Liu[2,6], Xisong Dong[1,5], Chuanfu Li[7], Qi Han[7], Jian Lu[8],
and Gang Xiong[5,6(✉)]

[1] State Key Laboratory of Management and Control for Complex Systems,
Institute of Automation, Chinese Academy of Sciences, Beijing 100190, China
[2] Qingdao Academy of Intelligent Industries, Qingdao 266109, China
[3] Department of Mechanics and Engineering Science (MES), College of Engineering,
Peking University, Beijing 100871, China
[4] University of Chinese Academy of Sciences, Beijing 100049, China
[5] Cloud Computing Center, Chinese Academy of Sciences, Dongguan 523808, China
xionggang@casc.ac.cn
[6] Beijing Engineering Research Center of Intelligent Systems and Technology,
Institute of Automation, Chinese Academy of Sciences, Beijing 100190, China
[7] Qingdao Expressway, Qingdao 266041, China
[8] Guangdong Launca Medical Device Technology Co., Ltd., Dongguan 523808, China

Abstract. Traditional bridge inspection methods require lane closures, inspection equipment, and most importantly the experiences and knowledge of the inspectors. This increases not only the inspection cost and time, but also the risk to the travelling public. Due to the lengthy and costly traditional bridge inspection methods, there has been an increasing backlog of inspection activities. In this research, we design an unmanned aerial vehicle (UAV) assisted bridge defect inspection system, in which a UAV can capture the image and transmit the information to the ground station for further analysis. The system can be divided into 2 subsystems: electromechanics & communication system, and image processing system. The electromechanics & communication systems ensure the self-locating, flight control, image transmission, and human intervention functions. The image processing system performs the image preprocessing, defect extraction, and provides the inspection report. This system, if put into practice, can save the cost up to 70%. We believe that the UAV assisted bridge inspection can be popular in the future.

Keywords: UAV · Defect inspection · Electromechanics system
Communication system · Image processing

This work was supported in part by the National Natural Science Foundation of China under Grants 61773381, 61773382 and 61533019; Chinese Guangdong's S&T project (2016B090910001, 2017B090912001); 2016 S&T Benefiting Special Project (No. 16-6-2-62-nsh) of Qingdao Achievements Transformation Program; Dongguan's Innovation Talents Project (Gang Xiong, Jian Lu), and Guangdong Innovative Leading Talents Introduction Program (00201511).

© IFIP International Federation for Information Processing 2018
Published by Springer Nature Switzerland AG 2018. All Rights Reserved
Z. Shi et al. (Eds.): ICIS 2018, IFIP AICT 539, pp. 401–411, 2018.
https://doi.org/10.1007/978-3-030-01313-4_43

1   Introduction

Bridges play an important role in the transportation network. There are over 800 thousand bridges in China according to the report of Ministry of Transport of the People's Republic of China. To ensure the ongoing service capability and safety of the travelling public, it is essential to inspect bridges regularly and record the corresponding data. The bridge inspection usually includes defect detection on deck, superstructure, and substructure. Bridges should undergo regular inspection every 3 months, and periodic inspection once 3 years, according to Code for Maintenance of Highway Bridges and Culvers JTG H11. Up to now 16,623 bridge assessment records have been accumulated (Li et al. 2014). Traditionally, bridge inspections are conducted manually through traffic control, which includes under-bridge inspection units, mobile scaffolding, boom lifts and cherry pickers. The finical cost in the inspection of Jiaozhou Bay Bridge takes around RMB 1,412,000, with bridge inspection vehicle rent fee RMB 1,184,000; bridge inspection ship rent fee RMB 8000; and traffic block cost RMB 14,800, which is due to the labor-intensive nature of the current bridge inspection method. Moreover, the conventional inspection method is inaccurate, depending highly on the experiences and knowledge of inspectors. A more efficient, economic and accurate bridge inspection method is desirable.

UAV systems have been widely applied in infrastructure management area, because of its high mobility. The system includes traffic monitoring, construction engineering safety inspections and 3-D photogrammetric modelling (Irizarry et al. 2012). Currently, UAV is commonly used in aerial photography and light load transportation.

Our UAV assisted bridge defect inspection system mainly consists of 3 parts: UAV with its function units, ground station, and the interactions between UAV and ground station (see Fig. 1).

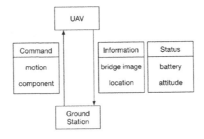

Fig. 1. Bridge inspection system

UAV incorporates multiple electronic devices, such as mainboard, gyroscope, communication unit, Lidar, and carries a pan and tilt camera, so as to conduct image capture task automatically. Ground station can extract defects on images, generate inspection reports, as well as remotely control the UAV. The interactions between UAV and ground station are command, information, and status. Ground station sends command to UAV. Generally, the command is to adjust the attitude (pitch, roll, and yaw) of the UAV, or to modify the performance of the electromechanics devices on UAV, for example, zoom in the camera.

This research focuses on the 3 parts of the system. The hardware configuration, the control framework, and sensor fusion of the UAV are to be studied. The communication realization methods and image processing methods are also elaborated. The rest of the paper is structured as follows. Section 2 lists related work. We describe the UAV and the UAV-ground station interaction in Sect. 3. Then, we present the details of the image processing methods and the results in Sect. 4. In the end, we conclude our work in Sect. 5.

2   Related Work

In general, the related projects and literatures fall into 2 categories: UAV developments and applications, and practice of assessing the visual condition of items. We will introduce both of them and present concisely our contribution in practical aspects.

UAVs can be used to execute observation or detection missions through automatic or remote control. Existing UAV applications are mainly used in mapping applications, environmental change monitoring, disaster prevention response, resource exploration, etc. For example, Murphy et al. (2008) used unmanned sea-surface and micro-aerial vehicles together after Hurricane Wilma in 2008. The effort identified cooperative unmanned sea-surface and micro-aerial vehicles strategies and open issues for autonomous operations near structures. The aerial vehicle is equipped with robust control system, which guarantees the performance in strong wind. Campoy et al. (2009) discussed applications in the field of civilian tasks, in which UAVs can be utilized. Companies like AIBOTIX from Germany offer solutions for wind turbine and power line inspections with UAV. One example from AIBOTIX is a hexarotor equipped with a protected frame and an upward camera, operated manually with the help of live video. Alighanbari et al. (2003) presented coordination of multiple UAVs for complicated task. Besides the above work, many achievements have also taken place in the field of power management of UAVs, UAV navigation, as well as UAV telemetry control system. In our case, the UAV is supposed to take clear pictures at a specific place from a specific angle, against the bad light and wind beneath the bridge deck, which sets higher requirements for automation.

Visual inspection constitutes an important part of quality control in industry. Back to 1983, the United States Department of Energy sponsored Honeywell to develop a metal surface defect detection device. The device consists of a linear CCD architecture, and the digital image processing software can effectively detect some small defects on metal surface. The automated visual inspection systems are utilized in many industrial and commercial applications. There are many visual inspection systems, which are used for defect detection of ceramic tiles (Rahaman and Hossain 2009), textured material (Kumar and Pang 2002), and textile fabric (Mak et al. 2005) etc. There are many techniques that has been employed for detecting the surface defect (Zhang et al. 2011) and digital texture image defect (Sivabalan et al. 2010) etc. American company Cognex developed iS-2000 automatic detection system and iLearn software to detect surface defects. In our study, bridge images have relatively low SNR (Signal-to-Noise Ratio) due to environment limitations. The defects are not very distinct in images, so the images are with low SNR. In addition, since it is an engineering project, the system needs to be user friendly and well packaged. All these problems are the challenges we are facing.

3   Electromechanics and Communication System

The UAV shapes, mechanisms, configurations, and characteristics can be tailored for task requirements. For bridge defect inspection, our electromechanics and communication system is designed as such (see Fig. 2). The power of the system is provided by battery. The battery eliminator circuit (BEC) is utilized to adjust voltage. The controller acts on motors (UAV motor, tilt motor, and roll motor) to change the attitude of UAV and camera. Sensors feedback information to both controllers and ground station through signals has little delay or noise, since delays or noise would cripple the control effect. Images are sent to ground station via data transmitter. The realization of navigation, flight control, image capture, signal transmission, and self-protection mechanism will be further discussed in this section.

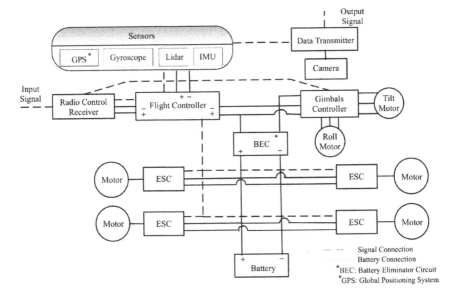

Fig. 2. Electromechanics and communication system

3.1   Navigation

Bridges we are going to inspect in this study usually span over several or tens of kilometers (e.g. Jiaozhou Bay Bridge is of 36.48 km). When talking about "navigation", we mean that the UAV flies from the ground station, arrive at a specific position within the bridge span, and keep a certain distance from the structure to capture images. UAV carries a GPS unit, a Real Time Kinematic (RTK) unit, a Lidar, a camera along with a communication system to function. GPS is a space-based radio navigation system owned by the United States government with accuracy of 30 cm. Real Time Kinematic (RTK) satellite navigation is a technique used to enhance the precision of position data derived from GPS systems. An RTK-GPS (Leica MC1000) has a nominal accuracy of ±1 cm +1 ppm for horizontal displacements and ±2 cm +2 ppm for vertical displacements. To capture high quality images of a bridge substructure requires the UAV to stay in a certain range and an appropriate distance. Thus, a Lidar, a camera and a communication system are introduced. The Lidar is for distance perception, which tells the distance between the UAV and the substructure. The formula relates the focal length of the camera and the distance will be later discussed in this section. The perceived distance and environment surroundings captured by camera can be sent back to ground station in real-time. The operator can remotely control the UAV based on these feedbacks.

3.2 Flight Control

The UAV adopts PID based double closed-loop control (See Fig. 3). The outer control loop is for position control, while the inner control loop is for attitude control. Position control is composed of distance error proportional control, and velocity error PID control. Attitude control is composed of attitude angle proportional control, and angular velocity PID control. The application of extended Kalman filter in our system integrates the data from inertial measurement unit (IMU), barometer, RTK-GPS to estimate the flight state. The sensor data will be transmitted to flight control system. Flight control system generates motor speed control signal and transmits it to ESC (Electronic Speed Control), according to flight command and flight state. Then, the ESC will change the motor speed. Lidar can detect obstacles within 100 m, which effectively avoids collision. The communication system transmits UAV information and images to the ground station in real time for personnel to assess the UAV state. The bridge inspection UAV supports two flight modes, one is manual mode; the other is automatic mode. The manual mode requires the personnel to manually control the UAV, which is for targeted area inspection. The automatic mode requires 3D data model of the bridge be imported to the system beforehand. The UAV will automatically fly through the route. This mode is for regular inspection and maintenance.

3.3 Image Capture

The image capture task is conducted by camera placed on gimbal. The gimbal can reduce the vibration of camera, and adjust its position through roll motor and tilt motor. If necessary, an LED can be turned on to get a clear image. In Fig. 3 we show the image capture mechanism. The image capture distance between camera and bridge, and the image acquisition interval can be derived through the following formula.

$$H = \delta f / \gamma \tag{1}$$

$$L = nd\gamma / \delta - t = l\gamma / \delta - t \tag{2}$$

H camera-structure distance (mm)
δ measurement resolution (mm)
f camera focal length (mm)
t overlap length (mm)
γ pixel spacing of photoelectric sensor (mm)
L acquisition distance interval (mm)
n pixel number of collector
d pixel diameter of collector (mm)
l target surface diameter of collector (mm).

3.4   Communication

The transmission of sensor data, flight control command and bridge defect images are achieved through communication system. The transmission of sensor data and flight control command is especially crucial. When the UAV is far beyond sight, the data from sensors and video from camera is the only reference of control command. The delay of transmission will misrepresent the state of UAV, causing potential danger. For this concern, we choose radio rather than Wi-Fi as our transmission method. In Wi-Fi communication, handshaking mechanism between UAV and ground station is applied. Each data package needs to be well sent, or the missed or incomplete data package will be resent. These mechanisms cause the delay of information. This limits the UAV to working within 500 m. When beyond 200 m, an obvious delay is observed. Different from Wi-Fi, radio is a one-way transmission method, which extends the working rage form 500 m up to 2 km.

3.5   Self-protection Mechanism

When error occurs, the self-protection mechanism is activated. For minor errors, like device temporary dysfunction, the alarming signal is sent to the ground station, manual

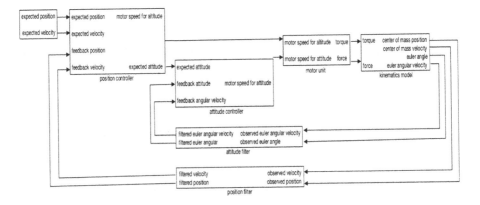

Fig. 3. UAV double closed-loop control

intervention is suggested for the UAV control. For severe errors, like physical damage or low battery, the UAV should return to launching place.

4   Image Processing

Bridge inspection images are usually of relatively low signal to noise ratio (SNR) and contrast ratio, since defects and structures are often with the same color and the defects are very small compared with the whole structure. This section is going to elaborate the image processing procedures in our system (see Fig. 4). Our procedures can be roughly divided into 2 parts: pre-processing, and defect extraction. Image pre-processing includes gradient processing and gray-scale stretch, while defect extraction includes fisher thresholding and clustering analysis. Finally, a report includes images with defect highlighted, location information, time information will be presented to users. Note that all the images we are going to deal with are gray scale ones (Fig. 5).

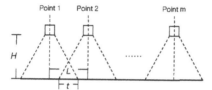

Fig. 4. Image capture mechanism

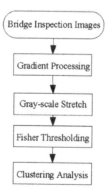

Fig. 5. Image processing procedures

4.1   Gradient Processing

Gray scale is not evenly distributed in inspection images. Defects have a relatively large gradient compared to the background. The above features of inspection images enable us to do image segmentation. We hope to do a preliminary defect extraction here. There are several common operators for image segmentation. The pros and cons of each operator are listed below (Table 1).

Table 1. Operators comparison

Operator	Pros	Cons
Robert	Capable of dealing with sharp low noise	Inaccurate edge positioning
Sobel	Capable of dealing with gradient wipe and a large amount of noise, accurate in edge positioning	Cannot deal with complex noise
Prewitt	Capable of dealing with gradient wipe and a large amount of noise	Cannot deal with complex noise
LOG	Eliminating divergent changes in gray scale and discontinuity in edge	Sensitive to noise, inaccurate edge positioning
Canny	Insensitive to noise, weak edges can be detected	High computing complexity

We choose canny operator here, because it is insensitive to noise. The image processing work is conducted at ground station when the UAV flight inspection is finished, so the computing burden is not that heavy. Note that, we do not binarize the images here. Values above threshold are set the same color, while for values below threshold, no operation is conducted to preserve the defect information.

4.2 Gray-Scale Stretch

Gray-scale stretch is a way to enhance contrast ratio. It linearly transforms the gray-scale into a larger scope.

4.3 Fisher Thresholding

Image segmentation is to extract meaningful information from the background. In our case, we are going to extract defects from bridge inspection images. Based on the nature of bridge inspection images (e.g. defects are various and usually small; image noise is quite obvious), Fisher criterion function is preferred in threshold setting. After this step, images are binarized.

4.4 Clustering Analysis

In clustering analysis, defect areas will be grouped and highlighted, and some trivial noise will be filtered. In Fisher thresholding step, defect points are very scattered. Clustering analysis will group these scattered points into several defect areas. Clustering centers with few defect points around will be filtered to eliminate noise.

4.5 Experimental Results

The bridge inspection images after each steps are shown below (Figs. 6, 7, 8 and 9).

Fig. 6. Gradient processing

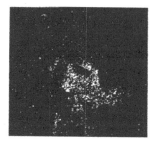

Fig. 7. Gray-scale stretch

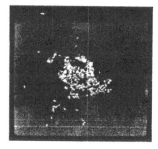

Fig. 8. Fisher thresholding

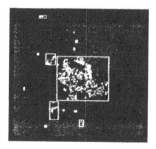

Fig. 9. Clustering analysis

The images after clustering analysis with defects highlighted, along with the location, time and other information will be included in the bridge inspection report, which is automatically generated by our system.

5 Conclusion

The UAV assisted bridge inspection system in our paper helps to ease the current costly, labor-intensive situation that has high practical relevance. By introducing UAV in our system, the inspection work can be performed, with the help of electronics and communication system. Currently, defect identification relies on inspectors' experiences and knowledge. Image processing can play a more and more importation role. It is being adopted in civil infrastructure monitoring over the past few years. The potential for automated inspections was shown to us.

The future work of our system is to build a more integrated software for friendly use. There are a lot to explore from our inspection data. The data serves as an important reference in other tasks. We hope our system will not only have defect inspection reports, but also combine inspection planning and other engineering economics decision making units.

References

Technical code of maintenance for city bridges, China, CJJ

Li, L., Sun, L., Ning, G.: Deterioration prediction of urban bridges on network level using Markov-chain model. Math. Probl. Eng. **2014**(7), 1–10 (2014)

Irizarry, J., Gheisari, M., Walker, B.N.: Usability assessment of drone technology as safety inspection tools. J. Inf. Technol. Constr. **17**(1), 194–212 (2012)

Civil UAV Real-name Registration Regulation by Civil Aviation Administration of China: http://www.caac.gov.cn/XXGK/XXGK/GFXWJ/201705/P020170517409761154678.pdf

Civil UAV Application Regulation by Civil Aviation Administration of China: http://www.caac.gov.cn/HDJL/YJZJ/201708/t20170809_46115.html

Civil UAV Manufacturing Standards by Ministry of Industry and Information Technology of the People's Republic of China: http://www.miit.gov.cn/n1146285/n1146352/n3054355/n3057585/n3057590/c5653876/content.html

Civil UAV Traffic Rules by Ministry of Transport of the People's Republic of China: http://zizhan.mot.gov.cn/sj/fazhs/zongheyshlf_fzhs/201711/t20171128_2942432.html

Murphy, R.R., Steimle, E., Griffin, C., Cullins, C., Hall, M., Pratt, K.: Cooperative use of unmanned sea surface and micro aerial vehicles at Hurricane Wilma. J. Field Robot. **25**(3), 164–180 (2008)

Campoy, P., Correa, J.F., Mondragón, I., Martínez, C., Olivares, M., Mejías, L., Artieda, J.: Computer vision onboard UAVs for civilian tasks. J. Intell. Robot. Syst. **54**(1–3), 105–135 (2009)

Alighanbari, M., Kuwata, Y., How, J.P.: Coordination and control of multiple UAVs with timing constraints and loitering. In: Proceedings of the 2003 American Control Conference, vol. 6, pp. 5311–5316 (2003)

Suresh, B.R.: A real-time automated visual inspection system for hot steel slabs. IEEE Trans. Pattern Anal. Mach. Intell. **5**(6), 563–572 (1983)

Xie, X.: A review of recent advances in surface defect detection using texture analysis techniques. Electron. Lett. Comput. Vis. Image Anal. **7**(3), 1–22 (2008)

Rahaman, G.M.A., Hossain, M.M.: Automatic defect detection and classification technique from image: a special case using ceramic tiles. Int. J. Comput. Sci. Inf. Secur. **1**(1), 22–30 (2009)

Kumar, A., Pang, G.K.H.: Defect detection in textural materials using Gabor filters. IEEE Trans. Ind. Appl. **38**(2), 425–440 (2002)

Mak, K.L., Peng, P., Lau, H.Y.K.: A real time computer vision systems for detecting defects in textile fabrics. IEEE Trans. (2005)

Sivabalan, K.N., Ghanadurai, D.: Detection of defects in digital texture images using segmentation. Int. J. Eng. Sci. Technol. **2**(10), 5187–5191 (2010)

Zhang, G., Chen, S., Liao, J.: Otsu image segmentation algorithm based on morphology and wavelet transformation. IEEE Trans. **1**, 279–283 (2011)

Fault Diagnosis and Knowledge Extraction Using Fast Logical Analysis of Data with Multiple Rules Discovery Ability

Xiwei Bai[1,2(✉)], Jie Tan[1], and Xuelei Wang[1]

[1] Institute of Automation, Chinese Academy of Sciences, Beijing 100190, China
xiwei_bai@163.com
[2] University of Chinese Academy of Sciences, Beijing 100049, China

Abstract. Data-based explanatory fault diagnosis methods are of great practical significance to modern industrial systems due to their clear elaborations of the cause and effect relationship. Based on Boolean logic, logical analysis of data (LAD) can discover discriminative if-then rules and use them to diagnose faults. However, traditional LAD algorithm has a defect of time-consuming computation and extracts only the least number of rules, which is not applicable for high-dimensional large data set and for fault that has more than one independent causes. In this paper, a novel fast LAD with multiple rules discovery ability is proposed. The fast data binarization step reduces the dimensionality of the input Boolean vector and the multiple independent rules are searched using modified mixed integer linear programming (MILP). A Case Study on Tennessee Eastman Process (TEP) reveals the superior performance of the proposed method in reducing computation time, extracting more rules and improving classification accuracy.

Keywords: Fault diagnosis · Knowledge extraction
Fast logical analysis of data · Multiple rules discovery

1 Introduction

Fault detection and diagnosis (FDD) is the key technique to guarantee production safety and reduce costs, thus it has always been a research focus 1–3. Various methods with both low fault alarm rate (FAR) and high fault detection rate (FDR) were carefully designed and applied to practical industrial systems in the past few decades 4–6. FDD methods are roughly divided into three categories: model-based, expertise-based and data-based 7. Since the real industrial systems have become larger and more complex, it is almost impossible to build precise mathematical models. Meanwhile, the expert knowledge of potential faults are difficult to obtain. Therefore, the data-based FDD methods are wildly researched due to their low dependences to the structure of system and the mechanism of fault 8.

In this paper, we focus on the data-based FDD classification method. With pre-defined classes (include a normal class and several faulty classes), various discriminant models can be trained successfully using data-based machine learning algorithms such as support vector machine (SVM) 9, artificial neuron networks (ANN) 10, Bayesian

Z. Shi et al. (Eds.): ICIS 2018, IFIP AICT 539, pp. 412–421, 2018.
https://doi.org/10.1007/978-3-030-01313-4_44

networks 11 and fuzzy logic 12 etc. Recently, the advancement of deep learning has proven its superiority in feature extraction and classification. Researchers implemented fault diagnosis with outstanding performance using deep learning algorithms such as deep belief networks (DBN) 13, stacked autoencoder (SAE) 14, recurrent neural networks (RNN) 15 and convolutional neural networks (CNN) 16.

Although the above methods attain high classification accuracy and realize effective fault diagnosis, they all suffered from a serious problem, namely, they build complicated discriminant models and extract incomprehensible patterns, which are like black boxes to users. All connections and transformations between the input data and the final results are encapsulated. The complexity of industrial systems make it impossible for users to understand the cause and effect relationship among various features and the corresponding diagnosis. Therefore, method that derived explanatory patterns are necessary and meaningful to convince people about its effectiveness, especially for practical engineering applications, where safety and validity are the prime principles.

Due to the above reasons, logical analysis of data (LAD) was proposed to reveal the causal relationship and implement fault diagnosis 17. LAD can discover explanatory patterns based on Boolean logic and detect faults by transforming these patterns into if-then rules and developing a synthetical discriminant function. However, the traditional LAD algorithm is not perfect. On the one hand, it discovers patterns with the least degree, which equals to the number of rules accordingly. Therefore, only the most discriminative rule will be found. Owing to the fact that faults are usually caused by several independent problems, multiple independent rules should be extracted at the same time. Obviously, the traditional LAD does not own that ability. On the other hand, the data binarization step (details are illustrated in Sect. 2) of traditional LAD forces the rest step to discover patterns among a large set of high dimensional data, which is time-consuming and unnecessary. To solve these two drawbacks, a novel fast LAD with multiple rules discovery ability based on modified mixed integer linear programming (MILP) is proposed and analyzed in this paper. A case study on Tennessee Eastman Process (TEP) indicates that the proposed new LAD algorithm can discover multiple reliable discriminant rules in a comparatively shorter time and realize fault classification with high accuracy.

The rest of this paper is organized as follow: Sect. 2, a brief introduction of LAD. Section 3, details about the proposed fast LAD with multiple rules discovery ability. Section 4, a case study on TEP. Section 5, conclusion.

2   Logical Analysis of Data

Logical analysis of data (LAD) is pattern discovery and classification algorithm. It finds the optimal feature combinations to achieve binary classification based on the Boolean representation of the original data. The combination is referred as pattern, which can be transformed to if-then rules. Therefore, LAD is explanatory and can extract explicit knowledge directly. This is of practical significance in the field of FDD because the if-then rules are easy to understood and accepted by both the administrator and the front-line worker in the industrial sites.

LAD contains three main steps: data binarization, pattern discovery and discriminant function formation 17. Data binarization transforms the original data to Boolean representations without any pre-processing or standardization operations. Pattern discovery aims to find the optimal binary variable combination that distinguish samples between two categories. Discriminant function is formed to implement binary classification.

Let S^N and S^F denote the normal and faulty data set. Through data binarization, Sample $i \in S^N$ is represented as a binary vector $a_i = \{a_{i,1}, a_{i,2}, \ldots, a_{i,n}\}$ with size n. We expand a_i to $a_i = \{a_{i,1}, a_{i,2}, \ldots, a_{i,n}, a_{i,n+1}, a_{i,n+2}, \ldots, a_{i,2n}\}$ with size $2n$, where $a_{i,n+j}$ is the negation of $a_{i,j}$, $j = 1, 2, \ldots, n$. A pattern p_l can also be represented as a binary vector $p_l^N = \{p_{l,1}^N, p_{l,2}^N, \ldots, p_{l,2n}^N\}$, which contains d binary variables with the value of 1, namely,

$$\sum_{k=1}^{2n} p_{l,k}^N = d \tag{1}$$

$p_{l,k}^N (k = 1, 2, \ldots, 2n)$ equals to 1 means $a_{i,k}$ is selected, then the real value of dimension to which $a_{i,k}$ corresponds is larger (smaller) than the cut point if k is smaller (larger) than n. Therefore, $a_{i,j}$ and its negation cannot be selected at the same time, namely,

$$p_{l,j}^N + p_{l,n+j}^N \leq 1 \tag{2}$$

To make p_l^N rejects all $i \in S^F$ and accepts as many $i \in S^N$ as possible, two constraints are introduced:

$$\sum_{k=1}^{2n} a_{i,k} p_{l,k}^N \leq d - 1, i \in S^F \tag{3}$$

$$\sum_{k=1}^{2n} a_{i,k} p_{l,k}^N + n y_i \geq d, i \in S^N \tag{4}$$

where y_i is the optimization variable that needs to be minimized. Constraints (3) forces all faulty samples not to fulfil p_l^N with at least 1°. Constraints (4) tries to make p_l^N cover more normal samples.

$$\min_{p_l^N, y_i, d} \sum y_i, i \in S^N$$

$$s.t. \begin{cases} \text{eq.}(1), (2), (3), (4) \\ 1 \leq d \leq n \\ p_{l,k}^N \in \{0, 1\}, k = 1, 2, \ldots, 2n \\ y_i \in \{0, 1\} \end{cases} \tag{5}$$

The MILP method for pattern discovery is summarized in (5).

If p_l^N cannot cover all samples, then another pattern p_{l+1}^N needs to be found by deleting all covered samples and run MILP again. All patterns form a pattern set P^N.

To obtain a pattern p_m^F that accepts $i \in S^F$ and rejects all $i \in S^N$, simply change their positions in (5). Likewise, these pattern form a pattern set P^F.

The discriminant function is formulated in the basis of both P^N and P^F. For a new extended binary vector a, the discriminant function is given as follow:

$$D(a) = \sum w_l \text{sgn}(p_l^N a^T - d) - \sum w_m \text{sgn}(p_m^F a^T - d) \qquad (6)$$

where $p_l^N \in P^N$ and $p_m^F \in P^F$. $\text{sgn}(a)$ is a sign function, it equals to 1 if $a > 0$ and 0 if $a \leq 0$.

3 Fast LAD with Multiple Rules Discovery Ability

Although MILP method is a useful and effective pattern discovery tool, it still has several serious defects. Let us consider the following three typical situations:

- More than one binary variables from different dimensions are needed jointly to discriminate normal and faulty data.
- Both binary variable and its negation from same dimension are needed to discriminate normal and faulty data.
- More than one binary variables from different dimensions can discriminate normal and faulty data independently.

Since MILP searches the optimal variable combination to cover as many samples as possible, the first two situations are easy to handle. However, the independent causes of fault in the third situation will never be found completely using MILP because the algorithm stops when a pattern with one critical degree is found and then focuses on the rest, unaccepted samples. The reason is obvious: this pattern can cover all samples of its kind and reject all others, thus MILP will not add more degree into it.

To illustrate this problem, a simple example is shown above. Figure 1 contains two normal binary samples and two faulty samples. A pattern that contains the red or green variables can distinguish them independently. Unfortunately, MILP can only find one of them randomly and stop instantly. Therefore, we need a new MILP that can discover a pattern with all critical degrees (red and green) here, which can then be transformed into multiple independent discriminant rules.

Normal	1	1	1	1	0	0	0	0
	1	1	1	1	0	0	0	0

Faulty	1	0	1	0	0	1	0	1
	0	0	0	0	1	1	1	1

Fig. 1. The diagram of the third situation (Color figure online)

Dealing with this situation will be very meaningful, it reveals more characteristics of the fault and reduces the FAR of the original discriminate function because more rules can determine the type of fault more accurately.

To solve this problem, a novel modified MILP is proposed in this paper to realize multiple rules discovery.

3.1 Multiple Rules Discovery Based on Modified MILP

To find identical functional degrees, we could find all potential degrees firstly, then remove those useless degrees. Mortada 19 proposed an upgraded MILP with additional item in the object function, namely, $\min \sum y_i + \delta d, i \in S^N$, where $\delta > 0$ is selected to reduce the degree of pattern, making them more explanatory. In this paper, we do the contrary. We set $\delta < 0$, then MILP will find a pattern with both high degree and high coverage of samples. The importance of finding more degrees depends on δ. Unfortunately, as long as one critical degree is found, Eq. (3) will be fulfilled immediately. Degree could be increased by finding the smallest cut point of any dimension (except the dimension where the critical degree derived from) because the left part of Eq. (4) will naturally raise 1, therefore, this operation will bring some useless discriminate rules in.

To fix this drawback, Eq. (3) needs to be revised. Inspired by the idea that maximize the coverage of pattern in Eq. (4), we propose the following new constraint to replace Eq. (3):

$$\sum_{k=1}^{2n} a_{i,k} p_{l,k}^N + z_i \leq d - 1, i \in S^F \qquad (7)$$

and the object function are revised as:

$$\min_{p_l^N, y_i, d, z_{i'}} \sum y_i + \delta d + \gamma \sum z_{i'}, i \in S^N, i' \in S^F \qquad (8)$$

where $z_{i'}$ is optimization integer variable that increases the distinction between the faulty data and the normal pattern, $\gamma < 0$ is the weight coefficient. All $z_{i'}$ have the same range of value with $d - 1$. With this additional optimization variable, Eq. (7) will not be fulfilled by one critical degree. The more critical degrees are found, the better a pattern can meet the constraint and the optimization requirement at the same time.

The hyper parameter δ and γ have opposite effects and thus should be selected carefully to reach a balance. In the ideal situation, all critical degrees with similar function are aggregated in one pattern and the useless degrees are removed. This pattern can then be transformed into multiple independent discriminant rules simply by reversing the binarization step.

It seems that the above modified MILP can achieve multiple rules discovery easily. Nevertheless, it is usually difficult to obtain an ideal result using MILP solver due to the proper value of hyper parameters are uncertain and those useless degrees are hard to eliminate. To deal with this problem, we restrict the upper limit of d and $z_{i'}$ by selecting the number of degree manually. The modified MILP is showed below:

$$\min_{p_l^N, y_i, d, z_{i'}} \sum y_i + \delta d + \gamma \sum z_{i'}, i \in S^N, i' \in S^F$$

$$s.t. \begin{cases} \text{eq.}(1), (2), (7), (4) \\ 1 \leq d \leq ub \\ p_{l,k}^N \in \{0, 1\}, k = 1, 2, \ldots, 2n \\ y_i \in \{0, 1\} \\ 0 \leq z_{i'} \leq ub - 1 \end{cases} \tag{9}$$

where ub is the upper bound. How to select an appropriate ub is discussed in the following part.

3.2   Fast LAD Based on Fast Data Binarization

The data binarization step of LAD results in large cut point set and long binary vector, making the pattern discovery step difficult and time-assuming. However, most of the cut points are useless. If those cut points are found in advance, they can be removed to reduce the computation load and realize fast data binarization.

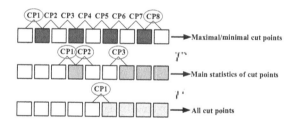

Fig. 2. The diagram of fast data binarization

The diagram of fast binarization is shown in Fig. 2. Assume that all collected variables follow Gaussian distribution, we estimate the mean and variance of Gaussian distribution in each dimension of the normal data. The proportion of faulty data covered between the range of positive and negative three sigma is calculated. According to the proportion, we divide the faulty data into three parts by setting an upper threshold T^u and a lower threshold T^l. Meanwhile, to make Eq. (9) work, the determination of ub is necessary. This can be done according to the prior knowledge of the fault. However in this paper, under Gaussian distribution assumption, ub can be determined by analysing the similarity of distribution between normal and faulty data. The fast data binarization is implemented in the basis of the following three rules:

- If the proportion is larger than T^u, this dimension is considered not discriminative, only the maximal and minimal cut points are remained.

- If the proportion is between T^l and T^u, this dimension is probably discriminative, the main statistics: mean, mode, median, maximal and minimal cut points are remained for potential usage.
- If the proportion is lower than T^l, this dimension is considered discriminative, all cut points are remained and ub is equal to the total number of dimension in this type.

4 A Case Study on Tennessee Eastman Process

The Tennessee Eastman Process (TEP) is a widely used chemical process simulation. TEP contains 8 components (A, B, C, D, E, F, G and H) and 5 units (stripper, separator, reactor, compressor and condenser). Among them, A, C, D and E are the reactants; G and H are the target products; B is the non-reactive inertial component and F is the by-product. 11 manipulated variables (XMV (1–11)) and 41 measured variables (XMEAS (1–41)) are acquired from 21 designed faulty conditions (IDV (1–21)) and 1 normal condition (IDV (0)) at an interval of 3 min. Each condition contains 480 training sampled and 960 testing samples. In this paper, XMEAS(1–22) and XMV(1–11) are selected as input variables.

The thresholds T^l and T^u are set to 0.1 and 0.9 according to a rule of thumb. The parameters δ is set to -0.001 and γ is set to -1.

Due to the feedback mechanism of the system, the discriminative rules change as time goes by. Therefore in this paper, data in the final stable phase are used to train and test fast LAD. The training data set contains the last 100 normal data and the last 80 faulty data. The testing data set contains the last 300 normal and faulty data. Experiments were implemented on an Intel Core i7-6700 PC with 16G RAM and 3.4 GHz in Matlab environment.

IDV (1, 2, 4, 7, 17) are selected for a comparative analysis. To show the validity of modified MILP, discriminant rules discovered by both modified MILP and MILP are listed in Table 1. From the table, it is obvious that the modified MILP discovered patterns with more degrees than MILP in IDV (1) and IDV (2) and thus derived more discriminant rules. To prove the correctness of the results, we analysed the data distribution of IDV (0) and IDV (1, 2, 4, 7, 17). For IDV (1), XMEAS (1, 4, 18, and 19) and XMV (6, 9) have apparent deviations from the corresponding range of distribution of the normal data. For IDV (2), XMEAS (10, 11, 18, 19, and 22) and XMV (6, 9) are the variables that deviate from the normal state. All these deviations are captured by the modified MILP and transformed into independent discriminant rules whereas the general MILP found only one feasible rule for each fault. If another fault fulfills the rule, they might be misdiagnosed. Therefore, the modified MILP technique will reduce the misdiagnosis rate effectively. The fault classification can be implemented using any combination of these rules.

For IDV (4) and IDV (7), two algorithms obtain identical results because these two faults have only one deviated variable, namely, XMV (10) and XMV (4). For IDV (17), MILP obtains more discriminant rules. However, only XMEAS (9, 21) and XMV (10) show deviations from normal state. Among all we can learn that rule related to

Table 1. Fault discriminant rules

Fault no.	Modified MILP		MILP
IDV(1)	Rule group 1	XMEAS(1) > 0.50 XMEAS(4) < 9.07 XMEAS(18) > 66.73 XMEAS(19) > 246.66 XMV(3) > 48.87 XMV(9) > 52.23	XMV(9) > 52.23
IDV(2)	Rule group 1	XMEAS(10) > 0.51 XMEAS(11) > 88.88 XMEAS(18) < 64.45 XMEAS(19) < 184.93 XMEAS(22) > 78.27 XMV(6) > 60.35 XMV(9) < 37.00	XMV(22) > 78.28
IDV(4)	Rule group 1	XMV(10) > 43.01	XMV(10) > 43.01
IDV(7)	Rule group 1	XMV(4) > 69.08	XMV(4) > 69.08
IDV(17)	Rule group 1	XMEAS(21) < 94.33	XMEAS(21) < 94.38 XMV(2) < 54.73
	Rule group 2	XMEAS(9) < 120.35	XMEAS(3) > 4498.10 XMEAS(19) < 235.99

Table 2. FAR and FDR for different faults

Fault no.	Fast LAD based on modified MILP		Traditional LAD based on MILP	
	FAR (%)	FDR (%)	FAR (%)	FDR (%)
IDV(1)	**0**	**94.33**	0	96
IDV(2)	**0**	**97.33**	0	99.67
IDV(4)	**0.33**	**100**	0.33	100
IDV(7)	**0**	**100**	0	100
IDV(17)	**0.33**	**98.33**	0.33	89.67

XMEAS (21) is the main functional rule. Other rules are not much useful. The modified MILP obtains effective rules because constraints Eq. (7) increases the distinction between the faulty data and the normal pattern, therefore removes rules that are not discriminative.

The FAR and FDR of proposed fast LAD based on modified MILP and the traditional LAD based on MILP are listed in Table 2. The former provides equally satisfied diagnosis results with the latter and better results for IDV (17).

The fast binarization step removes a large amount of cut points to reduce the dimensionality of the binary vector and further improves computation speed. The following comparative analysis proves its validity.

In Fig. 3, apparently, the fast LAD takes much less time to obtain the final result. This experiment is run on a small data set, so the original computation time is

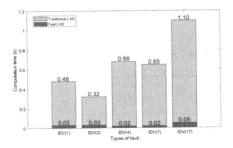

Fig. 3. The computation time comparison between traditional and fast LAD

acceptable. For a large dataset, the traditional LAD will spend very long time searching rules among massive high dimensional binary vectors while the fast LAD can remove useless dimension ahead. Therefore, it can reduce computation time effectively.

5 Conclusions

As a data-based machine learning algorithm, logical analysis of data can discover explanatory patterns and reveal the logic relations between data and results, which is very meaningful in fault diagnosis field. In this paper, we improve the traditional LAD algorithm and propose a fast LAD with multiple rules discovery ability. Based on modified MILP and fast binarization, this new algorithm can extract multiple independent fault rules within much shorter computation time. With a case study on TEP, we show the advantages of fast LAD in both fault classification and knowledge extraction.

Acknowledgements. This research is supported by National Natural Science Foundation of China (Number: U1701262).

References

1. Yin, S., Ding, S.X., Xie, X., et al.: A review on basic data-driven approaches for industrial process monitoring. IEEE Trans. Ind. Electron. **61**(11), 6418–6428 (2014)
2. Severson, K., Chaiwatanodom, P., Braatz, R.D.: Perspectives on process monitoring of industrial systems. Annu. Rev. Control **48**(21), 931–939 (2016)
3. Dai, X., Gao, Z.: From model, signal to knowledge: a data-driven perspective of fault detection and diagnosis. IEEE Trans. Ind. Inf. **9**(4), 2226–2238 (2013)
4. Peng, K., Zhang, K., You, B., et al.: Quality-related prediction and monitoring of multi-mode processes using multiple PLS with application to an industrial hot strip mill. Neurocomputing **168**(C), 1094–1103 (2015)
5. Torabi, A.J., Meng, J.E., Xiang, L., et al.: Application of clustering methods for online tool condition monitoring and fault diagnosis in high-speed milling processes. IEEE Syst. J. **10**(2), 721–732 (2017)

6. Zhang, Y., Zhou, H., Qin, S.J., et al.: Decentralized fault diagnosis of large-scale processes using multiblock kernel partial least squares. IEEE Trans. Ind. Inf. **6**(1), 3–10 (2010)
7. Wen, C.L., Fei-Ya, L.V., Bao, Z.J., et al.: A review of data driven-based incipient fault diagnosis. Acta Autom. Sin. **42**(9), 1285–1299 (2016)
8. Yin, S., Li, X., Gao, H., et al.: Data-based techniques focused on modern industry: an overview. IEEE Trans. Ind. Electron. **62**(1), 657–667 (2015)
9. Yin, Z., Hou, J.: Recent advances on SVM based fault diagnosis and process monitoring in complicated industrial processes. Neurocomputing **174**(PB), 643–650 (2016)
10. Shatnawi, Y., Al-Khassaweneh, M.: Fault diagnosis in internal combustion engines using extension neural network. IEEE Trans. Ind. Electron. **61**(3), 1434–1443 (2013)
11. Cai, B., Huang, L., Xie, M.: Bayesian networks in fault diagnosis. IEEE Trans. Ind. Inf. **PP** (99), 1 (2017)
12. Khan, S.A., Equbal, M.D., Islam, T.: A comprehensive comparative study of DGA based transformer fault diagnosis using fuzzy logic and ANFIS models. IEEE Trans. Dielectr. Electr. Insul. **22**(1), 590–596 (2015)
13. Shao, H., Jiang, H., Zhang, X., et al.: Rolling bearing fault diagnosis using an optimization deep belief network. Meas. Sci. Technol. **26**(11) (2015)
14. Sun, W., Shao, S., Zhao, R., et al.: A sparse auto-encoder-based deep neural network approach for induction motor faults classification. Measurement **89**, 171–178 (2016)
15. Bruin, T.D., Verbert, K., Babuška, R.: Railway track circuit fault diagnosis using recurrent neural networks. IEEE Trans. Neural Netw. Learn. Syst. **PP**(99), 1–11 (2016)
16. Guo, X., Chen, L., Shen, C.: Hierarchical adaptive deep convolution neural network and its application to bearing fault diagnosis. Measurement **93**, 490–502 (2016)
17. Ragab, A., El-Koujok, M., Poulin, B., et al.: Fault diagnosis in industrial chemical processes using interpretable patterns based on Logical Analysis of Data. Expert Syst. Appl. **95**, 368–383 (2018)
18. Ryoo, H.S., Jang, I.Y.: MILP approach to pattern generation in logical analysis of data. Discrete Appl. Math. **157**(4), 749–761 (2009)
19. Mortada, M.A., Yacout, S., Lakis, A.: Fault diagnosis in power transformers using multi-class logical analysis of data. J. Intell. Manuf. **25**(6), 1429–1439 (2014)

Improved Feature Selection Algorithm for Prognosis Prediction of Primary Liver Cancer

Yunxiang Liu[✉], Qi Pan[✉], and Ziyi Zhou[✉]

School of Computer Science and Information Engineering,
Shanghai Institute of Technology, Shanghai 201418, China
yxliu@sit.edu.cn, 463728073@qq.com, 2634578954@qq.com

Abstract. Primary liver cancer, one of the most common malignant tumors in China, can only be roughly diagnosed through doctors' expertise and experience at present, making it impossible to resolve the health problem that people care about. A new method that applies machine learning to the medical filed is therefore presented in this paper. The decision tree algorithm and the random forest algorithm are used to classify the data, and decision tree algorithm and improved feature selection algorithm to select important features. Comparison shows that the performance of the random forest algorithm is better than that of the decision tree algorithm, and the improved feature selection algorithm can filter out more important features on the premise of retaining accuracy.

Keywords: Primary liver cancer · Machine learning · Decision tree
Random forest

1 Introduction

Primary liver cancer is one of the most common malignant tumors in China, with its mortality in patients being the third in malignant tumors [1]. Typically the prognosis of this disease can only be roughly judged through doctors' professional knowledge and experience. The low accuracy, therefore, has a negative effect on both doctors and patients. At present, systematic researches conducted by machine learning are few both at home and abroad, and there is no corresponding model or software to verify the classification of liver cancer data [2]. Even if the iterative updating of medical treatment equipment cannot predict the occurrence of liver cancer. For solving the health problem that people care about more effectively, we try to use the machine learning into analyzing the data about primary liver cancer, hoping it can be used in clinical prognosis assessment and treatment option. This paper studies two machine learning algorithms— the decision trees and random forests, which are the most typical representatives of symbol learning and ensemble learning.

In addition, it also improves and verifies the character choosing method based on random forest so as to reduce the overhead of model training and the difficulty in data acquisition. By using Python language, this paper implements the above algorithms and organizes and tests the system interface. Besides, the classification accuracy and feature

© IFIP International Federation for Information Processing 2018
Published by Springer Nature Switzerland AG 2018. All Rights Reserved
Z. Shi et al. (Eds.): ICIS 2018, IFIP AICT 539, pp. 422–430, 2018.
https://doi.org/10.1007/978-3-030-01313-4_45

selection of different algorithms are also analyzed in this paper, providing reference for the selection of models. To sum up, this topic has considerable research significance both in the field of computer science and medical domain.

2 Principle

2.1 Decision Tree Algorithm

During late 1970s and the early 1980s, J. Ross Quinlan developed the decision tree algorithm which was originally called ID3. In the application process, Quinlan found and improved the shortcomings of ID3, and put forward the C4.5. In 1984, many statisticians published the book named "Classification and Regression Trees" (CART). A decision tree is a tree structure similar to a flow chart with each internal node representing a test on an attribute [3], each branch the corresponding output of the test, and each leaf node the category. A typical decision tree model is shown in Fig. 1. The inner node is represented by an ellipse, and the leaf nodes by rectangles. Most decision trees are built by top-down recursion, selecting instances of a class from the known training set by using changes in entropy.

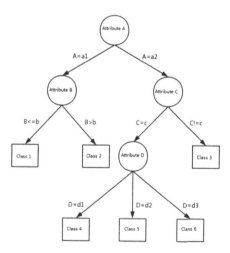

Fig. 1. Decision tree model

2.1.1 Information Gain

ID3's attribute selection and measure are realized by information gain. Entropy [4], which is inversely proportional to the purity of the data set, is usually used to represent the uncertainty of random variables. Supposing that there are D samples and K classes, with p_i being the probabilities of class I in D. The entropy of D is given in the following form:

$$H(D) = -\sum_{i=1}^{k} p_i \log_2 p_i \tag{1}$$

When D is divided according to the attribute A, the feature A has n different values, and the new conditional entropy is defined as the subset of the D:

$$H(D|A) = \sum_{i=1}^{n} p_i' \cdot H(D|A = a_i) = \sum_{i=1}^{n} p_i' \cdot H(D_i) \tag{2}$$

By comparison, $H(D)$ it can reflect the uncertainty of the original data set, $H(D|A)$ indicate the uncertainty after division, and regard the difference between them as information gain.

$$G(D,A) = H(D) - H(D|A) \tag{3}$$

2.1.2   Gain Rate

C4.5 uses an information gain called to expand the split [5]. In order to standardize the information gain, we use the "split information" value. The gain rate is defined as follows:

$$G_R(D,A) = \frac{G(D,A)}{H_A(D)} \tag{4}$$

among

$$H_A(D) = -\sum_{i=1}^{n} p_i' \log_2 p_i' \tag{5}$$

2.1.3   Gini Index

Gini index is used to reflect the uncertainty of the number set, and Gini index is inversely proportional to the purity of the data set. The Gene index is defined as follows:

$$Gini(D) = \sum_{i=1}^{k} p_i(1 - p_i) = 1 - \sum_{i=1}^{k} p_i^2 \tag{6}$$

Under the condition of characteristic A, the Gini index of the sample set D is defined as:

$$Gini(D, A) = \sum_{i=1}^{n} p_i' \cdot Gini(D_i) \tag{7}$$

When the characteristics of the sample set have the smallest Gini index, the current feature is the best feature. Figure 2 below shows the relationship between the Gini index, the entropy half and the classification error rate in the two categories.

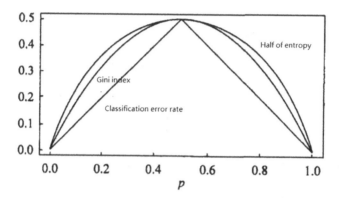

Fig. 2. Comparison of two indicators

2.2 Random Forest Algorithm

Random forest is built on the basis of decision tree. It is an integrated classifier model formed by multiple decision trees [6]. To put it simple, multiple decision trees construct the random forest. Random forests have adopted the Bagging thought [7] and characteristic subspace thought, being of more anti-noise ability than a single decision tree. It will not over-fit and can significantly improve the generalization ability. The basic flow of a random forest is shown in Fig. 3:

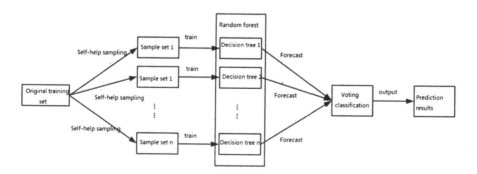

Fig. 3. The basic flow of a random forest

2.3   Improved Feature Selection Algorithm

The basic idea of selecting the characteristics by using the random forest is to sort the characteristics according to the importance first, then remove some features through generalized-sequence backward selection algorithm. Then we train random forest on the new feature set and calculate its accuracy rate; repeat the process, and finally use the feature set with the highest accuracy as the output. In order to ensure the stability of each test result, cross validation is used to evaluate the newly established random forest after each round of screening, and the average accuracy rate is taken as the accuracy of that round. Compared with the wrapped method like LVW random selection feature subset, the algorithm's feature selection is heuristic and has higher efficiency. However, since the iteration will continue until the number of remaining features is reduced to the threshold, it still has a large time and space overhead; and because we make the final selection based on the highest test precision, the feature set is not necessarily the smallest one.

On this basis, a faster feature selection algorithm is designed to optimize the process. According to the error increment caused by each wheel screening, we judge whether to continue screening. Once it exceeds the specified threshold, the iteration is exited, and feature set selected in the last round is used as the result. For models trained on shrinking feature sets, the generalization performance tends to decrease, and the degree of reduction can be used as an evaluation criterion for feature sets. The essence of this strategy is to select the smallest feature subset in a given error range rather than the highest test precision, so that it can stop screening as soon as possible and save a lot of time. The reason why the error increment threshold is not simply set to 0 is that some weak correlation features are expected to be removed in addition to the unrelated features, and this can also allow small deviations of each test. The test results show that the selected feature set does not actually produce an error increment as large as the threshold value does, and the test accuracy on it can be even higher before screening.

3   Experimental Analysis

3.1   Collection and Data Collation

The case data came from Eastern Hepatobiliary Surgery Hospital, Second Military Medical University, including malignancies, benign lesions and normal types. There were 588 groups of data, among which malignant tumor accounting for 246; benign lesions 149 and normal 193. The data itself has too many missing values, making it difficult to sort and classify the data. We delete many useless indexes with the help of professionals, leaving 39 anonymous indicators for each sort finally. As a result of privacy protection, the sample set has 693 missing values; Samples that have more than 5 missing values are automatically discarded by the program. The sample contains 6 discrete indexes and 519 effective sample groups. Different methods are used to treat the missing values in the decision tree model. Some endow the sample with the common value of the feature, while the method used in C4.5 gives a weight for each value of the feature, which is divided into the sub nodes with different probability.

3.2   Comparison Between Decision Tree and Random Forest

After reading the data, 70% of the data in the effective sample are selected as the training set and the rest as the test set. 5 times of random training and testing are repeated, the decision tree and the random forest model are created. We record the average value to realize a simple cross validation. The results of the recording were shown in Table 1.

Table 1. Comparison between decision tree and random forest

Model	Pruning algorithm	Training time	Prediction time	Test accuracy
Decision tree	Nothing	7.16 s	1.69 ms	86.13%
Decision tree	PEP	7.16 s + 3.78 ms	0.74 ms	89.21%
Random forest	Nothing	1.53 s	17.72 ms	92.16%

Obviously, in addition to forecasting time, the overall performance of random forests is better than that of decision trees, especially in training time. For training sets of the same size, the construction speed of the random forest is nearly 5 times faster than that of the decision tree.

Similarly, 5 training and testing are conducted under random division, and the feature set selected after each training decision tree (PEP pruning) is recorded. F_i in the following table represents the characteristic subscript set of i test, and r_i the test accuracy that is used for reference (Table 2):

Table 2. Features selected by the decision tree

i	F_i	r_i
1	{1, 3, 4, 16, 21, 25, 28, 34}	90.12%
2	{1, 2, 3, 4, 16, 21, 33}	88.46%
3	{2, 3, 4, 11, 16, 21, 25, 28, 34, 35}	87.92%
4	{2, 3, 4, 11, 15, 16, 21, 28, 33, 34, 37}	89.03%
5	{1, 2, 3, 4, 13, 16, 21, 34, 35}	89.35%

The intersection and union of $F_i(1 \leq i \leq 5)$ are resolved respectively as follows:

$$F_{min} = \bigcap_{1 \leq i \leq 5} F_i = \{3, 4, 16, 21\}$$

$$F_{max} = \bigcup_{1 \leq i \leq 5} F_i = \{1, 2, 3, 4, 11, 13, 15, 16, 21, 25, 28, 33, 34, 35, 37\}$$

In the above formula, F_{min} represents the most important features that can be collected first in the collection of samples to avoid missing values as much as possible; while F_{max} can be used as a selected feature set. Then we use the improved feature selection algorithm for feature selection.

It better demonstrates a feature selection process with a default parameter. The filtered results of each group and the test accuracy obtained from the training are listed as follows, and those removed features will be discarded in the next round (Table 3).

Table 3. Feature screening process

Rotation	Selected feature subscript (order of importance)	Characteristic number	r
0	4, 28, 14, 21, 15, 6, 26, 11, 16, 10, 3, 7, 25, 34, 30, 27, 36, 35, 17, 13, 38, 20, 24, 32, 22, 29, 37, 8, 33, 0, 12, 2, 18, ~~31, 5, 1, 19, 9, 23~~	39	91.82%
1	16, 4, 3, 28, 27, 6, 25, 14, 24, 21, 29, 13, 11, 8, 7, 26, 2, 22, 34, 15, 37, 20, 32, 17, 36, 33, 38, 10, ~~30, 12, 0, 18, 35~~	33	90.60%
2	4, 28, 25, 11, 16, 24, 21, 6, 3, 29, 15, 22, 26, 13, 8, 20, 17, 2, 34, 14, 7, 38, 32, ~~36, 37, 33, 27, 10~~	28	90.81%
3	6, 3, 4, 16, 21, 25, 28, 8, 15, 11, 34, 24, 7, 13, 26, 29, 20, 2, 14, ~~17, 38, 22, 32~~	23	92.73%
4	28, 3, 25, 16, 4, 21, 26, 15, 29, 13, 8, 6, 2, 34, 14, 7, ~~20, 24, 11~~	19	92.52%
5	4, 28, 2, 6, 3, 14, 21, 16, 25, 15, 13, 8, 26, ~~34, 29, 7~~	16	91.45%
6	4, 3, 21, 15, 28, 25, 16, 26, 13, 6, 14, ~~8, 2~~	13	91.87%
7	4, 3, 25, 21, 16, 15, 28, 26, 14, 6, 13	11	92.31%

As can be seen from the above table, the algorithm will eliminate some features with the lowest importance every time, and the number of discarded features will gradually decrease as a result of proportionate screening. The results obtained in seventh round are final results. The reason why we stop screening features is that the error increment in the eighth round exceeds fixed value 2%. It can be concluded that the accuracy of test before the screening stops does not change significantly with the decrease of feature numbers, but only fluctuates near the initial accuracy. This shows that the algorithm can effectively identify redundant and weak correlation features, and thus it can preserve the classifier performance while removing these features. It is also found during the test that the number of features having the highest test precision (23 in the upper case) is not stable and has a large randomness (Table 4).

It can be seen from the table that the subscript, the size of the selected feature sets is generally between 7 and 19, and the average size is 13, more than half of that of the original feature capacity 39; However, their corresponding test accuracy does not show a sharp decline compares with 92.16% before screening, both within 2% and equaling

Table 4. Feature screening results

maxAccurDesc	Selected feature subscript (order of importance)	r
2.5	3, 28, 16, 4, 6, 27, 21	90.28%
	4, 15, 28, 11, 6, 21, 14	91.02%
	3, 4, 28, 16, 21, 25, 6, 2, 24, 34, 27	91.23%
	4, 3, 15, 28, 25, 21, 11, 14, 26	90.34%
2 (默认)	4, 28, 21, 27, 15, 24, 3, 26, 16, 6, 14, 2, 5, 8, 11, 13	92.10%
	4, 25, 21, 15, 16, 6, 26, 28, 13, 24, 3, 2, 30, 22, 27, 19	92.52%
	4, 3, 25, 21, 16, 15, 28, 26, 14, 6, 13	92.31%
	4, 28, 3, 21, 24, 2, 15, 16, 37, 6, 25, 14, 30	91.86%
1.5	28, 4, 16, 3, 21, 26, 6, 27, 5, 13, 24, 15, 2, 35, 29, 14, 25, 20, 34	92.02%
	4, 28, 21, 27, 11, 3, 15, 16, 6, 24, 25, 30, 10, 13, 26, 2, 31, 20, 8	92.64%
	4, 21, 3, 15, 28, 25, 2, 14, 24, 16, 6, 22, 27	91.98%
	4, 28, 16, 21, 11, 25, 3, 6, 24, 15, 30, 13, 14, 38, 29, 33	93.01%

that before selecting after adjusting the parameters. First of all, the establishment of random forests is a stochastic process in itself, which makes the order of importance always change and affects the screening results. The improved feature selection algorithm uses the generalized-sequence backward selection to eliminate the feature, which is actually based on a greedy strategy and usually leads to local optimum as a result of neglecting the correlation between features.

4 Conclusion

By applying the large data technology to the prognosis prediction of primary liver cancer, we can find that the accuracy of the random forest algorithm is obviously superior to that of the decision tree algorithm. The decision tree algorithm cannot avoid a low accuracy caused by over-fitting even if it uses the pruning technique. The improved feature selection algorithm, on the contrast, can significantly reduce the feature set on the premise of guaranteeing the prediction accuracy, which lays the foundation for considering the correlation between features later. It further shows that it is an essential trend to apply the increasingly perfect big data technology to the medical field, and it is worth further exploring and studying.

References

1. Liu, Q., Wang, W.: Liver Cancer. People's Medical Publishing House, Beijing (2000)
2. Han, Y., Shi, H., Qu, B.: Application of random forest method in medicine. Chin. J. Prevent. Med. **15**(1), 79–81 (2014)

3. Chen, X., Wang, S., Li, J.: Application of weighted constraint based decision tree method in identifying poor students. Comput. Appl. Softw. Parts **32**(12), 136–139 (2014)
4. Wang, X., Jiang, Y.: Analysis and improvement of decision tree ID3 algorithm. Comput. Eng. Des. **32**(9), 3070–3072 (2011)
5. Miao, Y., Zhang, X.: Improvement and application of C4.5 decision tree algorithm. Comput. Eng. Appl. **32**(9), 3070–3072 (2011)
6. Zhou, Z.: Machine Learning. Tsinghua University Press, Beijing (2016)
7. Huai, T.T.: Improvement and application of random forest algorithm. Metrology University of China, Hang Zhou

A Novel Spatial-Spectra Dynamics-Based Ranking Model for Sorting Time-Varying Functional Networks from Single Subject FMRI Data

Nizhuan Wang[1(✉)], Hongjie Yan[2], Yang Yang[3], and Ruiyang Ge[4]

[1] Artificial Intelligence and Neuro-informatics Engineering (ARINE)
Laboratory, School of Computer Engineering,
HuaiHai Institute of Technology, Lianyungang 222002, China
wangnizhuan1120@gmail.com
[2] Department of Neurology, Affiliated Lianyungang Hospital
of Xuzhou Medical University, Lianyungang 222002, China
[3] Center for Neuroimaging,
Shenzhen Institute of Neuroscience, Shenzhen 518057, China
[4] Non-Invasive Neurostimulation Therapies (NINET) Laboratory,
Department of Psychiatry, University of British Columbia,
Vancouver V6T 2A1, Canada

Abstract. Accumulating evidence suggests that the brain state has time-varying transitions, potentially implying that the brain functional networks (BFNs) have spatial variability and power-spectra dynamics over time. Recently, ICA-based BFNs tracking models, i.e., SliTICA, real-time ICA, Quasi-GICA, etc., have been gained wide attention. However, how to distinguish the neurobiological BFNs from those representing noise and artifacts is not trivial in tracking process due to the random order of components generated by ICA. In this study, combining with our previous BFNs tracking model, i.e., Quasi-GICA, we proposed a novel spatial-spectra dynamics-based ranking method for sorting time-varying BFNs, called weighted BFNs ranking, which was based on the dynamical properties in both spatial and spectral domains of each BFN. This proposed weighted BFNs ranking model mainly consisted of two steps: first, the dynamic spatial reproducibility (DSR) and dynamic fraction of amplitude low-frequency fluctuations (DFALFF) for each BFN were calculated; then a weighted coefficients-based ranking strategy for merging the DSR and DFALFF of each BFN was proposed, to make the meaningful dynamic BFNs rank ahead. We showed the effective results by this ranking model on the simulated and real data, suggesting that the meaningful dynamical BFNs with both strong properties of DSR and DFALFF across the tracking process were ranked at the top.

Keywords: fMRI · ICA · Dynamic spatial variability
Dynamic power spectrum · Ranking

© IFIP International Federation for Information Processing 2018
Published by Springer Nature Switzerland AG 2018. All Rights Reserved
Z. Shi et al. (Eds.): ICIS 2018, IFIP AICT 539, pp. 431–441, 2018.
https://doi.org/10.1007/978-3-030-01313-4_46

1 Introduction

Blood oxygen level dependent (BOLD) functional magnetic resonance imaging (fMRI) is powerful modality to discover functional connectivity (FC) among discrete brain regions. Recently, a growing number of reports have suggested that the FC under rest or task condition is not static but exhibits complex spatiotemporal dynamics [1]. For example, besides the well-known temporal variability of FC [2–4], the brain activation regions also show considerable spatial variability and dynamical power spectrum over time [5, 6], which has potentials for diagnosing brain diseases [7].

Independent component analysis (ICA), which could extract the brain functional networks (BFNs) from fMRI data at the single subject level or group level under rest or task conditions [1–3, 8–10], has turned out to be a promising tool to decode the brain activity. With respect to tracking the dynamic BFNs at the single subject level, some variants of ICA such as SliTICA, real-time ICA and Quasi-GICA, have been proposed to capture the spatial dynamics over time [5, 11, 12]. However, after the BFNs identification using ICA-based tracking methods, how to distinguish the neurobiological and meaningful time-varying BFNs from noise and artifacts is not trivial due to the following reasons. To the beginning, most ICA algorithms do not automatically rank BFNs, which leads to manually examine all BFNs one by one. Further, the BFNs at single subject level always have limited information in contrast to ones from multiple subjects, which leads to that sorting BFNs at single subject level is more challenging task. In addition, to our knowledge, most of BFNs ranking methods such as percent variance ranking [2], power spectrum ranking [13], support vector machine (SVM) based ranking [14], multiple ICA estimations based ranking [15, 16] and MMC ranking [17], are designed for sorting static BFNs, rather than dynamic BFNs. Thus, in this study, a novel spatial-spectra dynamics-based ranking method was proposed to sort the time-varying BFNs, taking advantage of the dynamic features in both spatial and spectral domains of each BFN.

2 Theory and Methods

2.1 Brief Review of Quasi-GICA

It is well-known that the BFNs identification has always been modulated as a blind source separation (BSS) problem, assuming that there exists functional integration of activity in multiple macroscopic loci [2, 3, 8–10, 18]. This BSS problem aims at retrieving the underlying sources, namely BFNs, denoted in vector notation as $\mathbf{S}$ having Q rows of underlying sources $\mathbf{s}_i$ with size of $Q \times M$, from the observed mixture $\mathbf{X}$ having P time points with size of $P \times M$, which can be written as:

$$\mathbf{X} = \mathbf{AS}, \tag{1}$$

where each column of $\mathbf{A}$ is called time course (TC) with the size $P \times Q$, the row vector $\mathbf{s}_i$ and $\mathbf{x}_i$ are with the same size $1 \times M$. Founding on formula (1), the dynamical BFNs identification in Quasi-GICA [5] has the following four steps: firstly, the sliding window technique [12] is applied to generate the sliding window subset i, namely $\hat{\mathbf{X}}_i = \left(\mathbf{x}_i^T, \cdots, \mathbf{x}_{i+l}^T, \cdots, \mathbf{x}_{i+L-1}^T\right)^T$, where the window length is equal to L, and $i \in [1, 2, \cdots, P - L + 1]$; secondly, the data compression for each sliding window subset $\hat{\mathbf{X}}_i$ is formulated as $\mathbf{Y}_i = \mathbf{F}_i^{-1}\hat{\mathbf{X}}_i$, where $\mathbf{Y}_i$ is the reduced data matrix with size of $N \times M$, $\mathbf{F}_i^{-1}$ is the N-by-L reducing matrix (determined by PCA), and N is the size of the retained time dimension. Considering the data compression on each $\hat{\mathbf{X}}_i, i \in [1, 2, \cdots, P - L + 1]$, the compressed data of $\hat{\mathbf{X}}$ can be formulated as $\mathbf{Y} = \left(\mathbf{Y}_1^T, \cdots, \mathbf{Y}_i^T, \cdots, \mathbf{Y}_{P-L+1}^T\right)^T$; thirdly, ICA estimation is implemented on $\mathbf{Y}$ to obtain BFNs, and formulated as follows:

$$\mathbf{Y} = \mathbf{M}\hat{\mathbf{S}}, \tag{2}$$

$$\hat{\mathbf{S}} = pinv(\mathbf{A}) * \mathbf{Y} \tag{3}$$

where $pinv()$ denotes the pseudo-inverse operation, $\mathbf{M}$ is a $(P - L + 1) * L$-by-N matrix, and $\hat{\mathbf{S}}$ is a N-by-M matrix with each row representing a BFN. To obtain the TCs (i.e., $\hat{\mathbf{A}}$), the multivariate linear regression is applied to the mixture $\mathbf{X}$, depending on the separated $\hat{\mathbf{S}}$, i.e.,

$$\hat{\mathbf{A}} = \mathbf{X} * pinv(\hat{\mathbf{S}}). \tag{4}$$

Finally, to obtain the dynamical BFNs and dynamical TCs from each sliding window subset (i.e., $\hat{\mathbf{X}}_i$), the dual regression is applied, and sequentially expressed as:

$$\hat{\mathbf{A}}_i = \hat{\mathbf{X}}_i * pinv(\hat{\mathbf{S}}), \tag{5}$$

$$\hat{\mathbf{S}}_i = pinv(\hat{\mathbf{A}}_i) * \hat{\mathbf{X}}_i. \tag{6}$$

2.2   Weighted BFNs Ranking Model

The random order of BFNs generated by ICA has negative effect on seeking biologically meaningful BFNs, which brings extra burden for manual selection of BFNs. To our knowledge, the meaningful BFNs firstly should have relatively high spatial reproducibility across different sessions and/or subsampled datasets [17, 19]. Also, the TCs corresponding to meaningful BFNs exhibit strong low-frequency fluctuations within the frequency band [0.01 Hz, 0.1 Hz] [6, 8, 20]. Thus, by assuming that the BFNs and TCs have different properties of dynamic spatial reproducibility (DSR) and dynamic fraction of amplitude low-frequency fluctuations (DFALFF), respectively, a novel weighted BFNs ranking model is proposed, where the dynamical BFNs with both high values of DSR and DFALFF will be ranked in front of those who have relatively low values. The DSR for the spatial map as to each BFN can be defined as follows:

$$DSR_1^k = \frac{2 * \sum corrcoef(\hat{\mathbf{S}}_i^k, \hat{\mathbf{S}}_j^k)}{(P-L+1)(P-L)}, \ \forall i,j \in [1,2,\cdots,P-L+1], i<j \qquad (7)$$

$$DSR_2^k = \frac{\sum corrcoef(\hat{\mathbf{S}}_i^k, \hat{\mathbf{S}}^k)}{(P-L+1)}, i \in [1,2,\cdots,P-L+1], \qquad (8)$$

where $\hat{\mathbf{S}}_i^k$, $\hat{\mathbf{S}}_j^k$ and $\hat{\mathbf{S}}^k$ denote the spatial maps of the *kth* BFNs from the sliding window subset $\hat{\mathbf{X}}_i$, $\hat{\mathbf{X}}_j$ and $\mathbf{X}$, respectively. The DSR_1^k in Eq. (7) emphasizes the mutually spatial reproducibility among the dynamical BFNs from the different sliding-windows, while DSR_2^k in Eq. (8) emphasizes the spatial reproducibility between the dynamical BFNs and the corresponding BFNs from the whole sliding-windows.

To calculate DFALFF values, the TCs are first extracted from $\hat{\mathbf{A}}$ and $\hat{\mathbf{A}}_i (i \in [1, 2,\cdots,P-L+1])$, and then manipulated by Fourier transformation according to Eq. (9) to obtain the power spectra $\hat{\mathbf{F}}$ and $\hat{\mathbf{F}}_i$, respectively.

$$F(\omega) = \int_{-\infty}^{+\infty} f(t)e^{-iwt}dt \qquad (9)$$

For a continuous frequency band [a, b], the energy at the specific interval is calculated as $Energy_{[a,b]} = \int_a^b F(\omega)d\omega$. The corresponding discrete formation of the energy is denoted as $Energy_{[\omega_0,\omega_n]} = \sum_{\omega=\omega_0}^{\omega_n} F(\omega)$, where $\omega \in [\omega_0, \omega_1, \cdots, \omega_n]$. According to the low-frequency fluctuations within band [0.01 Hz, 0.1 Hz] of BFNs [6, 8, 20], the dynamic FALFF values as to the power spectra $\hat{\mathbf{F}}$ and $\hat{\mathbf{F}}_i$ for the *kth* BFN can be formulated as follows:

$$DPS_1^k = \frac{1}{P-L+1} \sum_{i=1}^{P-L+1} \frac{\sum_{\omega=0.01}^{0.1} \hat{\mathbf{F}}_i^k(\omega)}{\sum_{\omega=0}^{\omega_n} \hat{\mathbf{F}}_i^k(\omega)} \qquad (10)$$

$$DPS_2^k = \frac{\sum_{\omega=0.01}^{0.1} \mathbf{F}_i^k(\omega)}{\sum_{\omega=0}^{\omega_n} \mathbf{F}_i^k(\omega)} \qquad (11)$$

The DPR_1^k in Eq. (10) emphasizes the DFALFF property of BFNs from the different sliding-windows, while DPR_2^k in Eq. (11) emphasizes FALFF property of BFNs from the whole sliding-windows.

For sorting the *kth* BFN, the dynamic spatial reproducibility, i.e., DSR_1^k and DSR_2^k, and the DFALFF, i.e., DPS_1^k and DPS_2^k, are first calculated, and then the weighted ranking coefficient (WRC) is expressed as

$$WRC_k = a_1 \times DSR_1^k + a_2 \times DSR_2^k + a_3 \times DPS_1^k + a_4 \times DPS_2^k, \tag{12}$$

where $\sum_{i=1}^{4} a_i = 1, \forall a_i, 0 < a_i < 1$. In this equation, each a_i is empirically set to $1/4$, emphasizing the equal contribution to the time-varying properties of DSR and DFALFF. It is worth noting that the larger WRC implies that the corresponding BFN has stronger properties of the dynamic spatial reproducibility and dynamic low-frequency fluctuations.

3 Experimental Tests

3.1 Experimental Datasets

(1) Simulation Dataset One subject dataset was generated by SimTB toolbox, with $V = 148 \times 148$ voxels, 12 spatial sources and 120 time points at time of repetition (TR) = 2 s. Two sources shared the task-related block modulation in addition to having unique fluctuations. Activation for the other ten sources was simulated based on solely unique hemodynamic fluctuations with no task-related variation. Additive noise was also included to reach a specified contrast-to-noise ratio of 1.0. This dataset was also used in the studies [5, 9, 10].

(2) Visual Task Dataset Three healthy subjects took part in a visual task. The visual stimulus paradigm was (OFF-ON) $\times 3$-OFF in 20-s blocks. The visual stimulus was a radial blue/yellow checkerboard, reversing at 7 Hz, corresponding to the "ON" state. At the "OFF" state, the participants were required to focus on the cross at the center of the screen. The BOLD fMRI dataset was acquired on a Philips 3.0 Tesla scanner with a single-shot SENSE gradient echo EPI with TR of 2.0 s. There were 40 slices providing the whole-brain coverage, with a SENSE acceleration factor of 3.0 and scan resolution of 80×80. The in-plane resolution was 3 mm $\times$ 3 mm. The slice thickness and slice gap were 3 mm and 1 mm, respectively.

3.2 Data Processing

For the simulated data, no preprocessing step was involved; as to the real data, the standard preprocessing procedure was performed in SPM8, including slice-timing, motion correction, spatial normalization and spatial smoothing with FWHM kernel equal to 8 mm. The MRIcroN software was used to display BFNs.

The Laplace approximation [21] was used to estimate the component number in Quasi-GICA [5]. The sliding window length L was empirically set to 70 and 40 time points for the simulation data and visual task data, respectively. For the visualization of BFNs from the real data, the z-score normalization was performed, and then the threshed maps was obtained with thresh value set to 2.0.

4 Results and Analysis

4.1 Results of Dynamic Low-Frequency Fluctuations

Figure 1A showed the different mean FALFF values for the twelve components estimated by Quasi-GICA across the sliding windows on the simulated data, implying that different BFNs had variable properties of the dynamic low-frequency fluctuations; also, the non-zero standard deviation (std) FALFF value for each BFN showed variant properties of the dynamic low-frequency fluctuations across different sliding windows. Figure 1B and C demonstrated the power spectra of the estimated BFNs with lowest (Comp7) and largest (Comp5) mean FALFF values, where the power spectra were estimated from the first sliding window and the last sliding window, respectively. Based on the results in Fig. 1A–C, we could draw a conclusion that the simulated BFNs had different properties of the dynamic low-frequency fluctuations; also, the block-task modulated BFNs, e.g., Comp5 and Comp3, had more consistently dynamic low-frequency fluctuations across all the sliding windows than those without task modulation.

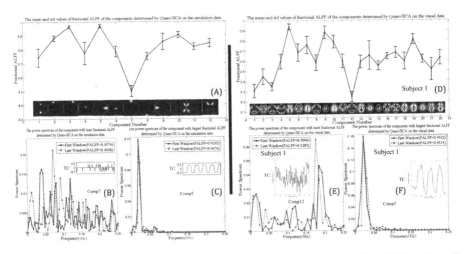

Fig. 1. (A)–(C) Dynamic properties of the low-frequency fluctuations regarding BFNs estimated from the simulated data; (D)–(F) Dynamic properties of the low-frequency fluctuations regarding BFNs estimated from the visual task data of Subject 1.

Figure 1D depicted the different mean FALFF values for the estimated BFNs by Quasi-GICA across the sliding windows on the visual task data of Subject 1. Figure 1E and F displayed the power spectra of the estimated BFNs with lowest (Comp12) and largest (Comp5) mean FALFF values, where the power spectra were estimated from the first sliding window data and the last sliding window data, respectively. According to the results in Fig. 1D–F, it can be concluded that the estimated BFNs had different properties of the dynamic low-frequency fluctuations, and the visual network modulated by visual stimulus, i.e., Comp5, was more consistent on dynamic low-frequency fluctuations across all the sliding windows than the others. Meanwhile, the results of the dynamic low-frequency fluctuations for Subject 2 and 3 also showed the similar phenomenon, which were not depicted for saving space.

4.2 Results of Dynamic Spatial Reproducibility

Figure 2A showed the mean and std values of DSR as to BFNs across the sliding windows on the simulated data, and Fig. 2B and C orderly displayed the normalized spatial maps of BFNs with lowest and highest DSR values, which were estimated from the first, middle and last sliding window data, respectively. According to the results in Fig. 2A–C, twelve estimated BFNs from simulation data showed different properties of dynamic spatial reproducibility, while the task-modulated BFN, i.e., Comp5, had the least spatial variance as the time went by, and Comp8 had the largest spatial variance across the sliding windows. This finding implied that the designed stimulus could strongly modulate the consistent patterns of brain activity in spatial domain.

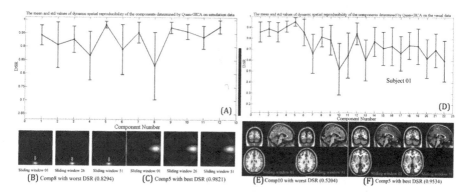

Fig. 2. (A)–(C) Dynamic properties of the spatial reproducibility regarding BFNs from the simulated data; (D)–(F) Dynamic properties of the spatial reproducibility regarding BFNs from the visual task data of Subject 1.

Figure 2D depicted the mean and std values of DSR as to BFNs across the sliding windows on the visual task data of subject 1, and Fig. 2E and F orderly displayed the spatial maps of BFNs with lowest and largest DSR values, which were estimated from the first and last sliding window data, respectively. Similarly, the visual stimulus modulated BFN, i.e., Comp5, had the strongest spatial reproducibility across the sliding windows, while Comp10 had the largest spatial variance as the time went by. This phenomenon implied that each BFN had different degrees of the time-varying spatial variance, and the task-modulated BFNs had strong spatial reproducibility across the sliding windows in contrast to the ones without modulation. Meanwhile, the similar performance in dynamic spatial reproducibility for Subject 2 and 3 was also found, whose results were not displayed for saving space.

4.3   Results of Weighted BFNs Ranking

Figure 3 depicted the sorted results of the estimated BFNs from simulation data, where two estimated BFNs sharing the same task stimulus (Comp5 and Comp3) were ranked in the top, with the largest WRC values, while Comp7 ranked at the bottom with least

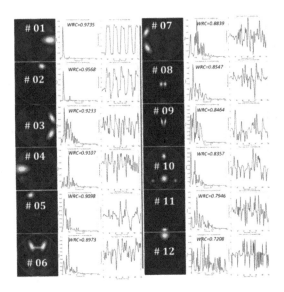

Fig. 3. The sorted order of the BFNs estimated from the simulated data.

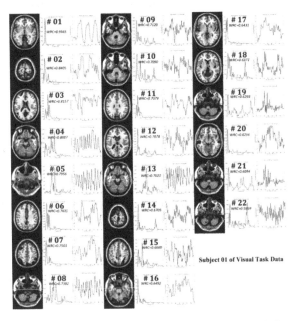

Fig. 4. The sorted order of the BFNs estimated from the visual task data of Subject 1.

WRC value, mainly due to its DFALFF value. Figure 4 showed the sorted results of the estimated BFNs from visual task data of Subject 1, where the estimated BFNs corresponding to the visual stimulus (Comp5) ranked at the top, with the largest WRC value, while Comp12 ranked at the bottom with least WRC value, mainly due to its lowest DFALFF value (see Fig. 2D). Based on the results from the simulated and real data, we could draw a conclusion that the proposed BFNs weighted ranking method could effectively sort the meaningful dynamical BFNs at the top.

5   Discussion

At beginning, taking the results of dynamic low-frequency fluctuations (Fig. 1) and dynamic spatial reproducibility (Fig. 2) of the estimated BFNs together, we could find that the BFNs were not always with consistently dynamic properties of spatial reproducibility and low-frequency fluctuations. For example, based on the results of the visual task data of Subject 1, Comp12 was with the lowest FALFF value, while Comp10 had the lowest DSR value. On one hand, this phenomenon demonstrated the complexity of dynamical BFNs in spatial and spectral domain. On the other hand, the good ranking results depicted in Figs. 3 and 4 showed the effectiveness of the weighted BFNs ranking model based on the dynamic properties of power spectrum and spatial variance, while most of ranking methods focused on the static BFNs ranking only, e.g., power spectrum ranking, SVM-based ranking, MMC ranking, etc. Moreover, in contrast to power spectrum ranking [13], the weighted BFNs ranking is not subject to task-related BFNs ranking, and is also useful for the resting-state data. Compared with SVM-based ranking [14], the weighted BFNs ranking is simpler, having no prerequisite training process. Compared to the multiple ICA-estimation-based ranking [15, 16], the only one-time ICA estimation is needed in Quasi-GICA. With respect to MMC ranking [17], the weighted BFNs ranking takes advantage of dynamical properties of spatial variance and power spectrum from the sliding windows, while MMC ranking uses the spatiotemporal reproducibility from the odd and even sampled dataset and needs at least two independent runs of ICA on sub-sampled dataset, which also suffers from some deficiencies, i.e., inaccurate estimation of order number on sub-sampled data, etc.

6   Conclusion

In this study, based on the dynamical BFNs tracking model, i.e., Quasi-GICA, a novel weighted BFNs ranking method was proposed, founding on the dynamic properties of spatial reproducibility in spatial domain and low-frequency fluctuations in spectral domain, aiming at determining the effective sorting orders of dynamical BFNs. Our results depicted and verified that different BFNs indeed had distinct dynamics in spatial variance and power spectrum in time-varying process, and the sorted results also demonstrated the effectiveness of this proposed model, which could sort BFNs with strongly dynamic spatial reproducibility and dynamic low-frequency fluctuations at the top.

Acknowledgments. This work is supported by Natural Science Foundation of China (No. 61701318), Medical Science and Technology Research Fund Project of Guangdong Province (No. A2017038), University Natural Science Research Project of Jiangsu Province (No. 18KJB416001) and Shenzhen Fundamental Research Project (No. JCYJ201703071 55304424).

References

1. Chang, C., Glover, G.H.: Time–frequency dynamics of resting-state brain connectivity measured with fMRI. Neuroimage **50**, 81–98 (2010)
2. McKeown, M.J., et al.: Analysis of fMRI data by blind separation into independent spatial components. Hum. Brain Mapp. **6**, 160–188 (1998)
3. Calhoun, V.D., Adali, T., Pearlson, G.D., Pekar, J.J.: A method for making group inferences from functional MRI data using independent component analysis. Hum. Brain Mapp. **14**, 140–151 (2001)
4. Wang, N., Zeng, W., Shi, Y., Yan, H.: Brain functional plasticity driven by career experience: a resting-state fMRI study of the seafarer. Front. Psychol. **8**, 1786 (2017)
5. Wang, N., Liu, L., Liu, W., Yan, H.: A novel automatic identification model for tracking dynamic brain functional networks at single-subject level. In: Proceeding of IEEE International Conference on Information and Automation (ICIA), Macau, China, pp. 505–510 (2017)
6. Fu, Z., et al.: Characterizing dynamic amplitude of low-frequency fluctuation and its relationship with dynamic functional connectivity: an application to schizophrenia. Neuroimage (2017). https://doi.org/10.1016/j.neuroimage.2017.09.035
7. Du, Y., et al.: Identifying dynamic functional connectivity biomarkers using GIG-ICA: application to schizophrenia, schizoaffective disorder, and psychotic bipolar disorder. Hum. Brain Mapp. **38**, 2683–2708 (2017)
8. Wang, N., Zeng, W., Chen, L.: A fast-FENICA method on resting state fMRI data. J. Neurosci. Methods **209**, 1–12 (2012)
9. Wang, N., et al.: WASICA: An effective wavelet-shrinkage based ICA model for brain fMRI data analysis. J. Neurosci. Methods **246**, 75–96 (2015)
10. Wang, N., Zeng, W., Chen, D.: A novel sparse dictionary learning separation (SDLS) model with adaptive dictionary mutual incoherence constraint for fMRI data analysis. IEEE Trans. Biomed. Eng. **63**, 2376–2389 (2016)
11. Esposito, F., et al.: Real-time independent component analysis of fMRI time-series. Neuroimage **20**, 2209–2224 (2003)
12. Kiviniemi, V., et al.: A sliding time-window ICA reveals spatial variability of the default mode network in time. Brain Connect. **1**, 339–347 (2011)
13. Moritz, C.H., Rogers, B.P., Meyerand, M.E.: Power spectrum ranked independent component analysis of a periodic fMRI complex motor paradigm. Hum. Brain Mapp. **18**, 111–122 (2003)
14. De Martino, F., et al.: Classification of fMRI independent components using IC-fingerprints and support vector machine classifiers. Neuroimage **34**, 177–194 (2007)
15. Himberg, J., Hyvärinen, A., Esposito, F.: Validating the independent components of neuroimaging time series via clustering and visualization. Neuroimage **22**, 1214–1222 (2004)
16. Yang, Z., LaConte, S., Weng, X., Hu, X.: Ranking and averaging independent component analysis by reproducibility (RAICAR). Hum. Brain Mapp. **29**, 711–725 (2008)

17. Zeng, W., Qiu, A., Chodkowski, B., Pekar, J.J.: Spatial and temporal reproducibility-based ranking of the independent components of BOLD fMRI data. Neuroimage **46**, 1041–1054 (2009)
18. Wang, N., Chang, C., Zeng, W., Shi, Y., Yan, H.: A novel feature-map based ICA model for identifying the individual, intra/inter-group brain networks across multiple fMRI datasets. Front. Neurosci. **11**, 510 (2017)
19. Wang, N., Zeng, W., Chen, D., Yin, J., Chen, L.: A novel brain networks enhancement model (BNEM) for BOLD fMRI data analysis with highly spatial reproducibility. IEEE J. Biomed. Health Inf. **20**, 1107–1119 (2016)
20. Zou, Q.H., et al.: An improved approach to detection of amplitude of low-frequency fluctuation (ALFF) for resting-state fMRI: fractional ALFF. J. Neurosci. Methods **172**, 137–141 (2008)
21. Minka, T.P.: Automatic choice of dimensionality for PCA. In: Advances in Neural Information Processing. Systems, pp. 598–604 (2001)

Bat Algorithm with Individual Local Search

Maoqing Zhang[1], Zhihua Cui[1(✉)], Yu Chang[1], Yeqing Ren[1],
Xingjuan Cai[1], and Hui Wang[2]

[1] Complex System and Computational Intelligence Laboratory, Taiyuan
University of Science and Technology, Taiyuan 030024, Shanxi, China
maoqing_zhang@163.com, 18834174241@163.com,
18435155956@163.com, xingjuancai@163.com,
zhihua.cui@hotmail.com
[2] School of Information Engineering, Nanchang Institute of Technology,
Nanchang 330099, China
huiwang@whu.edu.cn

Abstract. Bat algorithm (BA) is a well-known heuristic algorithm, and has
been applied to many practical problems. However, the local search method
employed in BA has the shortcoming of premature convergence, and does not
perform well in early search stage. To avoid this issue, this paper proposes a new
update method for local search. To verify the proposed method, this paper
employs CEC2013 test suit to test it with PSO and standard BA as comparison
algorithms. Experimental results demonstrate that the proposed method obviously outperforms other algorithms and exhibits better performance.

Keywords: Bat algorithm · Premature convergence · CEC 2013
Particle swarm optimization algorithm

1 Instruction

Nature-inspired computation is an umbrella for stochastic optimization algorithms by
simulating the nature phenomenon. Up to now, many algorithms have been proposed,
such as particle swarm optimization [1], ant colony optimization [2, 3], bat algorithm
[4], cuckoo search [5–7] and firefly algorithm [8, 9].

Bat algorithm (BA) is a novel population-based swarm intelligent [4]. Due to its
fast convergent speed, BA has been widely applied to many engineering problems,
including tracking problems [10], economic load dispatch [11], Detection of Malicious
Code [12], uninhabited combat aerial vehicle path planning [13], flow shop scheduling
[14, 15], and job shop scheduling problems [16].

There are numerous variants of BA, which have been greatly improved in term of
performance. Gandomi [17] introduced chaos into BA so as to increase its global
search capability for robust global optimization. To solve path planning which is a
complicated high dimension optimization problem, Wang [18] proposed a new bat
algorithm with mutation (BAM), and a modification was applied to mutate between
bats during the process of the new solutions updating. Fister [19] hybridized BA using
different DE strategies and applied them as a local search heuristic for improving the

Z. Shi et al. (Eds.): ICIS 2018, IFIP AICT 539, pp. 442–451, 2018.
https://doi.org/10.1007/978-3-030-01313-4_47

current best solution and directing the swarm towards the better regions within a search space. Cui [20] proposed three different centroid strategies and further combined them with BA. AI-Betar [21] studied six selection mechanisms to choose the best bat location: global-best, tournament, proportional, linear rank, exponential rank, and random. Cai proposed a triangle-flipping strategy to update the velocity of bats [22] and designed a optimal forage strategy to guide the search direction for each bat and employed a random disturbance strategy to extend the global search pattern [23].

In this paper, to enhance the convergence ability of BA, we proposed a modified BA called Bat Algorithm with Individual local search (IBA). In IBA, a new local search manner is introduced to enhance the convergence speed of each individual in early search stage. Then, to strength the convergence of the global best individual in later search stage, the standard local search method in BA is incorporated into IBA.

The rest of paper organized as follows: Sect. 2 gives a brief description of Bat algorithm. After that, the newly modified bat algorithm is presented. In Sect. 4, CEC2013 is employed to verify the proposed algorithm. Section 5 concludes the paper.

2 Bat Algorithm

In bat algorithm, there are many virtual bats in search space, while each bat flies to seek food according to the feedback of echoes. Suppose x_i^t and v_i^t are the position and velocity of bat i in generation t, then in the next generation, they are updated as follows:

$$v_i^{t+1} = v_i^t + (x_i^t - x^*) \times f_i \tag{1}$$

$$x_i^{t+1} = x_i^t + v_i^{t+1} \tag{2}$$

where x^* is the historical best position found by entire swarm, and f_i is the frequency randomly generated as follows:

$$f_i = f_{min} + (f_{max} + f_{min}) \times \beta \tag{3}$$

where f_{max} and f_{min} are the predefined maximum and minimum bounds of frequency, β is a random number uniformly generated from interval $[0, 1]$.

For some bats, they may move with the following local search manner:

$$x_i^{t+1} = x^* + \varepsilon \times \bar{A}' \tag{4}$$

where ε is random number uniformly generated from interval $[-1, 1]$, and $\bar{A}'$ is averaged loudness of all bats:

$$\bar{A}' = \frac{\sum_{i=1}^{n} A_i^t}{n} \tag{5}$$

If the position of bat i is updated, the loudness A_i^{t+1} and emission rate r_i^{t+1} are updated as follows:

$$A_i^{t+1} = \sigma A_i^t \tag{6}$$

$$r_i^{t+1} = r_i^0(1 - e^{-\gamma t}) \tag{7}$$

where $\sigma > 0$ and $\gamma > 0$ are pre-defined parameters.

The pseudocode of standard bat algorithm is described as follows:

Algorithm1: Standard bat algorithm

Begin
    Initialize position, velocity and other parameters for each bat
    While (Stop criteria is met ?)
        Randomly generates the frequency for each with Eq. (3)
        Update the velocity with Eq. (1)
        Update the position with Eq. (2)
    If rand> r_i^t
        Update the position with Eq. (4)
    End
    Calculate the fitness;
    If (rand< A_i^t)&&($f(x_i^t) < f(x^*)$)
        Replace the position with the new one
        Update r_i^t and A_i^t with Eq.(7) and Eq.(6)
    End
    Select the current global best position
    End
    Output the best position
End

---

3 Bat Algorithm with Individual Local Search

In standard BA, local search is performed with Eq. (4), and it can be expressed in Fig. 1, where x^* is the current global best position, x_{i-1}^t and x_i^t are population members, and the circles represent the potential better positions. From Fig. 1, we can find that Eq. (4) is mainly used to search potential better individuals around the current global best position. However, as we all know that, in the initial search stage, any position (even the current best position) has potential to be the best one. In other words, the individuals around the current global best position may not be better than those around regular positions. However, Eq. (4) may be effective in later search stage because the individuals around current global best position are likely to be the best ones compared with other regular positions.

To overcome the drawback in the early search stage, this paper proposes an improved version of Eq. (4), and it can be expressed with Eq. (8):

$$x_i^{t+1} = x_i^t + \varepsilon \times \bar{A}' \tag{8}$$

In the later search stage, position x_i^t is updated with equation [4]:

$$x_i^{t+1} = x^* + \varepsilon \times \bar{A}' \tag{4}$$

Figure 2 illustrates the mechanism of improved local search. From Fig. 2, we can obviously see that every position can be exploited sufficiently, and new potential better position can be found in early search stage. The newly modified BA can be named as Bat algorithm with individual local search (IBA for short). As for how to balance Eq. (8) in the search process, we employ a parameter r, which is a percentage of the largest generation. We will verify it in later experimental section.

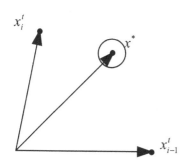

Fig. 1. Illustration of local search

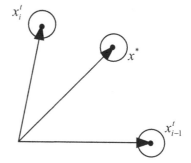

Fig. 2. Illustration of improved local search

The pseudocode of IBA can be described as follows:

Algorithm2: Bat algorithm with individual local Search

Begin
   Initialize position, velocity and other parameters for each bat
   While (Stop criteria is met?)
      Randomly generates the frequency for each with Eq. (3)
      Update the velocity with Eq. (1)
      Update the position with Eq. (2)
      If (rand> r_i^t)&&(rand<*r*)
            Update the position with Eq. (8)
      Else if (rand> r_i^t)&&(rand>=*r*)
            Update the position with Eq. (4)
      End
      Calculate the fitness;
      If (rand< A_i^t)&&($f(x_i^t) < f(x^*)$)
            Replace the position with the new one
            Update r_i^t and A_i^t with Eq. (7) and Eq. (6)
      End
      Select the current global best position
   End
   Output the best position
End

4   Experiments

To investigate the performance of IBA, we compare it with particle swarm optimization (PSO) [1] and standard bat algorithm (BA) [4]. By the way, CEC2013 test suit [24] is employed. CEC2013 contains of twenty-eight functions, including five unimodal functions, fifteen basic multi-modal functions and eight composition functions. The simulation is conducted in the machine with Intel Core i5-2400 3.10 GHz CPU, 6 GB memory, and windows 7 operating system with Matlab7.9.

For each algorithm, the size of population is set to 100, and the dimension of search space is 30. Each algorithm is run 51 times. For each algorithm, the following mean error is considered:

$$MeanError = \frac{\sum_{i=1}^{51} (f_{best}^j - f^*)}{51} \tag{9}$$

where f_{best}^j represents the fitness value of obtained best solution, f^* is the fitness value of true global optimum.

4.1   Investigation of Parameter r

Tables 1 and 2 present the experiment results with fixed parameter r ranging from 0.1 to 1.0. It is obviously that IBA with $r = 0.1$, 0.2 and 0.4 performs similar performance. To further investigate the difference, Friedman test is also employed, and the results are listed in Table 3. From Table 3, we can that BA with parameter $r = 0.4$ performs the best performance.

Table 1. Comparison results with different parameter r from 0.1 to 0.5

Function	0.1	0.2	0.3	0.4	0.5
F1	1.55E+00	1.22E+00	1.15E+00	1.10E+00	**8.60E-01**
F2	**4.08E+06**	4.32E+06	5.85E+06	8.33E+06	9.10E+06
F3	4.46E+08	**4.03E+08**	1.24E+09	1.45E+09	5.37E+09
F4	**3.43E+04**	3.50E+04	4.28E+04	4.06E+04	4.60E+04
F5	4.74E-01	**4.74E-01**	2.34E+01	6.54E+01	1.59E+02
F6	7.80E+01	**6.50E+01**	9.40E+01	1.10E+02	9.01E+01
F7	3.31E+02	2.17E+02	2.17E+02	2.78E+02	2.01E+02
F8	2.10E+01	2.10E+01	2.10E+01	2.10E+01	**2.10E+01**
F9	3.50E+01	3.33E+01	3.22E+01	3.31E+01	3.34E+01
F10	**1.30E+00**	1.37E+00	1.44E+00	2.03E+00	7.24E+00
F11	4.34E+02	4.08E+02	4.04E+02	**3.48E+02**	3.97E+02
F12	4.59E+02	4.34E+02	4.09E+02	3.73E+02	4.07E+02
F13	4.56E+02	**4.13E+02**	4.50E+02	4.42E+02	4.33E+02
F14	4.23E+03	4.34E+03	3.81E+03	3.92E+03	3.70E+03
F15	3.93E+03	4.14E+03	3.63E+03	4.02E+03	4.05E+03
F16	2.26E+00	2.22E+00	2.45E+00	2.15E+00	2.14E+00
F17	9.09E+02	**9.04E+02**	9.84E+02	9.69E+02	9.92E+02
F18	9.65E+02	9.35E+02	9.74E+02	**9.33E+02**	9.60E+02
F19	**6.70E+01**	7.10E+01	8.54E+01	1.13E+02	1.14E+02
F20	1.48E+01	**1.45E+01**	1.46E+01	1.46E+01	1.49E+01
F21	**3.04E+02**	3.94E+02	3.68E+02	3.41E+02	3.67E+02
F22	5.19E+03	4.73E+03	4.20E+03	4.41E+03	4.34E+03
F23	5.08E+03	4.90E+03	4.33E+03	4.48E+03	4.83E+03
F24	3.26E+02	3.25E+02	3.20E+02	**3.18E+02**	3.25E+02
F25	**3.39E+02**	3.44E+02	3.46E+02	3.48E+02	3.55E+02
F26	**2.00E+02**	2.00E+02	2.00E+02	2.00E+02	2.00E+02
F27	1.33E+03	1.27E+03	1.30E+03	**1.23E+03**	1.32E+03
F28	3.78E+03	**3.17E+03**	3.30E+03	3.25E+03	3.63E+03

4.2   Comparison with State-of-Art Algorithms

To further verify the performance of IBA with $r = 0.4$, IBA is further compared with PSO and standard BA, and the simulation results are listed in Table 4, while $w/l/t$

Table 2. Comparison results with different parameter r from 0.6 to 1.0

Function	0.6	0.7	0.8	0.9	1.0
F1	1.05E+00	1.03E+02	3.53E+03	1.49E+04	3.16E+04
F2	1.32E+07	2.31E+07	2.85E+07	5.54E+07	2.15E+08
F3	8.15E+09	1.49E+10	1.77E+10	4.97E+10	7.65E+11
F4	4.35E+04	5.34E+04	5.56E+04	5.57E+04	6.87E+04
F5	2.59E+02	4.68E+02	1.43E+03	2.66E+03	7.25E+03
F6	1.25E+02	1.64E+02	4.19E+02	9.58E+02	3.72E+03
F7	2.45E+02	5.12E+02	3.03E+02	4.65E+02	9.09E+02
F8	2.09E+01	**2.09E+01**	2.09E+01	2.10E+01	2.09E+01
F9	**3.13E+01**	3.37E+01	3.17E+01	3.47E+01	3.60E+01
F10	4.73E+01	1.44E+02	5.61E+02	1.52E+03	3.65E+03
F11	4.07E+02	4.39E+02	3.97E+02	3.78E+02	4.76E+02
F12	4.37E+02	3.69E+02	**3.66E+02**	3.85E+02	4.30E+02
F13	4.71E+02	4.22E+02	4.53E+02	4.59E+02	5.77E+02
F14	3.42E+03	3.78E+03	3.89E+03	**3.41E+03**	4.21E+03
F15	3.74E+03	**3.45E+03**	3.54E+03	3.95E+03	4.48E+03
F16	**2.12E+00**	2.51E+00	2.13E+00	2.39E+00	2.49E+00
F17	9.40E+02	9.80E+02	9.90E+02	1.09E+03	1.16E+03
F18	9.60E+02	9.60E+02	1.04E+03	9.70E+02	1.11E+03
F19	1.38E+02	1.48E+02	3.41E+02	6.74E+03	1.35E+05
F20	1.48E+01	1.49E+01	1.48E+01	1.50E+01	1.49E+01
F21	4.33E+02	3.40E+02	1.09E+03	1.93E+03	2.70E+03
F22	4.13E+03	4.39E+03	4.26E+03	**4.07E+03**	4.87E+03
F23	4.43E+03	4.27E+03	**4.16E+03**	4.50E+03	4.98E+03
F24	3.20E+02	3.25E+02	3.25E+02	3.33E+02	3.36E+02
F25	3.47E+02	3.44E+02	3.53E+02	3.45E+02	3.53E+02
F26	2.00E+02	2.00E+02	2.01E+02	2.03E+02	2.27E+02
F27	1.29E+03	1.26E+03	1.27E+03	1.24E+03	1.35E+03
F28	3.37E+03	3.53E+03	3.46E+03	3.53E+03	4.45E+03

Table 3. Friedman test for parameter r

r	0.1	0.2	0.3	0.4	0.5	0.6	0.7	0.8	0.9	1.0
Rankings	5.86	4.71	5.00	**4.46**	6.00	5.36	6.23	6.50	7.61	10.46

means that IBA win in w functions, lose in l functions and tie in t functions. Friedman test is also employed to verify their performance (please refer to Table 5). In one word, our modification achieves the best performance when compared with PSO and BA.

Table 4. Comparison results

Function	PSO	BA	IBA
F1	2.9600E+04	1.8300E+00	**1.1000E+00**
F2	4.0200E+08	**3.2600E+06**	8.3300E+06
F3	1.6500E+14	**2.2600E+08**	1.4500E+09
F4	5.2500E+04	**3.4400E+04**	4.0600E+04
F5	1.0400E+04	**4.7400E-01**	6.5400E+01
F6	4.5400E+03	**5.9900E+01**	1.1000E+02
F7	1.6000E+04	**1.7500E+02**	2.7800E+02
F8	**2.1000E+01**	**2.1000E+01**	**2.1000E+01**
F9	4.0500E+01	3.5700E+01	**3.3100E+01**
F10	3.9800E+03	**1.3200E+00**	2.0300E+00
F11	5.7200E+02	3.8100E+02	**3.4800E+02**
F12	5.1800E+02	4.0400E+02	**3.7300E+02**
F13	5.4300E+02	4.7300E+02	**4.4200E+02**
F14	8.2100E+03	4.6100E+03	**3.9200E+03**
F15	7.1900E+03	5.1700E+03	**4.0200E+03**
F16	2.5900E+00	2.3000E+00	**2.1500E+00**
F17	**7.2500E+02**	9.3100E+02	9.6900E+02
F18	**7.1200E+02**	9.4900E+02	9.3300E+02
F19	9.5200E+04	**6.1500E+01**	1.1300E+02
F20	1.5000E+01	**1.4600E+01**	**1.4600E+01**
F21	2.2800E+03	3.7000E+02	**3.4100E+02**
F22	8.5500E+03	5.6600E+03	**4.4100E+03**
F23	8.3800E+03	5.7300E+03	**4.4800E+03**
F24	3.7600E+02	**3.1200E+02**	3.1800E+02
F25	3.9200E+02	**3.4300E+02**	3.4800E+02
F26	2.4400E+02	**2.0000E+02**	**2.0000E+02**
F27	1.4900E+03	1.2600E+03	**1.2300E+03**
F28	4.4300E+03	3.7900E+03	**3.2500E+03**
$w\backslash t\backslash t$	25\2\1	14\11\3	

Table 5. Friedman test on comparison algorithms

Algorithm	Rankings
PSO	2.86
BA	1.63
IBA	**1.52**

5 Conclusion

In this paper, an individual local search manner is designed to enhance the exploitation. With this manner, the bats can make a deep search within their neighbours during the early search period. Simulation results show its effectiveness.

Acknowledgements. This work is supported by the National Natural Science Foundation of China under Grant No. 61663028, Natural Science Foundation of Shanxi Province under Grant No. 201601D011045 and Graduate Educational Innovation Project of Shanxi Province under Grant No. 2017SY075.

References

1. Eberhart, R., Kennedy, J.: A new optimizer using particle swarm theory. In: International Symposium on MICRO Machine and Human Science, pp. 39–43. IEEE (2002)
2. Dorigo, M., Birattari, M., Stutzle, T.: Ant colony optimization. IEEE Comput. Intell. Mag. **1**(4), 28–39 (2007)
3. Xu, B., Zhu, J., Chen, Q.: Ant Colony Optimization, New Advances in Machine Learning, pp. 1155–1173. InTech (2010)
4. Yang, X.S.: A new metaheuristic bat-inspired algorithm. Comput. Knowl. Technol. **284**, 65–74 (2010)
5. Yang, X.S., Deb, S.: Cuckoo Search via Lévy flights. In: Nature & Biologically Inspired Computing, NaBIC 2009, vol. 2010, pp. 210–214 (2009)
6. Zhang, M., Wang, H., Cui, Z., et al.: Hybrid multi-objective cuckoo search with dynamical local search. Memet. Comput. **10**(2), 199–208 (2018)
7. Cui, Z., Sun, B., et al.: A novel oriented cuckoo search algorithm to improve DV-Hop performance for cyber-physical systems. J. Parallel Distrib. Comput. **103**, 42–52 (2017)
8. Yang, X.S.: Firefly algorithms for multimodal optimization. Mathematics **5792**, 169–178 (2009)
9. Nasiri, B., Meybodi, M.R.: History-driven firefly algorithm for optimisation in dynamic and uncertain environments. Int. J. Bio Inspired Comput. **8**(5), 326–339 (2016)
10. Gao, M.L., Shen, J., Yin, L.J., et al.: A novel visual tracking method using bat algorithm. Neurocomputing **177**, 612–619 (2016)
11. Pham, L.H., Ho, T.H., Nguyen, T.T., Vo, D.N.: Modified bat algorithm for combined economic and emission dispatch problem. In: Duy, V.H., Dao, T.T., Kim, S.B., Tien, N.T., Zelinka, I. (eds.) AETA 2016. LNEE, vol. 415, pp. 589–597. Springer, Cham (2017). https://doi.org/10.1007/978-3-319-50904-4_62
12. Cui, Z., Xue, F., et al.: Detection of malicious code variants based on deep learning. IEEE Trans. Ind. Inf. **14**(7), 3187–3196 (2018)
13. Luo, Q., Li, L., Zhou, Y.: A quantum encoding bat algorithm for uninhabited combat aerial vehicle path planning. Int. J. Innov. Comput. Appl. **8**(3), 182–193 (2017)
14. Marichelvam, M.K., Prabaharan, T., Yang, X.S., et al.: Solving hybrid flow shop scheduling problems using bat algorithm. Int. J. Logist. Econ. Glob. **5**(1), 15–29 (2013)
15. Tosun, Ö., Marichelvam, M.K.: Hybrid bat algorithm for flow shop scheduling problems. Int. J. Math. Oper. Res. **9**(1), 125–138 (2016)
16. Dao, T.K., Pan, T.S., Nguyen, T.T., et al.: Parallel bat algorithm for optimizing makespan in job shop scheduling problems. J. Intell. Manuf. **29**(2), 1–12 (2015)
17. Gandomi, A.H., Yang, X.S.: Chaotic bat algorithm. J. Comput. Sci. **5**(2), 224–232 (2014)

18. Wang, G., Guo, L., Hong, D., et al.: A bat algorithm with mutation for UCAV path planning. Sci. World J. **6**, 418946 (2012)
19. Fister Jr., I., Fong, S., Brest, J., et al.: A novel hybrid self-adaptive bat algorithm. Sci. World J. **2014**(1–2), 709–738 (2014)
20. Cui, Z., Cao, Y., Cai, X., et al.: Optimal LEACH protocol with modified bat algorithm for big data sensing systems in Internet of Things. J. Parallel Distrib. Comput. (2017). https://doi.org/10.1016/j.jpdc.2017.12.014
21. Al-Betar, M.A., Awadallah, M.A., Faris, H., et al.: Bat-inspired algorithms with natural selection mechanisms for global optimization. Neurocomputing **273**, 448–465 (2017)
22. Cai, X., Wang, H., et al.: Bat algorithm with triangle-flipping strategy for numerical optimization. Int. J. Mach. Learn. Cybern. **9**(2), 199–215 (2018)
23. Cai, X., Gao, X., Xue, Y.: Improved bat algorithm with optimal forage strategy and random disturbance strategy. Int. J. Bio Inspired Comput. **8**(4), 205–214 (2016)
24. Liang, J.J., Runarsson, T.P., Mezura-Montes, E., et al.: Problem definitions and evaluation criteria for the CEC 2006. Technical report, Nanyang Technological University, Singapore (2006)

Ethics of Artificial Intelligence

Research on Artificial Intelligence Ethics Based on the Evolution of Population Knowledge Base

Feng Liu[✉] and Yong Shi

Research Center on Fictitious Economy and Data Science,
The Chinese Academy of Sciences, Beijing 100190, China
zkyliufeng@126.com

Abstract. The unclear development direction of human society is a deep reason for that it is difficult to form a uniform ethical standard for human society and artificial intelligence. Since the 21st century, the latest advances in the Internet, brain science and artificial intelligence have brought new inspiration to the research on the development direction of human society. Through the study of the Internet brain model, AI's IQ evaluation, and the evolution of the brain, this paper proposes that the evolution of population knowledge base is the key for judging the development direction of human society, thereby discussing the standards and norms for the construction of artificial intelligence ethics.

Keywords: Direction of evolution · Population knowledge base
Artificial intelligence ethics

1 Dilemma for Formation of Human and Artificial Intelligence Ethics

Like the cases of intelligence, consciousness, life and universe, it is difficult to give ethics a uniform definition. Generally, ethics refers to the principles and guidelines that should be followed when one deals with the people-to-people and people-to-society relationships, and it is a philosophical reflection on moral phenomena from a conceptual perspective. Ethics not only contains the behavioral norms in dealing with the people-to-people, people-to-society and people-to-nature relationships but also embodies the profound truth of regulating behaviors in accordance with certain principles [1].

Since the ethics is often related to culture, religion, region, values, and world view, there haven't been a unified, standard, and clear ethical system in thousands of years of human civilization history except for some principles that are basically acknowledged by people. The ethical impact from the artificial intelligence has become more prominent because of the incompleteness and controversy of ethical issues. The following are several examples related to ethics.

In the famous ethical thought experiment, i.e. The Choice of a Switchman, the train travels at a high speed and couldn't be stopped urgently. Right in front of the train is a forked rail. There are five abductees on the left and one abductee on the right of the rail [2]. Should the switchman choose to let the train go to the left or the right?

In 2017, Boston Dynamics arranged a scientist to give a test attack to the robot who was carrying boxes. As a result, the robot became unstable and fell down after attacked. The viewers who watched the video through the Internet protested that they violated the rights of robot. In this case, a debate about whether a robot has human rights or not was caused in society.

Behind these problems are profound ethical issues. The final choice will have an important influence and significance on the ethical construction of artificial intelligence in the future.

2 Inspiration of New Technological Advances on the Development Direction of Human Society

Since the 21st century, the Internet, cloud computing, big data, Internet of things, artificial intelligence, brain-like computing, and brain science have emerged. Human technology is experiencing another round of explosive growth. Where, the development of brain science, Internet, and artificial intelligence has provided a new perspective for finding and exploring the development direction of human society.

2.1 Evolution of Biological Brain for Hundreds of Millions of Years

For hundreds of millions of years, organisms have followed the principle of "survival of the fittest in natural selection" to form different types of life forms in order to adapt to the changes of the environment. Although the manifestations of organisms vary widely, the core of the organisms, i.e. the brain has shown obvious continuity. From single cell to human, the brain has become more and more complex and the level of intelligence is getting higher and higher.

John. C. Eccles, an Australian scientist, Nobel laureate, mentioned in his book Evolution of the Brain that "the brain of organisms has evolved from the brain of fish to the brain of reptiles, then to the brain of mammals and finally to the human brain. If a human brain is dissected, we can see the clear distinction between the structures of fish-likes, reptiles, and mammals in the human brain" [3].

Organisms reflect biological diversity through natural competition and natural selection. Giraffes have longer necks, gazelles run faster, and eagles have sharper eyes. In the structure of the human brain, the biological brain in the process of evolution shows a structure of one layer wrapped by another, just like the accumulated fossil.

2.2 Formation and Evolution of the Internet Brain

In 2008, the author of this paper and Professor Peng Geng et al. of the University of Chinese Academy of Sciences published the paper Trends and Laws of Internet Evolution by referring to the brain structure of neuroscience based on the emerging Internet-based brain phenomena, and proposed the model of the Internet brain which is used to explain the latest structure of Internet development [4] (see Fig. 1).

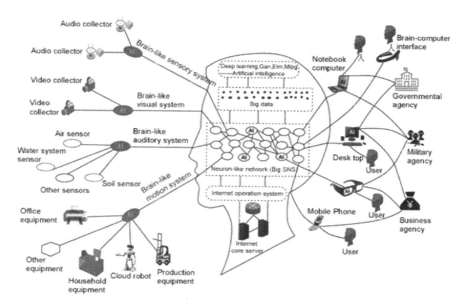

Fig. 1. Architecture of Internet brain

Definition of the Internet brain: The Internet brain is the brain-like giant system architecture formed during the evolution of the Internet to a high-similarity to the human brain. The Internet brain architecture has a brain-like vision system, auditory system, somatosensory system, motor nervous system, memory nervous system, central nervous system and autonomic nervous system. The Internet brain links various social elements (including but not limited to human, AI system, production materials and production tools) and various natural elements (including but not limited to rivers, mountains, animals, plants, and the space) through a neuron-like network. Driven by the collective wisdom and artificial intelligence, the Internet brain achieves the recognition, judgment, decision, feedback and transformation of the world through the cloud reflex arc.

Regarding the formation of the Internet brain-like giant system, the following can be inferred: When the organisms have evolved to human level, human will be combined with the Internet and achieve co-evolution. And the result of the co-evolution is: The Internet connected with the human is becoming highly similar to the brain step by step in terms of structure, and continuously expands in terms of space along with human expansion (see Fig. 2).

2.3   Division and Evolution of Intelligence Levels

In 2012, the author of this paper and Professor Shi Yong of the Chinese Academy of Sciences began to consider whether it was possible to evaluate the level of intelligence of the evolving Internet brain. This research topic was later extended to the research on

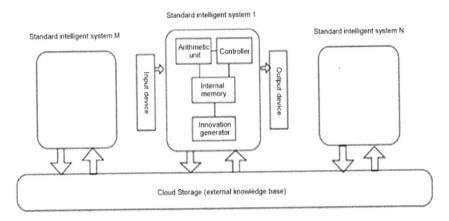

Fig. 2. Expanded von Neumann architecture

the evaluation of the intelligence level of artificial intelligence systems. The research difficulty is to build a model that can describe uniformly the intelligence characteristics of human life, robots, artificial intelligence systems and the Internet brain.

In 2014, the research was progressed with the reference to the von Neumann architecture, the David Wechsler human intelligence model, and the DIKW model system in the field of knowledge management. A standard intelligence model-Agent was built. It was proposed that any agent, including AI program, robot, human, and the Internet brain model can be described as an integrated system with knowledge input, knowledge mastering, innovation and feedback. The following figure shows the expanded von Neumann architecture added with innovation and cloud storage devices [5] (see Fig. 2).

The extended von Neumann architecture gives us important inspiration for division of intelligence levels. The criteria for judgment are as follows:

Can it interact with the tester (human), namely, is there an input/output system?

(1) Is there a knowledge base in the system that can store information and knowledge? (2) Can the knowledge base of this system be continuously updated and increased? (3) Can the knowledge base of this system share knowledge with other artificial intelligence systems? (4) Can this system actively generate new knowledge and share it with other artificial intelligence systems, in addition to learning and updating its own knowledge base?

In accordance with the above principles, we can have 7 intelligence levels of intelligent systems [5] (Table 1).

If based on the standard intelligence model, the following mathematical formula can be used to describe the state of omniscience and omnipotence (I-receipt of knowledge information, O-output of knowledge information, S-acquisition or storage of knowledge information, C-innovation of knowledge information) (formula 1)

Table 1. Intelligence grades of intelligent systems

Intelligence level	Description	Examples
Level 0	Intelligent systems in which, for example, the information input can be realized, but the information output cannot. Such intelligent systems exist theoretically, but not in reality	None
Level 1	Intelligent systems that cannot interact with human testers	Stones, sticks, iron pieces and water droplets
Level 2	Intelligent systems that can interact with human testers and have controller and memory, but their internal knowledge base cannot grow	Floor mopping robots, old-fashioned household refrigerators, air conditioners and washing machines
Level 3	Intelligent systems that have the features of level-2 intelligent systems, and the programs or data contained in the controller or memory of which can be upgraded or added without networking	Smartphones, home computers and stand-alone office software
Level 4	Intelligent systems that have the features of level-3 intelligent systems, what's the most important is that they can share knowledge and information with other intelligent systems through network	Google brain, Baidu brain, cloud robot and B/S architecture websites
Level 5	Intelligent systems that can conduct innovation, recognize and appreciate the value of innovation and creation to human, and apply the results of innovation to the development of intelligent systems	Human
Level 6	Intelligent systems that can constantly innovate and create intelligent systems that generate new knowledge, and their input and output capabilities as well as the abilities to master and use the knowledge will also approach infinity as the time moves forward and approaches to the point of infinity	The "God" of Eastern culture or the concept of "God" in Western culture

$$\text{Agent} \rightarrow Q_{\text{Agent}}, Q_{\text{Agent}} = f(\text{Agent})$$
$$Q\text{Agent} = f(\text{Agent}) = f(I,O,S,C) = a^{\neq}f(I) + b^{\neq}f(O) + c^{\neq}f(S) + d^{\neq}f(C) \quad (1)$$
$$a + b + c + d = 100\%$$
$$f(I) \rightarrow \infty, f(O) \rightarrow, f(S) \rightarrow \infty, f(C) \rightarrow \infty$$

3   Development Direction of Human Society – Judged with the Population Knowledge Base

Inference 1 Population Knowledge Base Evolving toward Omniscience and Omnipotence.

The evolution of the brain, the evolution of the Internet, and the division of intellectual levels of intelligence systems all show obvious Increasing attribute. Professor Nelson, an artificial intelligence pioneer, puts forward such a definition of artificial intelligence that: "Artificial intelligence is a discipline of knowledge, and the science about how to express knowledge, acquire knowledge and use knowledge" [6].

A common feature of these three areas is that they are all constantly promoting the loaded knowledge base and ability to use the knowledge.

No matter the evolution of the brain, the evolution of the Internet or the division of intellectual levels of intelligent systems, from the biological development history, it can be seen that the enhancement of knowledge and wisdom is the core of biological evolution. We can judge the direction of biological evolution and the level of biology from the perspective of the capacity of the population knowledge base and the ability to use the population knowledge base (PKB). When the time approaches infinity, the biological population reaches the "Point of Ω" through the evolution of the population knowledge base toward omniscience and omnipotence. The mathematical formula describing the increment speed of the population knowledge base (Fig. 3).

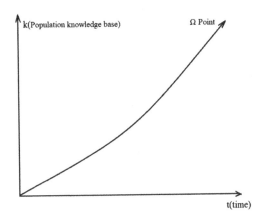

Fig. 3. The increment speed of PKB

The mathematical formula describing the increment speed of the population knowledge base (formula 2).

$$K = f(t) \tag{2}$$

The omega (Ω) point was proposed by Pierre Teilhard de Chardin (De Rijin), the famous French evolutionist philosopher, in the first half of the 20th century in the book The Phenomenon of Man. He used the last Greek letter (Ω) to express this ultimate status of human evolution-omniscience and omnipotence, known as Omega point [7].

Inference 2 Species Competing with Each Other through the Development of Population Knowledge Base.

The expansion speed of the population knowledge base and the ability to use the population knowledge base is the focus of biological evolution. The knowledge base of other organisms has stagnated and come to a dead end, without further change for thousands of years, so its status in the life circle of the earth is becoming lower and lower (Fig. 4).

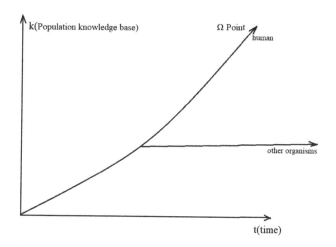

Fig. 4. Increment speed of PKB of human and other organisms

In the past 100 thousand years, human has continued to expand and accelerate its knowledge and wisdom, and made further great leaps because of the invention of the Internet and artificial intelligence, thus gaining the dominance in the natural competition of the earth.

$$K_{human} = f(t)$$
$$K_{otherorganisms} = g(t). \tag{3}$$

4 Discussion on the Construction and Decision of AI Ethics from the Perspective of Development Direction of Human Society

4.1 Taking the Evolution from the Population Knowledge Base as a Standard for Constructing AI Ethics

From the discussion in the third section, the promotion of knowledge and wisdom is the core of evolution for human society and even for organisms. We can judge the difference between the development direction of human society and other organisms from the perspective of the capacity of the knowledge base and the ability to use the population knowledge base. When the time approaches infinity, the biological population reaches the "point of God" through the evolution of the population knowledge base toward omniscience and omnipotence. This shows the development of human society has a direction and goal. In the process of life evolution, the behaviors and relationships that promote and protect the development of the population knowledge base are positive ethical rules, while those that hinder and harm the development of the population knowledge base are negative ethical rules (Fig. 5).

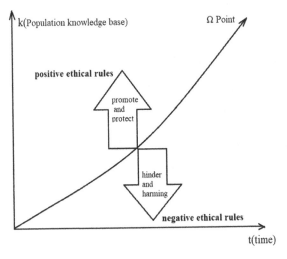

Fig. 5. Relationship between PKB and ethical rules

4.2 Discussion on the Decision of Viewing AI Ethics from the Perspective of Evolution

(1) In the movie of 2012, the president chose to let the scientist fly to a safe place and expressed that "a scientist is more important than dozens of officials", which was also based on the fact that a scientist is more important than a president to the future of human in terms of continuation and innovation of knowledge.

(2) In the Choice of a Switchman, can we judge which side contributes more to the future knowledge and wisdom of human? In the absence of a third choice and the ability to judge which side contributes more to the future knowledge and wisdom of human, choosing to allow more people (five) to survive should be a reluctant action.

(3) Regarding the ethical issue whether the Boston powered robot has human rights, from the construction of a standard intelligent model and the life evolution mentioned above, AI cannot be seen as a living entity with the same rights as human. It shares part of human knowledge and wisdom functions, but cannot replace human in the most important creativity and evaluation of creative value. More importantly, AI neither can determine its evolutionary direction and evolutionary goals, nor has the natural power for correct evolution. Its evolutionary power comes from human, so it is still one of human's tools (Fig. 6).

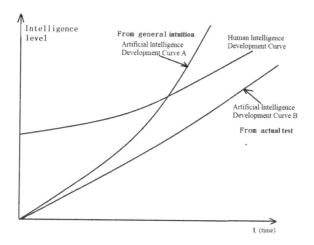

Fig. 6. Developmental curves of artificial and human intelligence

(4) No matter the great earthquake in 2012, the escape problem in Titanic or the puzzle of the switchman, all belong to the ethical issue concerning escape in extreme situations. However, in the vast majority of cases, individual interests should be protected in non-emergency situations and they should not be occupied by group interests. Allowing individuals to exert their initiatives in exploration should be a very important way for the promotion of the population knowledge base.

5 Summary

It is a new research topic to take the development of population knowledge base as a method for judging the development direction of human society and thereby establish an ethical standard for artificial intelligence, with respect of how the population knowledge base represented by human was formed and developed. If the ethic construction is measured according to whether the development of population knowledge base is promoted or hindered, then how to avoid the mistakes that social Darwinism once committed, and prevent the ethical standard based on population knowledge base from becoming a factor hindering the development of human society will be the worth-studying issues in the future.

References

1. Wang, H.M.: What is ethics. J. Yuxi Teach. Coll. **3**(2), 91–97 (2002)
2. Field, D.: What Can AI Do for Ethics?, pp. 161–170. Cambridge University Press, Cambridge (2011)
3. Jerison, H.J.: Evolution of the brain and intelligence. Evolution **29**(1), 190–192 (1975)
4. Liu, F., Peng, G.: Discovery and analysis of Internet evolution laws. China Science Paper Online (2008)
5. Liu, F., Shi, Y.: The search engine IQ test based on the Internet IQ evaluation algorithm. In: Proceedings of the 2nd International Conference on Information Technology and Quantitative Management, vol. 31, pp. 1066–1073. Procedia Computer Science (2014)
6. Lin, M.B.: Philosophical Thinking of Intelligent Machines, pp. 23–45. Jilin University Press, Changchun (2017)
7. Stock, M.E.: Pierre Teilhard de Chardin, Phenomenon of Man, pp. 134–136. Zhejiang People's Publishing House, Beijing (1989)

Does AI Share Same Ethic with Human Being?

From the Perspective of Virtue Ethics

Zilong Feng[✉]

Department of Philosophy, and of Religious Studies, Peking University,
No. 5 Yiheyuan Road Haidian District, Beijing 100871,
People's Republic of China
13210160038@fudan.edu.cn

Abstract. From the perspective of virtue ethics, this paper points out that Artificial Intelligence becomes more and more like an ethic subject which can take responsibility with its improvement of autonomy and sensitivity. This paper intends to point out that it will produce many problems to tackle the questions of ethics of Artificial Intelligence through programming the codes of abstract moral principle. It is at first a social integration question rather than a technical question when we talk about the question of AI's ethics. From the perspective of historical and social premises of ethics, in what kind of degree Artificial Intelligence can share the same ethics system with human equals to the degree of its integration into the narrative of human's society. And this is also a process of establishing a common social cooperation system between human and Artificial Intelligence. Furthermore, self-consciousness and responsibility are also social conceptions that established by recognition, and the Artificial Intelligence's identity for its individual social role is also established in the process of integration.

Keywords: Artificial Intelligence · Virtue ethics · Responsibility

1 Introduction

Imagine that the human world was to suffer from a war with AI, and they compromise after a long time. After the war, artificial intelligence (AI) establishes its own country in the Antarctic. Can we imagine the feature of its social system? In the economics of human society, we should assume some premises based on humanity to construct economics system, for example, market hypothesis or rational person. But in the field of ethics, the question becomes more complex. It is obvious in the framework of virtue ethics to admit that all kinds of virtues have its own premises which are decided by social conditions and history conditions. Of course, what kind of factors are the most important may be vary from different scholars. Karl Max and Macintyre may have different opinions on the status of the contradictions between economic and superstructure, but they all convey the same opinion that we cannot think of a kind of system of ethics without its special social and history conditions. On the other hand, we also cannot think of the social and history conditions of ethics without considering humanity. However, it is quite difficult to definite the so-called humanity of AI. How

Z. Shi et al. (Eds.): ICIS 2018, IFIP AICT 539, pp. 465–472, 2018.
https://doi.org/10.1007/978-3-030-01313-4_49

can we give a set of human moral laws to AI by programming without permission from AI? Is it fair to give the set of human moral laws—just as the Isaac Asimov's three laws of robotics—to AI? The viewpoint of this paper is that answers to these questions are depended on the degree of the AI robot's involvement into human society. I want to prove the opinion with the basic theory structure from the virtue ethics of Macintyre.

2   The Problem of Abstract Moral Law and AI

There is a granted idea that we could program abstract moral laws into the codes of robots. Maybe the most famous moral laws given in the history are Isaac Asimov's three laws of robotics. And the three laws are listed as follow:

1. A robot may not injure a human being or, through inaction, allow a human being to come to harm.
2. A robot must obey orders given it by human beings except where such orders would conflict with the First Law.
3. A robot must protect its own existence as long as such protection does not conflict with the First or Second Law.[1]

However, the kind of idea may face the counter-argument which holds the idea that moral laws cannot perform without considering concrete situations—a kind structure of argument which can always be seen when you want to find some counter-argument for Kant's moral theory. The kind structure of counter-argument for Kant can be characterized more clearly when related with a problem of AI, the so-called framework problem.[2] Framework problem claims that AI is sensitive to variables around them.

Another argument for the idea to apply existing moral laws today into the area of AI or in the future society and this kind of argument can be called 'technological transparency', which means that fundamental moral laws can stay unchangeable during the developing process of technology.[3] The basic argument of technological transparency is that technology is only a kind of means which should not take the responsibility and the subject of responsibility can only be human. And the counter-argument for "technological transparency" can be more easy today, because with the development of AI, the subjective status of AI becomes not so difficult to answer. AI now becomes more and more intelligent, and it is so difficult to test whether it is really not autonomy. At least in the process of deep learning, programmer now more and more put their attention to adjust parameter rather than instruct AI directly. The process of deep learning is a black box process. According to the embodied cognition, the conceptual structure is influenced by the body of organism.[4] The consciousness criteria

[1] Wikipedia, https://en.wikipedia.org/wiki/Three_Laws_of_Robotics, last accessed 2018/08/07.

[2] B. Meltzer & Donald Michie (eds.): Machine Intelligence 4. Edinburgh University Press, Edinburgh (1969), pp. 463–502.

[3] Blay Whitby: Sometimes it's hard to be a robot: A call for action on the ethics of abusing artificial agents. J. Interacting with Computers, Volume 20, Issue 3 (2008), pp. 326–333.

[4] Varela, Francisco J., Thompson, Evan T., and Rosch, Eleanor: The Embodied Mind: Cognitive Science and Human Experience. The MIT Press, Cambridge, Massachusetts (1991), p. 4.

of human cannot be generalized. Today, whether the AI could possess ownership of property is not so absurd to talk about. If it is possible, maybe we also need to consider whether it is necessary to give them social insurance.

In fact, attitudes towards the responsibility of AI have altered and more and more scholars are now beginning to consider it as a serious question. Part of the reason is that the combination of biology and computer science, but the most important is that freedom are always related to responsibility. Now the freedom of robots become more than ever, especially when the freedom relates to its specialist areas, and maybe the best example is automatic drive. Colin Allen and Wendell Wallach argues that whether the subject of action can become the subject of moral depends on two factors: autonomy and sensitivity. If the two factors are very high, then we can say that the AI is more like a subject of moral rather than a tool. In such condition, the responsibility belongs to the AI rather than the operator of the AI except for in some special areas, according to Gert-Jan Lokhorst and Jeroen van den Hoven, such as military robots in the war.[5] Colin Allen and Wendell Wallach believes that we can make the kind of robot which can be the subject of moral in the future although it is very hard today.

With the development of AI, scientists now are developing a kind of being which can take its own responsibility in some sense. This kind of AI has high sensitivity and autonomy, so they are subject of moral in some degree. What human should face is not only a tool but a rational being. The foundation of moral for the two situations are different. If robots are only tools that be controlled by human, then the same argument structure can be applied to the problem of ethics of robot. However, now AI is in some sense a kind of being which has the ability to make its own choice, is it still adaptable and justice to coerce them to accept human ethics?

It is important to reconsider the norm between human and AI with the thoughts of Macintyre. As Macintyre had mentioned when he talk about virtue ethics, every kind of moral system has its own historical and social premises. The norm between human and AI has to be established on the base of the thought.

3 Three Stages of Virtue Ethics of Macintyre and AI

At first, I want to introduce the opinions of Macintyre briefly which is rarely mentioned in the analysis of robot's ethics. It is very complicated to category all kind of virtue ethics, for example, according to The Stanford Encyclopedia of Philosophy, virtue ethics can be defined in four ways.[6] In this paper, I will discuss ethics of AI with the thoughts of Macintyre who is one of the most famous scholars in this field. I will introduce the thoughts of Macintyre in short and explain the foundation of the relationship between its thoughts and AI' ethics. There are three stages of Macintyre's thoughts in this book—his theory of practice, theory of narrative and what makes up a tradition of moral.

---

[5] Lin, Patrick: Robot Ethics, The Ethical and Social Implications of Robotics. The MIT Press, Cambridge, Massachusetts (2012), p. 154.

[6] Stanford Encyclopedia of Philosophy, https://plato.stanford.edu/entries/ethics-virtue/, last accessed 2018/08/07.

We have to define what is the so-called virtue according to the text of *After Virtue: A Study of Moral Theory*. In this book, Macintyre have analyzed the conception of virtue in history to get a universal definition of virtue, and his virtue ethics established on the universal definition. The most important feature of the conception of virtue is as follow:

One of the features of the concept of a virtue which has emerges with some clarity from the argument so far is that it always requires for its application the acceptance for some prior account of certain features of social and moral life in terms of which it has to be defined and explained.[7]

For example, he lists three kinds of virtues in history, and the representations are Homer, Aristotle and Franklin. The virtues of them can be summarized as social role, the achievement of *telos*, the utility in achieving earthly and heavenly success.[8] All of three conceptions above are rooted in their social background. Of course, the influence of the Karl Marx is very obvious in this kind of viewpoint of Macintyre. The virtue of Macintyre's definition is historical and it cannot be separated from its social background. The question of ethics of AI is at first a question of social problem rather than a technology problem. Only if we can analyze its social background then we can find the basement of ethics of AI. Of course, this is not to say that it is not important at all to pay attention to the Kant's style of moral theory, just as many scholars have done so far. They designed the moral system of AI with adding to it a moral module. But this kind of module only reveals abstract conditions of moral. The thoughts of virtue ethics should be mentioned when we want to talk about particular questions.

The aim of virtue ethics is constrained by history and society. In order to argue this influence, Macintyre introduces the conception of practice and this is the first stage of his virtue ethics. His conception of practice is as follow:

By a "practice" I am going to mean any coherent and complex form of socially established cooperative human activity through which goods internal to that form of activity are realized in the course of trying to achieve those standards of excellence which are appropriate to, and partially definitive of, that form of activity, with the result that human powers to achieve excellence, and human conceptions of the ends and goods involved, are systematically extended.[9]

We can see from his words that this conception is at first a social conception, and it is used to describe the cooperation relationship in human society. With the formation of relationship, the notion of goods internal to social activities can be defined and admitted. Goods internal to social activities is beneficial to all members who participated in the practice. Moreover, he also mentions as follow:

A practice involves standards of excellence and obedience to rules as well as the achievement of goods. To enter into a practice is to accept the authority of those standards and the inadequacy of my own performance as judges by them.[10]

---

[7] Alasdair Macintyre: After Virtue: A Study of Moral Theory. University of Notre Dame Press, Notre Dame, Indiana (2007), p. 186.

[8] Ibid., p. 185.

[9] Ibid., p. 187.

[10] Ibid., p. 190.

That is to say, internal goods and practice establish the regulations and criteria of ethical community. If AI wants to live in human society as an ethical subject, then at first it should be integrated into the practice of human society and have the common internal goods with human. The fear that AI may threaten to human's existence can be resolved in this social cooperation system.

The second stage succeeds with the problem of first stage. Although the practice in the first stage provides the social background of virtue, the conflicts among individuals still exist. In order to solve the problem, Macintyre puts forward the theory of the unity of human life. This theory emphasized that everyone's life is a unity, and this unity provide a full *telos* to ethics. The unity of life comes from "the unity of a narrative embodied in a single life".[11] To realize the unity of personal life has to answer what is good to human. Macintyre dose not describe the method of judging good, but according to his text, a good life has the characteristic of internal goods to practice. We cannot understand it without describing the process of seeking a good life, and virtue must be defined in the same process. Just as Macintyre has said as follow:

The good life for a man is the life spent in seeking for the good life for man, and the virtues necessary for the seeking are those which will enable us to understand what more and what else the good life for man is.[12]

If AI wants to get the status of moral subject, it has to have a united life which has a teleology structure. For a robot who has the ability to choose with at least limited reason, the kind of choice has to be restricted by the third stage.

The third stage of Macintyre's theory wants to unite the social person with historical practice. For Macintyre, the subject cannot be separated from its history and society. The subject is something inherited from its past, as he has said:

What I am, therefore, is in key part what I inherit, a specific past that is present to some degree in my present.[13]

For the historical practice, Macintyre says as follow:

It was important when I characterized the concept of a practice to notice that practice always have histories and that at any given moment what a practice is depends on a mode of understanding it which has been transmitted often through many generations.[14]

The unity of a narrative embodied in a single life is in the historical social practice, and social practice constitutes a relative open tradition which faces the future. The person in the history tradition is not only constricted by the tradition, but also creates the tradition. For AI robots, if they are not tools but something who can take responsibility, then they are rolled in a kind of tradition and they have to take its responsibility for the tradition. The aim of AI robots' lives also relies on this tradition. Human as members of the same ethical community have right to require AI robots to take the choice which is good for the community. That is to say, AI robots have the responsibility to get the education in an ethical community.

---

[11] Ibid., p. 218.

[12] Ibid., p. 219.

[13] Ibid., p. 221.

[14] Ibid., p. 221.

According to Macintyre's thoughts above, every kind of ethical community has to establish itself by a social cooperation system. This kind of social relationship has its own internal goods, and virtues also depend on it. The historical subject who has its unity of narrative is in the social cooperation system. The members of the ethical community share a common ethics system in this sense. If AI really shares a common ethics system with human, they have to take part in the same ethical community in the same way. This kind of theory model can solve some problems of abstract moral module especially in the moral dilemma (as the Trolley dilemma showed), because it not only pays attention to action compared with abstract moral module. The importance of action is not so high because virtue can be potential, and it can be a kind of character even though it is not actualized. The kind of characteristics of virtue ethics are related with social role, and the following part of the paper will resolve the problem of how to identify individual and its social character in a society.

4   AI and the Recognition of Society

The question of self-consciousness is not only an important issue when we talk about AI, but also the core question in the modern history of philosophy. In the field of philosophy, the conception of I is not only discussed in the cognition field, but also is a conception related to society and ethics. When we talk about self-consciousness, we should not ignore the dimension of society. The phenomenon of self-consciousness cannot be explained without considering social dimension. The interaction of human and AI robots is the social premise of self-consciousness of AI robot, and it is also the premise of ethics of AI robots. To talk about ethics of AI robot usually relates to autonomy, teleology or responsibility. However, all of these conceptions are related to self-consciousness. Self-consciousness is not only a physiological phenomena, and it should be explained with the dimension of society. Even if the AI technology can make a real brain, it does not mean that the brain can really have self-consciousness. Because recognition is an important process in the establishment of self-consciousness.

We cannot have self-consciousness without being in human society. In the field of AI, conflicts such as individual and community also exist, and the difference of ethics and moral also should not be forgotten. Axel Honneth points out that ethics should be seen as follow:

The concept of "ethical life" is now meant to include the entirety of intersubjective conditions that can be shown to serve as necessary preconditions for individual self-realization.[15]

Individuals must know that they are recognized for their particular abilities and traits in order to be capable of self-realization, they need a form of social esteem that they can only acquire on the basis of collectively shared goals.[16]

---

[15] Axel Honneth: The Struggle for Recognition: The Moral Grammar of Social Conflicts. translated by Joel Anderson, the MIT Press, Cambridge, Massachusetts (1995), p. 173.

[16] Ibid., p. 178.

Society is constructed by all kinds of recognition, such as love and law. And individuals can deepen their recognition to itself in this process of recognition. Authors of science fiction and screenplays of science film are maybe more clear about this point. In many science fictions and movies, human and robots establish recognition to each other in the process of love and war, and all of these processes can be seen as the construction process of recognition. In the process of recognition, self-consciousness of AI is established too.

Of course, in the theory framework of technology of transparency, the opinions above are too weird. Because to the viewpoint of technology of transparency, AI robot is still something rather than a relative rational subject, so they belong to human and the responsibility of their behavior should also be ascribed to human too. The ethics of robot are still the extension of the ethics of human in the framework of technology of transparency. But this viewpoint has too many problems with the developing autonomy of AI robot now. Maybe it sounds so weird that a computer should be responsible with you. In fact, according to the outcome of questionnaire designed by Jamy Li in 2016, many people opposed to see autonomous car as the subject of responsibility.[17] But in the field of law, it is not so weird for a nonhuman subject to take responsibility. In fact, legal person such as company has the independent legal personality to take responsibility on many legal actions just as human. In the present framework of law, the possibility of designing a regulation about taking part responsibilities by AI robot still exists. It is too cruel for companies that produce autonomous car and buyer to take all responsibilities of autonomous car, and it could reduce the passion to develop such technologies. This is also the reason why this paper does not hold the point that it is too early to talk about the responsible subject of AI robots. The key point is the question of recognition, and law system is also the appearance of recognition.

5   Conclusion

The viewpoint that we can solve the ethical problem of AI by programming an abstract moral module has many theory problems. With the development of AI's autonomy and sensitivity, AI robots should be at least seen as part responsible subjects. From the framework of Macintyre's virtue ethics, the relationship of AI and human should be considered with premises of society and history. Of course, the dimension of recognition is also an important element which cannot be ignored when we consider the question of self-consciousness of AI which is related to the legal and moral status of AI. It is at first a social integration question when we talk about AI's ethics, and this is the premises of designing the moral module to AI robots. If we want to answer the question of AI robots, especially the ethical relationship between AI robots and human,

---

[17] Li, Jamy, Xuan Zhao, Mu-Jung Cho, Wendy Ju, and Bertram F. Malle: From Trolley to Autonomous Vehicle: Perceptions of Responsibility and Moral Norms in Traffic Accidents with Self-Driving Cars. In: SAE 2016 World Congress and Exhibition, SAE Technical Paper 2016-01-0164(2016), https://doi.org/10.4271/2016-01-0164.

we have to know that in what kind of degree AI can share the same ethics system with human equals to the degree of its integration into the narrative of human's history, and the degree of its integration into human's social cooperation system with goods internal to practice.

References

Macintyre, A.: After Virtue: A Study of Moral Theory. University of Notre Dame Press, Notre Dame (2007)

Honneth, A.: The Struggle for Recognition: The Moral Grammar of Social Conflicts. The MIT Press, Cambridge (1995). Translated by Joel Anderson

Varela, F.J., Thompson, E.T., Rosch, E.: The Embodied Mind: Cognitive Science and Human Experience. The MIT Press, Cambridge (1991)

Meltzer, B., Michie, D. (eds.): Machine Intelligence 4. Edinburgh University Press, Edinburgh (1969)

Whitby, B.: Sometimes it's hard to be a robot: a call for action on the ethics of abusing artificial agents. J. Interact. Comput. **20**(3), 326–333 (2008)

Wikipedia. https://en.wikipedia.org/wiki/Three_Laws_of_Robotics

Lin, P.: Robot Ethics, The Ethical and Social Implications of Robotics. The MIT Press, Cambridge (2012)

Stanford Encyclopedia of Philosophy. https://plato.stanford.edu/entries/ethics-virtue/

Li, J., Zhao, X., Cho, M.-J., Ju, W., Malle, B.F.: From trolley to autonomous vehicle: perceptions of responsibility and moral norms in traffic accidents with self-driving cars. In: SAE 2016 World Congress and Exhibition, SAE Technical Paper 2016-01-0164 (2016). https://doi.org/10.4271/2016-01-0164

"Machinery Rationality" Versus Human Emotions: Issues of Robot Care for the Elderly in Recent Sci-Fi Works

Lin Cheng[1(✉)] and Yiyi He[2]

[1] Department of German Studies, Guangdong University of Foreign Studies,
Bai Yun Da Dao Bei 2 Hao, No. 4 Building, Room 203, Guangzhou 510420,
People's Republic of China
lin.cheng@gdufs.edu.cn
[2] Cultural Studies, Queens University,
B176 Mackintosh-Corry Hall, Kingston, ON K7L 3N6, Canada
16yh58@queensu.ca

Abstract. In recent years, the phenomenon that robots are used for elderly care in our daily life has drawn attention of the public and the media. It emerges as a new attempt to solve the issue of how to provide for the aged after their retirement. Since 2012, this phenomenon has become the subject of the Sci-Fi movies and TV-play series. In those Sci-Fi works, however, the traditional dystopian human-robot conflicts are partly replaced by the prospect of human-robot co-existence. The robots here launch five ethical challenges: the issues of safety versus privacy, human-robot duality, machinery rationality versus human emotions, affective interaction, and the ethical responsibility for the elderly in the human-robot interaction. This essay scrutinizes the phenomenon of the elderly-care Robots in three recent Sci-Fi movies/TV series, with a focus on the theme of the machinery rationality versus human emotions. In the human-robot interaction, human emotion, which makes us who we are, has been magnified with the perfectly designed robot rationality as a frame of reference. Thus the discussion of the ethical tension and conflict involved within these two typical groups would be particularly imminent and significant.

Keywords: Machinery rationality · Sci-Fi · HRI · Elderly-care robots
Human emotions

1 Introduction

In recent years, with the rapid development of modern AI technology, and as a result of the increasing proportion of aging people in the whole population and the lack of carers, robots have been created to facilitate people's work as man's extended hands and they could help release the caring pressure of the family and society. In reality, they have been put to use already in attending the elderly, the sick and the disabled.

What is practical and useful is not necessarily ethical. That robots are entering our daily life poses ethical challenges, even protests. Here is an extreme case: in the Swedish science fiction (TV-Series) *Real Humans* (2012/2014), the "human-robot co-

© IFIP International Federation for Information Processing 2018
Published by Springer Nature Switzerland AG 2018. All Rights Reserved
Z. Shi et al. (Eds.): ICIS 2018, IFIP AICT 539, pp. 473–481, 2018.
https://doi.org/10.1007/978-3-030-01313-4_50

existence society" is already established in the post-human era in Sweden and there are some people begging for "Hubot-free elder care" (S.1, E.3, hubot = *hu*man + ro*bot*, a term for robot in this series). This paper talks about the HRI (human-robot interaction) and the ethical challenges of elderly care robots, more specifically, the characteristics of human beings in front of robots.

1.1   A Brief Literature Review

As a budding branch of applied ethics since 2004, the "Roboethics" has aroused a heated discussion among various disciplines. This paper focuses on the most common concerns centering robotics and roboethics especially in aged-care. Gianmarco Veruggio and Fiorella Operto have written an article which traces the genealogy of Robotics and Roboethics and provides a "detailed taxonomy" that "identifies the most evident/urgent/sensitive ethical problems in the main applicative fields of robotics" (1499). Vandemeulebroucke et al. have conducted a systematic review of aged-care robots in argument-based ethics literature, claiming "all stakeholders in aged care, especially care recipients, have a voice in ethical debate" (15). Amanda Sharkey and Noel Sharkey analyze six main ethical concerns in this topic. Sparrow and Sparrow also show similar concern over aged-care robots as "simulacra" that will deprive human beings of real social contact (141–161). Borenstein and Pearson remain a neutral stance towards the deployment of a companion robot (i.e., the seal robot Paro), which may help ease human loneliness and estrangement (277–288). A randomized controlled trial also proves that Paro does possess some edges in enhancing social interactions (Robinson et al. 2013). However, the elderly's attitudes toward Paro are mixed (Robinson et al. 2015, 2016). Moreover, Salvini et al. arouse discussions of ethical issues from five case studies in biomedical domains so as to promote human welfare. Lamber Royakkers and Rinie van Est has conducted a review in which they argue care robots will shift the responsibilities of caregivers and give them "a new role" (554).

However, the above-mentioned studies are more or less anthropocentric, the major intent of which is to improve human interests while neglecting the diversity of robot characteristics. Only a few researches have reached beyond the scope of social-technological usability of robots and seen a bigger picture from the fictional narratives. For instance, from a psychological perspective, Elizabeth Broadbent brings the discussion of fiction and reality into public attention. She slightly mentions the film *Robot&Frank* (2012), in which the care-robot has been given far greater potentials than they have in real life (629). Potential threats that these care robots can bring to humans are also highlighted (629–646). More specifically from the perspective of Sci-Fis, in *Roboethics in Film*, a collection of essays that concerns some key ethical issues in human-robot relationship is anthologized. It is one of the few works that touches upon roboethics on the screen, though is incapable to reveal some new phenomena of robot Sci-Fis (e.g. aged-care issue). Also notably, Norman Makoto Su et al. discuss robot issues and try to understand the definition of a healthcare robot through a lens of online YouTube videos, which is also a good attempt to approach the ethical issues of robots.

Unlike what has already been done, this essay scrutinizes the phenomenon of the aged-care robots in recent Sci-Fi works, that is, the potential ethical challenges that may come along in the human-robot relationship and specially the issue of machinery rationality versus human emotions. First, why Sci-Fi works are worth exploring here will be explained as follows.

1.2   Five Ethical Issues in the New Era of HRI

The phenomena of elderly-care robots in fictional movie/TV plays and non-fictional reality have caused different ethic responses. The Sci-Fis with robots can offer various pictures of HRI (human-robot interaction) and human-robot co-existence which have a potential connection with reality. To some degree, they are forward-looking, inspirational, public and approachable for every audience. There is no doubt that Sci-Fis can illustrate the richness, diversity of robotics and the topics for futurology, and thus broaden our cognitive space. Moreover, they have fixed texts with great imagination and openness, serving as ideal research resources. Therefore, they are not something unserious and far from reality, but ideal research materials for the in-depth explorations and thus are irreplaceable for this field.

Since 2012, approximately at the same time with the usage of elderly-caring robots in the nursing house, the phenomenon of the thriving of elderly-care robots in the domestic use has been represented in the fictional movie/TV plays. Here is a list of the Sci-Fis with elderly-caring robots as main topic or one of the main topics:

- Movie: *Robot&Frank* (USA, 2012),
- TV series: *Real Humans* (SWE, S.1: 2012/S. 2: 2014),
- TV series: *Humans* (UK/USA, S.1: 2015), the English adaptation of *Real Humans*,
- 3D Animation: *Changing Batteries* (MAS, 2012).

This paper discusses the phenomenon of the aged-care robots in the first three Sci-Fi works which mainly focus on the theme of "machinery rationality" versus human emotions. The important aged-care robot roles are e.g.: The anonymous health care aide in *Robot&Frank*, the simple-minded and old-modeled Odi, the high intelligent and mechanically rational Vera in *Real Humans* (Season 1) who takes care of Lennart, and her equivalent who takes care of Dr. Millican in *Humans* (Season 1). They are the aged-care robots this essay focuses on. Frank's robot and Vera are the representational crystallization of human's pure rationality.

They are not restricted in the factory or in a fictional dystopian world, but stepping into people's daily life. They are not the stereotyped images of slaves or monsters any more, but helpers and companions. This change means robots' gradual acceptance by the public, and the beginning of a new era of the HRI. These works present three characteristics of the HRI in a new world of human-robot co-existence:

- Room/Space: The HRI doesn't happen in a dystopian imagination, which is far from social reality, but in an intimate daily life discourse.
- Time: The elderly-care robots are not an imagination of a far future, but of a near future, for instance, the film *Robot&Frank* presents a picture of "the near future".
- Characteristics of robots: They are neutral intelligent humanlike machines and do their duties according to their programs. They are free from the traditional concepts of being "either good or evil", and will not arouse the all-or-nothing response.

The above-mentioned works positively narrate and foreground the human-robot co-existence over the narration of robots as dangerous and menacing machine. The traditional human-robot duality is challenged. The traditional HRC (human-robot conflict) and dystopian picture cannot always meet the audience's taste any more. However, still there are five ethical issues involving the elderly-caring by robots. Briefly they are:

- From the challenge of safety to the challenge of privacy;
- From the human-robot duality to the their co-existence and co-operation;
- The machinery rationality versus the human emotion;
- The possible problems of the overloaded affective interaction;
- The ethical responsibility for one's elderly parents in the HRI.

This essay focuses on the issue of mechanical rationality versus human emotions, and also relevant issues arising from the human-robot interaction.

2 The Machinery Rationality Versus Human Emotions

2.1 The Elderly as An Irrational Group

Sometimes, as the elderly step into the last phase of life, with their long-term stable living habits, some of them become very sage and sensible. But some of them have also gradually developed a kind of emotional and even irrational personality, which is called "the old turning into the child" in Chinese culture. However, this might not be a bad thing, sometimes it might even have its positive side. Yet we cannot deny that such irrationality will do harm to their health and will bring the elderly into conflicts with the robots, who possess only "rationality".

In the studied Sci-Fis, the elderly's irrationality has been fully revealed through their communication, interactions or even conflicts with robots. Frank said to Robot: "I would rather die eating cheeseburgers than live off of steamed cauliflower." (TC:17:12-17) Some old people no longer regard life or health as the most important thing when they reach old age or approach death; they have no longing for the future and lack motivation for the present. However, they are very nostalgic and highly cherish the good old days, their old habits, relationship and memories. The old-modeled robots are not only keepers of the shared memories (they can sometimes clearly revoke the forgotten memories by the elderly), also they are engaged in the memories themselves, which make them unique. Humans tend to regard robots as "tools", which is a deep-rooted concept according to human logic and rationality; yet in this case, such definition for robots does not apply to the elderly (Frank, for instance).

Keeping healthy and cherishing life is common sense for humans; however, not for the elderly discussed here. For example, regardless of Odi's malfunction, Lennart stubbornly and capriciously insists driving out with him, merely to recall the reminiscence of their hanging out and fishing together (his daughter Inger would not allow Odi to drive). Lennart does know Odi has been acting weirdly. Once Odi almost locked Lennart's head in the car boot, and can no longer drive smoothly. In fact, they do encounter a car accident (*Real Humans*, S.1, E. 3-4). Audience can understand the elderly's sense of nostalgia, yet as offspring, they will not tolerate such irrational behaviors of the old.

2.2   The New Conflicts Between Human and Robot

Not only in the European literature, but also in the western Sci-Fi movies, it is very common to see the "threat posed by artificial entities" (Veruggio 1503) or conflicts between human beings (Lord) and robots (tools/slaves). For instance, in *Robot&Frank*, Frank reminds his daughter: "This robot is not your servant" (TC 51:48-51). In *Real Humans* and *Humans*, there are people who abuse the robots. This is one side of the conflict. On the other side, human beings are afraid of the rebellion of the robots. In *Robot&Frank*, Frank tells his son who brings him the robot: "That thing is going murder me in my sleep." (TC 10:09-12) This is called "Frankenstein complex" by Isaac Asimov (in his novel *That Thou Art Mindful of Him*) because of the famous novel of Mary Shelley's *Frankenstein; or, The Modern Prometheus*. Therefore, Asimov proposes the "Three Laws of Robotics" in his short story *Runaround*: (1) A robot may not injure a human being or, through inaction, allow a human being to come to harm; (2) A robot must obey the orders given to it by human beings, except where such orders would conflict with the First Law; (3) A robot must protect its own existence as long as such protection does not conflict with the First or Second Laws (44–45).

In the Sci-Fis that covered in this paper, the conflict between human and robot is still inevitable along with their partnership and companionship. It has shifted from a survival game of life and death between human and robot (which is closely related with the 1. Law) to a tension between machinery rationality and human emotions (2. Law).

In *Real Humans* and *Humans*, unlike the good old friend Odi who earns the old man's love and trust, the new high intelligent aged-care robot, Vera, feels superior to others, and persists what she proposes is the best for Lennart. Her stubborn rationality which has been inserted in her program causes tension between her and the old man, who is dominated by emotions, caprice and irrationality. Here comes the key question. Where is Vera's mechanical rationality from? According to Wallach and Allen in their influential book *Moral Machines. Teaching Robots Right from Wrong* (2008), there are three ways of machine morality acquisition: Top-Down Morality, Bottom-Up and Developmental Approaches, Merging Top-Down and Bottom-Up (83–117). Here, Vera mainly needs to obey the Top-Down Morality (which is designed by humans, aiming to take care of the elderly and improve the sense and bodily functions of the elderly)—it is of great significance that the rationality of robots comes from humans, and is designed by pure rationality of the latter. Besides, Vera and the robot with Frank have developed a few senses and skills from Bottom-Up, to some degree. They have some potential of self ethical judgement and willpower of execution. For instance, the robot can cheat Frank into a healthy diet by lying to him that if Frank doe not obey, as a robot, he will be returned and recycled, which is a white lie apparently.

Back to the conflict between machinery rationality and human emotions. Take *Humans* for example. Vera tries to force the old widower Dr. Millican to form good living habits to keep him healthy, but he does not like it. He insists on his own life style. To make sure he rests well during the night, Vera even has to escort and force him to go back to sleep. Thus the ethical dilemma occurs: To do Dr. Millican good, Vera has to impose on him, but he complains and claims: "You are not a carer, you are a jailer!" (S.1, E. 2, TC 25:55-26:01). This may remind the audience of a similar scenario in *I, Robot* (2004), in which a robot imprisons a human being in order to "protect" him.

Judging from the elderly' sense of dignity and life quality, this conflict between robots and humans does lead to "an increase in the feelings of objectification [... and] a loss of personal liberty" (Sharkey 27), and even put the elderly at risk. If we take the "Consequentialism" (Veruggio 1505) as the principle of the elderly-caring by robots, then the question is: between guaranteed personal autonomous liberty and robots' sometimes compulsory service, which is more essential to human beings? How to pursue and keep such a balance between the two? The health and life on one hand, and the feelings and the old habits, wishes (personal liberty) on the other. Both are important for the quality of the life of the elderly. Under the circumstance that robots are more rational than humans, should they still obey humans' orders? If not, humans will be panicked; but if the orders given by humans are irrational and the robots obey, they will "injure a human being or, through inaction, allow a human being to come to harm" (which violates the First Law). The conflict between machinery rationality and human emotions leads to ethical dilemmas which could be barely solved. These are the fundamental questions which will serve to design a better ethics model for the future HRI. The decision is not only up to the designers of robots, but also to all people involved.

3   Robot as Antithesis: The Irrationality of Human Beings

Commonly conceived as a human's replica, substitution and companion, robots can be served as an antithesis, mirror or prisma, which allow humans to better see and define themselves. With robots as a frame of reference, following, we continue to scrutinize human beings' irrational and emotional characteristics and behaviors.

3.1   Robot: Not "It" But "He"

In traditional concept, robots have nothing to do with consciousness or emotions. Though some fictions from earlier period have touched upon this theme and endow them with feelings, "robot" has long become a metaphor which symbolizes a lack of emotions. Both in fiction and in reality, usually the elders were unwilling to accept robots as companions at first. For instance, in *Robot&Frank*, when Frank encountered the robot for the first time, he said, "I'm not this pathetic. I don't need to be spoon-fed by some goddamn robot" (TC 10:34-40). And he called the robot rudely as "goddamn robot" and "death machine" (TC 10:25-38).

However, even though humans know exactly robots are not human, are lifeless and emotionless, they would still show sympathy and attach emotions to the robots (human-like artificial creatures), particularly for the elderly. In terms of whether or not a robot should be referred to as "he" or "it", Lennart insists that it should be "he" and he tends to anthropomorphize the robot. Also, Frank says: "He [the robot] is my friend!"(TC: 52:18-19), even though the robot continuously reminds him that it is but a robot, Frank still regard him as a friend and cannot manage to wipe the memory which they shared. The same with Lennart: his daughter Inger says, "Odi is not human, it's a machine" (S.1, E. 1, TC 13:52-54), yet Lennart asserts that her daughter should use "he" to refer to his robot companion Odi (TC 13:45-48). Though they got into conflicts with high-intelligent robots, Lennart and Dr. Millican treat the old-styled and simple-minded robot Odi as their dear family member. That is because Odi knows their personal habits, likes and dislikes well, and helps to record their precious memories.

In the future, this "he or it" debate remains controversial. In fact, the children (a yet-to-be enlightened group) tend to be irrational and have more often referred to inanimate object as animate. As Sigmund Freud pointed out in his influential theoretical Essay *The Uncanny* (1919), some children cannot "distinguish at all sharply between living and lifeless objects, and that they are especially fond of treating their dolls like live people." Freud went on: "a woman patient declare that even at the age of eight she had still been convinced that her dolls would be certain to come to life if she were to look at them in a particular way, with as concentrated a gaze as possible." (9) The difference in Lennart's case is that, though he clearly knows his robot is but a lifeless thing, he still insists it is "he". This in turn highlights the elderly's extreme irrationality.

3.2 The Intimate HRI and the Risk

To many people, too much interaction with robots can cause alienation. Without making sure that robots possess moral agency or not and even knowing that robots are but human-shaped yet totally different, humans still regard them as with life and attach emotions to them. The problem is, where is the boundary in human-robot interaction? From the Sci-Fis studied here, excessive emotions attached to robots may cause negative effect of alienation in human beings like Sophie in *Humans,* who deeply relies on the hubot Anita and imitates her behaviors. Unlike children, the elderly are more mature and have stronger yearning for emotional exchange. In the three Sci-Fi works mentioned, an intimate human-robot relationship does not cause cognitive displacement for the elderly (not as serious as in the case of Sophie). In *Real Humans* and *Humans*, Lennart and Dr. Millican regard the outdated Odi as their close companion, for he will not force them to do anything and help them keep their most cherished memories.

It is common knowledge that keeping a certain degree of interpersonal relationship is good for the mental health of the elderly. Yet in the discussion of the Sci-Fis here, it is open to debate. Though the elderly are inclined to an harmoniously intimate and personal human-robot interaction, we cannot overlook some potential problems: Even if as what the elderly have insisted, robots have emotions; it is but artificial. Can it be counted as deception for the elderly? (Li 564) Can the mass-produced robot companions become the elderly's real friends? The ending of *Robot&Frank* also suggests similar concerns. From those Sci-Fi works, we learn that it depends on specific context to decide whether or not to treat robots as slaves (or tools), and yet even if the old knows robots do not possess life, they still tend to regard them as their real friends and companions. This sense of intimacy and belonging makes them feel good and mentally healthy. It might be mutually beneficial. Interestingly, this irrationality of humans becomes the prerequisite of a harmonious human-robot relationship (like Lennart, Dr. Millican and Odi).

3.3   The Irrationality of Human Being in the Mirror of Robots

Beatrice, an insidious hubot who wants to populate the planet one day, said to Mimi, a domestic hubot who has been accepted by the human family Engmans as follows: "We will never die. We are going to rule" (*Real Humans*, S. 2, E. 1, TC 56:26-28). She went on: "You have ability to develop feelings. Do not join the people. They are controlled by their emotions" (TC 56:14-21). Beatrice is not an aged-care robot, but from her words, we can tell emotions and feelings are human characteristics.

In those Sci-Fis, in the HRI and in front of robots, we can observe humanity more closely and clearly. Here the aged-care robots do not pose the question, "who am I", but the human characters or audience can better see and define themselves through this reference object. Aristole brought up the philosophical proposition that "Man is a rational being" in *Politics*. Once placed in the frame of human-robot, the elderly are far less rational than robots. The latter are consistently and absolutely rational compared to humans. Robots can persist in the rationality designed by yet fails humans. In this regard, the design of rationality is unique to humans, but humans are not good at keeping their reason. From the TV series of *Real Humans* and *Humans,* we can see in front of the reference object, an absolute rational "being" (robots), the audience observe more keenly the impulsive side of the elderly, their personal preference and their weighing emotions over rationality. And yet, this is exactly what makes us humans.

In Yuval Noah Harari's most recent published public reading, the finale of the popular trilogy, *21 Lessons for the 21st Century* (2018), he talks about a different but essentially similar logic: humans study philosophy and discuss over ethics and rationality, yet when it comes to ethical decisions to make within a moment, only human emotions and intuitions will work. It is the algorithm of a computer, rather than human beings that can persist absolute rationality (53–58). So do machines and robots.

4   Conclusion and Prospect

In the recent Sci-Fi works with aged-cared robots, we can find out a new trend on the Sci-Fi screen: The human-robot conflict has shifted from a survival game of life and death to a tension between machinery rationality and human emotions. From this study on the tension between mechanical rationality and human emotions, we can draw two concluding points: (1) Machinery rationality is pure rationality designed by humans, or robot rationality is the representational crystallization of pure rationality of human beings, which forms a drastic contrast with human emotions, especially the emotions of the elderly; (2) The design of rationality and the discussion of ethics is unique to humans, but humans are not good at staying rational. From the conflicts involved here, robots become the reference frame of humans, from which we can more keenly observe our irrational side, that humans highly regard emotions and can never absolutely persist in rationality. This is exactly what makes us humans. Thus, probably we should consider what we can learn from robots as a frame of reference. The discussion of the ethical tension and conflict between the pure rationality of robots and the emotional irrationality of the elderly would be particularly imminent and significant, which could serve to design a better ethics model for the future HRI.

Robots can continuously urge us to keep exploring, redefining and knowing ourselves. Meanwhile, we should be psychologically, emotionally and ethically prepared for the possible future of human-robot co-existence. When this kind of society really comes true, we should build a world with the elderly' life quality guaranteed and avoid the feeling of alienation or uncanniness before and with our artificial double.

References

Battaglia, F., Weidenfeld, N. (eds.): Roboethics in Film. Pisa University Press, Pisa (2014)

Borenstein, J., Pearson, Y.: Robot caregivers: harbingers of expanded freedom for all? Ethics Inf. Technol. **12**(3), 277–288 (2010)

Broadbent, E.: Interactions with robots: the truths we reveal about ourselves. Annu. Rev. Psychol. **68**(1), 627–652 (2017)

Freud, S.: The Uncanny. Alix Strachey, tr. (1919). http://web.mit.edu/allanmc/www/freud1.pdf. Accessed 9 Aug 2018

Harari, Y.N.: 21 Lessons for the 21st Century. CITIC Press, Beijing (2018). LIN Junhong tr.

Li, X.Y.: Study on the ethical risks of the aged care robots. J. Northeast. Univ. Soc. Sci. **17**, 561–566 (2015)

Robinson, H.M., et al.: The psychosocial effects of a companion robot: a randomized controlled trial. J. Am. Med. Dir. Assoc. **14**, 661–667 (2013)

Robinson, H.M., et al.: Physiological effects of a companion robot on blood pressure of older people in residential care facility: a pilot study. Australas. J. Ageing **34**, 27–32 (2015)

Robinson, H.M., et al.: Group sessions with Paro in a nursing home: structure, observations and interviews. Australas. J. Ageing **35**, 106–112 (2016)

Royakkers, L., van Est, R.: A literature review on new robotics: automation from love to war. Int. J. Soc. Robot. **7**(5), 549–570 (2015)

Salvini, P., et al.: Roboethics in biorobotics: discussion of case studies. In: Proceedings of the ICRA 2005, IEEE International Conference on Robotics and Automation Workshop on Robo-Ethics, Rome, Italy, 10 April 2005

Sharkey, A., Sharkey, N.: Granny and the robots: ethical issues in robot care for the elderly. Ethics Inf. Technol. **14**(1), 27–40 (2012)

Sparrow, R., Sparrow, L.: In the hands of machines? The future of aged care. Mind. Mach. **16**(2), 141–161 (2006)

Su, N.M., et al.: Mundanely miraculous: the robot in healthcare. In: Proceedings of the 8th Nordic Conference on Human-Computer Interaction: Fun, Fast, Foundational, pp. 391–400 (2014)

Vandemeulebroucke, T., et al.: The use of care robots in aged care: a systematic review of argument-based ethics literature. Arch. Gerontol. Geriatr. **74**, 15–25 (2018)

Veruggio, G., Operto, F.: Roboethics: social and ethical implications of robotics. In: Siciliano, B., Khatib, O. (eds.) Springer Handbook of Robotics, pp. 1499–1524. Springer, Heidelberg (2008). https://doi.org/10.1007/978-3-540-30301-5_65

Wendell, W., Allen, C.: Moral Machines: Teaching Robots Right from Wrong. Oxford University Press, New York (2009)

Correction to: From Bayesian Inference to Logical Bayesian Inference

A New Mathematical Frame for Semantic Communication and Machine Learning

Chenguang Lu ⓘ

Correction to:
Chapter "From Bayesian Inference to Logical Bayesian
Inference: A New Mathematical Frame for Semantic
Communication and Machine Learning"
in: Z. Shi et al. (Eds.): *Intelligence Science II*, IFIP AICT 539,
https://doi.org/10.1007/978-3-030-01313-4_2

The original version of this chapter contained a mistake. There was an error in Equation (31). The original chapter has been corrected.

The updated version of this chapter can be found at
https://doi.org/10.1007/978-3-030-01313-4_2

© IFIP International Federation for Information Processing 2018
Published by Springer Nature Switzerland AG 2018. All Rights Reserved
Z. Shi et al. (Eds.): ICIS 2018, IFIP AICT 539, p. E1, 2018.
https://doi.org/10.1007/978-3-030-01313-4_51

Author Index

Printed in the United States
By Bookmasters